From the Calculus to Set Theory, 1630–1910

An Introductory History

From the Calculus to Set Theory, 1630–1910

An Introductory History

Edited by

I. Grattan-Guinness

with chapters by

H. J. M. Bos
R. Bunn
J. W. Dauben
I. Grattan-Guinness
T. W. Hawkins
K. Møller Pedersen

Duckworth

First published in 1980 by
Gerald Duckworth & Co. Ltd.
The Old Piano Factory
43 Gloucester Crescent, London NW1

ISBN 0 7156 1295 6

Printed by

Taylor and Francis (Printers) Ltd,
Rankine Road, Basingstoke, Hants.

Contents

Contents

Chapter 0

Introductions and Explanations

I. Grattan-Guinness

Nowadays the Philosophy of Mathematics is considered in connection with more general notions such as consistency, predicativity, or constructivity. So it might seem to us that the evolution of the foundations of the calculus amounts merely to the development of certain mathematical techniques, and the rejection or modification of such techniques on the grounds of obvious inconsistency. The picture is incomplete. It ignores the fact that, from the seventeenth to the nineteenth century, the history of the Philosophy of Mathematics is largely identical with the history of the foundations of the Calculus.

Abraham Robinson (*1966a*, 280)[1]

Most of my time [as an undergraduate at Cambridge] was taken up by mathematics, and mathematics largely dominated my attempts at philosophical thinking. . . . Those who taught me the infinitesimal Calculus did not know the valid proofs of its fundamental theorems and tried to persuade me to accept the official sophistries as an act of faith.

Bertrand Russell (*1959a*, 28, 35–36)

If I compare arithmetic with a tree that unfolds upwards in a multitude of techniques and theorems whilst the root drives into the depths . . . .

Gottlob Frege (*1893a*, xiii)

That which is provable, ought not to be believed in science without proof.

Richard Dedekind (*1888a*, preface)

[1] The manner of giving references is described in section 0.4 on page 8.

0.1. *Possible uses of history in mathematical education*

This book recounts the development of the differential and integral calculus from the early 17th to the late 18th centuries, and their subsumption under the broader subject of mathematical analysis in the 19th century. It describes the progress up to the early 20th century, and also records the introduction and progress of set theory and mathematical logic during the forty years beginning around 1870. Special attention is paid to the close relationships of these topics, and their unfolding one into the other.

All these topics have been treated by historians of mathematics before, but the kind of treatment given here is unusual ; for while my co-authors and I have discussed the mathematics involved in some detail, we also aimed at *introducing* the reader to its historical development. We hope that the book will be useful in mathematical education, and in ways which I shall now discuss.

During their undergraduate courses students learn in great detail substantial quantities of mathematics ; but usually they are given little historical information on its genesis, or on the motivations which led to its creation. This tradition of teaching has its own history. As with many sciences, mathematics began to enter its modern professionalised state about two hundred years ago, and mathematical education became correspondingly institutionalised a little later. The establishment of the *Ecole Polytechnique* and the re-forming of the *Ecole Normale* in Paris in 1795 were particularly important ' pace-setters ' in educational practice for science throughout Europe,[1] and the writing of textbooks based on courses at such institutions became a standard procedure. Indeed, some branches of mathematics were stimulated in their development by educational needs ; as we shall see, mathematical analysis was one of them.

The professionalisation of mathematics led to a vast increase in the number of research mathematicians and therefore in the amount of published work. In order to present to students the basic components of this expanding world in an intelligible form, teachers and textbook writers, who were not always prominent in research work, tried to present as well as they could the essentials of the particular branch of mathematics in question in an economical and rigorous form. This

[1] I wonder how the modern student would react to the teaching programmes meted out by these institutions. For example, at the *Ecole Polytechnique* around 1818 students were up for prayers at 5.30 a.m. and studied from 6.00 to 8.00, then after breakfast from 8.30 to 2.30 p.m., and after lunch and recreations from 4.30 (6.30 on Tuesdays) to 8.30. Bedtime was at 10.00. Sunday was the only day off this routine, though there were still studies of various kinds. (See *Ecole Polytechnique 1819a*, 55–63.)

style of presentation had the advantages of cramming the maximum amount of reliable mathematics into the given space, and of preventing already large books from becoming unreadably long. However, it meant that the mathematics seemed to fall complete and ' perfect ' onto the printed page ; while perhaps clear in form, it was largely unmotivated and therefore difficult for the student to appreciate from that point of view. The same approach to education became standard in lecture courses, with similar strengths and weaknesses in the consequences for the student.

Put another way, in this tradition of mathematical education emphasis is usually laid on the *accumulation* of mathematical *knowledge*, on the amassing of the ' facts ' which comprise a mathematical theory ; it does not much consider the *growth* of mathematical *understanding*, the appreciation of why a mathematical theory developed and took its *form*, and not merely that it does have its *content*. Now the history of mathematics may be of benefit here ; for after all, any mathematics is the result of human endeavour in the past, and strands at least of its original motivations may well still be relevant or have educational value. *How much* use can be made of historical aspects, and the manner in which they can most profitably be employed, are difficult questions which will not be pursued here. But it should be stressed that, while students should obtain a better understanding of mathematics (in the sense of ' understanding ' mentioned above), they cannot expect the mathematics to be ' easier ' to do. For example, if this book is read carefully, then it will be quite taxing in places, and pencil and paper will be needed. The rewards of understanding will, it is hoped, form sufficient compensation.

0.2. *The chapters and their authors*

The branches of mathematics whose historical development is described in this book were chosen because they are of great importance, are widely taught, pose interesting educational as well as historical and mathematical questions, and lead in some way to a more general appreciation of mathematics. In order to keep the book to a reasonable length I decided to concentrate on the ' pure ' and foundational aspects of the subject. But the technical aspects and applications have not been passed over in silence ; perhaps someone might find them worthy of similar treatment. In particular, the development of Newtonian mechanics and its broadening in the 19th century into mathematical physics (which included topics like electro-magnetism and potential theory) has a history closely related to the development of the calculus and its broadening into mathematical analysis.

The chapters lie more or less in chronological order of the historical development, and as the story unfolds the mathematics usually becomes more advanced in character while the historical context moves closer to the modern situation. In addition, the historical record is roughly the reverse of the formal presentation of mathematics ; it starts with calculus techniques and ends with set theory, whereas an axiomatic treatment of analysis could start with set theory and end with the calculus.

Let me now introduce the authors and the chapters which they have prepared.

The first chapter was written by Kirsti Møller Pedersen, of the Department of the History of Science at the University of Aarhus, Denmark. Her special interest is the infinitesimal calculus in the 17th century, and her chapter discusses the development of its techniques in the period 1630–1660. The emphasis has to be laid on techniques, for there was a lack of general theory in the calculus at that time. However, the adoption of algebraic symbolism in the 16th century led to a gradual synthesis of algebraic and (traditional) geometrical methods in problems such as the rectification and quadrature of curves, the construction of tangents, and the determination of maxima and minima. The algebraic techniques began to reveal their potential, although their level of rigour was low.

The second chapter was written by Henk Bos, of the Mathematical Institute of the University of Utrecht, The Netherlands. He has studied various aspects of the calculus in the 17th and 18th centuries, and in his chapter he continues the story of the development of the calculus up to about 1780. Two great mathematicians qualify for the claim of having ' invented ' the calculus : Newton and Leibniz. Both created a coherent system of concepts and methods to solve problems concerning tangents and quadratures of curves in general. Leibniz's differential and integral calculus and Newton's fluxional calculus, though different in many aspects, each involve a clear recognition of what we now call the inverse relationship between differentiation and integration. Moreover, both men worked out a system of notations, symbols and rules through which their methods could be applied in the form of algorithms performed on formulae, rather than in the form of geometrical arguments presented in prose with reference to figures. Of the two systems, Leibniz's was to have much the greater influence ; hence the developments achieved by Newton's followers have not been described. Jakob and Johann Bernoulli, l'Hôpital and many other mathematicians worked out the Leibnizian calculus and applied it with great virtuosity. Later, Euler cast analysis into a form which it was to retain for nearly a century. Meanwhile the foundations of analysis, especially the use of infinitesimal quantities, remained unclear. They were criticised by the philosopher George Berkeley and defended by mathematicians in various ways, none of them satisfactory from the logical point of view.

The third chapter was written by myself, in between working at Middlesex Polytechnic, England, and editing the history of science journal *Annals of science*. The chapter describes the development of the calculus on from Euler, and the emergence of mathematical analysis with Cauchy in 1820s and its further development to about 1880. Lagrange proposed a foundation for the calculus to replace Euler's use of infinitesimals, in the expansion of any function by a Taylor series ; but Cauchy demonstrated the falsehood of its assumption that all functions could be so expanded. Cauchy's alternative foundation used a theory of limits to a far greater extent than had been done previously, largely creating mathematical analysis as we now understand it by incorporating the convergence of series and the continuity of functions under the same technique. However, his system did not explicitly include a theory of multiple limits and related distinctions, such as those between the uniform and non-uniform convergence of series of functions. An important example of such series were Fourier series, introduced by Fourier in 1807. Dirichlet produced in 1829 a successful convergence proof for these series which motivated many later studies in limits, convergence, functions and integrals. Gradually the techniques of multiple-limit analysis emerged, especially from Weierstrass and his followers in the last third of the century, and the principal features of this kind of analysis are discussed. A final section reviews the progress described so far in this book, and looks forward to the following chapters.

The fourth chapter was written by Tom Hawkins, of the Department of Mathematics at Boston University, Massachusetts, U.S.A. His initial interest in the history of mathematics was with Lebesgue integration, and he treats here the development of the integral during the 19th and early 20th centuries. In the 18th century the integral was normally defined as the inverse of the differential, but the ever-widening class of functions being used provided so many exceptional cases to this definition that a return to the (pre-)Leibnizian interpretation of the integral primarily as an area or a sum was unavoidable. Cauchy developed the basic theory of the integral as the limiting value of a sequence of partition sums in the 1820s. His results were developed by others (especially Riemann) over the next forty years, but the types of function being used expanded still further and required the use of set-theoretic techniques to define the integral, or ' measure ', of a set function. Early definitions proved to be either inconsistent or inadequate in scope, but in the early 1900s Lebesgue (and others) formulated a definition which could provide the generality of definition then needed.

The fifth chapter was written by Joe Dauben, of the Department of History at Lehman College in the City University of New York, U.S.A. It encapsulates his special interest in the history of mathematics : the life and work of Georg Cantor. Certain set-theoretic techniques were perforce taken for granted historically in the previous chapter, but their development is described here in detail. Cantor was not the only or even the first mathematician to use such techniques, but after his theorems of the early 1870s on the uniqueness of the trigonometric representation of a function, he far surpassed all rivals to produce ‘ Cantorian set theory ’. This theory has twin components, excitingly unified in Cantor’s later conceptions ; the topology of the line and plane as expressed in the theory of sets of points, and the infinitely large cardinals and ordinals and their arithmetic, which were associated with a more general conception of a set. Cantor’s affirmation of the actual infinite produced strong reactions, both for and against, among his contemporaries.

The sixth and last chapter was written by Bob Bunn, who studied in the Department of Philosophy at the University of British Columbia, Vancouver, Canada, and now holds a Killam Postdoctoral Fellowship there. His special interest, in the development of abstract set theory (and related topics such as mathematical logic) during the half-century from 1880, has been distilled into this chapter, which surveys progress in most areas of the foundations of mathematics from the 1870s to about 1910. It draws together strands from previous chapters, for foundational studies became motivated by the formulation of mathematical analysis in the Weierstrassian era, by the paradoxes and other features of Cantorian set theory, by the efforts especially of Frege and Russell to find new foundations for arithmetic, by the espousal of mathematical logic by Peano and his school, and by the introduction of metamathematics by Hilbert. The complexity of inter-relations and the unavoidability of philosophical questions led to various differing views about the scope and limits of mathematical proof and the relationship between mathematics and logic. Although the foundations of mathematics and the nature of logic were studied before 1870, the achievements were relatively slight as compared with the developments described here. The only significant missing components are the introduction of Boolean algebra, and the development of the foundations of geometry in the 19th century.

0.3. *The book and its readers*

Having now described the book in some detail, I shall point out some other general features and indicate its readership. It is hoped that undergraduates can use it, especially in the later years of their course; but this is *not* the place from which the mathematics can be *learnt*. Such training must be obtained from books and courses on the calculus, mathematical analysis, set theory or mathematical logic, to which this book would serve as collateral reading. Naturally students may not be taking courses in *all* these subjects; they will then pick out the chapters of closest relevance to their studies, and possibly look at the others out of general interest. The book should also prove useful to post-graduates, teachers and research mathematicians in these branches of mathematics or in other related subjects, and to all who are acquainted with these fields and wish to obtain some impression of their historical development.

It may be that the reader will want to write something himself on a topic which we have discussed or mentioned, and that he has not previously done any historical research. This problem cannot be pursued in detail here, but a practical discussion of the collection and storage of references and the planning of a paper is given in May *1973a*, 3–34, a book which also contains many bibliographies for the history of mathematics.

I shall now point out some limitations of the book. Firstly, specialists in the history of mathematics will find some omissions of detail and over-simplifications in historical interpretation, which they must accept as consequences of the policy followed in designing the chapters. In particular, the discussions usually concentrate on the principal mathematicians and their main (usually published) writings, consigning to passing mention or even silence their relevant manuscripts and the work of lesser figures whose work was apparently, if undeservedly, neglected in its time. However, minor work is sometimes discussed when it provides interesting historical insights.

There is also a problem about languages. We have translated all quotations into English where necessary, but unfortunately no translations (or only unreliable ones) exist for many of the original sources. Further, we have cited worthwhile historical literature as locations of further information on a particular topic irrespective of the language in which it happens to be written. In fact, for the areas of mathematics with which this volume deals, much of the original material and the best historical literature was and is written in Latin, Russian, French and German. The bibliography, though substantial, is not intended to be exhaustive for either the primary or secondary literature; but it should enable the reader to work out from this book into related topics.

We have not described the lives of our historical figures in much detail. More biographical information can be found in several of the secondary sources which we have cited, in other works on the history of mathematics, and in encyclopedias and reference works. In this latter category the *Dictionary of scientific biography*, edited by C. C. Gillispie and in progress of publication since 1970 from Scribner's at New York, is particularly useful.

Finally, let me say that, while the chapters are designed to flow one into the other with as little repetition or discontinuity as possible, the complicated nature of the historical record and the use of several authors has inevitably made each chapter stand a little independently from the others. For this I crave the reader's indulgence—and mention the name and subject indexes at the end of the book, where will be found all locations of a person (and also, where known, his full name and dates) or topic.

0.4. *References and bibliography*

The citation of literature caused a number of difficulties. A wide range of primary and secondary literature is cited and the references include reprintings and translations. In order to avoid cluttering the text with lengthy footnotes providing all these details, works are cited in the text by dating codes or catchwords, with the full details of publication provided in a comprehensive bibliography after the chapters. The dating code takes the form '*1867a*'. The year is that of the first publication of the work (or circulation of a manuscript), although in one or two cases the year of composition has been used if publication occurred long after. The suffix '*a*' is added and the dating code italicised even if only one work of that year by that author is cited, in order to make clear that '*1867a*' cites a publication whereas '1867' refers to a year. (If more than one work for the year is cited, then the suffices '*b*', '*c*', . . . are used.) However, editions of works, which usually appeared long after the original publication of material and then in many volumes over several years, are cited by the catchword '*Works*', '*Writings*' or '*Papers*'. In addition, a few books which appeared in several editions at different times, and manuscripts of uncertain date of composition, are given a catchword appropriate to their title.

There was also the difficulty of concisely citing a specific passage in a work which was, for example, published more than once in the author's life-time, then republished in editions of his works or in other collections, and possibly also translated into English. In these cases we cited where possible the paragraph, section, chapter or even theorem number, for they would normally be preserved in republications whereas the page numbers would usually change each time. However, sometimes it was impossible to avoid page numbers, in which case we have cited them from one, or at most two, relatively convenient sources.

Each chapter of the book is divided into sections numbered in the style 2.4 (section 4 of chapter 2). These numbers and the word ' section ' are used to make cross-references.

0.5. *Mathematical notations*

On the whole, the notations used are standard today; many were introduced as part of the mathematical developments which we describe. But a variety of notations is used in mathematical logic and set theory, and I indicate here which ones have been adopted in chapters 3–6 (apart from in quotations from original sources, of course):

Set theory

$\in$	is a member of[1]	$\notin$	is not a member of
$\cap$ *or* $\bigcap$	intersection	$\cup$ *or* $\bigcup$	union
$\subset$	proper inclusion	$\subseteq$	improper inclusion
$\{a, b, \ldots\}$	unordered set of objects $a, b, \ldots$		
$\varnothing$	empty set	$\mid$	condition bar
$[a, b]$	closed interval	(a, b)	open interval
$\overline{}$	complement	$\sim$	in one–one correspondence

Mathematical logic

&	and	$\vee$	or
$\rightarrow$	if . . . then	$\leftrightarrow$	if and only if
$\neg$	not	ϕx	propositional function[2]
$(\forall x)(\ldots x \ldots)$	universal quantifier (' for all x, . . . ')		
$(\exists x)(\ldots x \ldots)$	existential quantifier (' there exists an x . . . ')		

November, 1976.

[1] This symbol ' $\in$ ' is to be distinguished from ' ϵ ', ' epsilon ', used in the ' (ϵ, δ)' style of mathematical analysis.

[2] Mathematical functions in the calculus are symbolised by, for example, ' $f(x)$ ', where brackets are placed around the argument variable.

Chapter 1

Techniques of the Calculus, 1630–1660

Kirsti Møller Pedersen

1.1. *Introduction*

During the first six decades of the 17th century mathematics was in a state of rapid development. In this period ideas were born and developed which were to be taken up later by Isaac Newton and G. W. Leibniz. Many methods were developed to solve calculus problems ; common to most of them was their *ad hoc* character. It is possible to find examples from the time before Newton and Leibniz which, when translated into modern mathematical language, show that differentiation and integration are inverse procedures ; however, these examples are all related to specific problems and not to general theories. The special merit of Newton and Leibniz was that they both worked out a general theory of the infinitesimal calculus. However, it cannot be said that either Newton or Leibniz gave to his calculus a higher degree of mathematical rigour than their predecessors had done.

As the ideas which were the basis of the methods preceding the work of Newton and Leibniz came to bear fruit, the methods themselves fell into oblivion. In this chapter, therefore, great importance will be attached to the earlier ideas, and the methods will be illustrated by simple examples. The picture of what the mathematicians of the time achieved may thus appear somewhat distorted, but a rendering of the more complicated examples would be all too easily submerged in calculations. That it is possible to find simple problems is due to the fact that it was the practice of the mathematicians of the time to verify their methods by applying them to problems of which the solutions were known beforehand. Then the next step was to find new results by means of these methods.

It is impossible to deal comprehensively with this topic in a single chapter. My approach will be to exemplify the calculus of the period by relatively few methods, which are described in some detail. This implies that the methods of many important mathematicians will have to be left unmentioned. A more general survey giving a more profound impression of the development of the calculus from 1630 to 1660 may be found in the rich literature on this subject.[1] I have made my choice on the assumption that to give even a tolerably satisfactory general survey in a single chapter would mean listing names and outlining techniques in a way which could not possibly give a proper impression of the methods and style of the time to a reader who is not acquainted with the period.

One criterion for the selection of methods has been that they should render a picture of the way in which the mathematicians of the time did actually solve the problems with which they were most heavily engaged; another has been that they should inform the reader of the ideas which were to become sources of inspiration for later methods. Where different methods are based on similar ideas, I have tried to select the writer who first formulated the idea.

Of the period 1630–1660, no less than of all other periods, it holds true that if you really want to set its mathematics into relief then you must know the mathematics which preceded it. The mathematics of the period in question were greatly influenced by classical Greek mathematics[2] and also by that of the previous period. The reason for the importance of Greek mathematics was that during the 16th century it had become usual for the mathematicians to acquire a knowledge of this discipline, and it formed a basic element in the mathematical equipment of most of them. Greek mathematics was especially admired for its great stringency. But its methods were not heuristic; they were not well-fitted to suggest ideas as to how to attack a new problem, a fact which will be illustrated later in connection with quadratures and cubatures.

It was natural, therefore, to search for other methods which, if they could not live up to the Greek requirement of exactness, were at least able to suggest ideas as to the solution of problems. The seeds of such methods are to be found in the previous period, the end of the 16th and the beginning of the 17th centuries, which was a fertile time for the exact sciences as a whole. Astronomy made great progress through the work of Johannes Kepler; Simon Stevin contributed much to statics with his treatise *De Beghinselen der Weeghconst* ('The elements of the art of weighing': *1586a*). In mechanics Galileo Galilei's deduction of the laws of freely falling bodies and of the parabolic paths of projectiles meant a break with Aristotelian physics and the beginning of a new epoch, where mathematics was to be extensively used in physics.

[1] See, for example, Baron *1969a*, Boyer *1939a* and Whiteside *1961a*, and their bibliographies.

[2] There are excellent bibliographies of Greek mathematics in Boyer *1968a* and Kline *1972a*.

Kepler made use of infinitesimal methods in his works. The interest he took in estimating the volumes of wine casks resulted in the book *Nova stereometria doliorum vinariorum* (' New measurement of large wine casks ' : *1615a*). There he considered solids of revolution as composed in various ways of infinitely many constituent solids. For example, he regarded a sphere as made up of an infinite number of cones with vertices at the centre and bases on the surface of the sphere. This led to the result that the sphere is equal in volume to the cone which has the radius of the sphere as altitude and as base a circle equal to the surface of the sphere, that is, a circle with the diameter of the sphere as radius (Kepler *1615a*, Prima Pars, Theorem 11 ; *Works*$_1$, vol. 4, 563, or *Works*$_2$, vol. 9, 23 f.).

Galileo planned to write a book on indivisibles, but this book never appeared ; however, his ideas had a great influence on his pupil Cavalieri, with whose work we shall deal later.

1.2. *Mathematicians and their society*

A great many mathematicians of the 17th century were not mathematicians by profession. This tendency was especially noticeable in France ; there only Gilles Personne de Roberval occupied a chair of mathematics, while great mathematicians like Pierre de Fermat, René Descartes and Blaise Pascal worked without any official connection with their discipline. Like the mathematician who inspired him, François Viète, Fermat was a lawyer, and worked as such in Toulouse for most of his career. Descartes and Pascal were men of private means and, apart from mathematics, were also occupied with physics and philosophy. Descartes spent a large part of his time outside France, living for long periods in Holland and elsewhere.

This stay of Descartes in Holland served to inspire several Dutch mathematicians, among whom was Frans van Schooten. He was a member of the School of Engineering at Leyden, while his more important pupils, whose treatises he published along with his own, mostly worked professionally outside mathematics. However, the most illustrious of his pupils, Christiaan Huygens, devoted his whole life to mathematics and physics. In 1666 the *Académie des Sciences* was founded in Paris, and Huygens was offered a membership which he accepted. As a member of the *Académie* he received an ample stipend. In Italy, the most outstanding mathematicians and physicists, such as Galileo Galilei, Bonaventura Cavalieri and Evangelista Torricelli, held offices within their own fields, partly at universities and partly as court mathematicians.

The development of that part of mathematics with which this chapter is concerned started later in England than on the Continent. Hence the only English mathematician with whom we shall deal in this chapter is John Wallis, who was Savilian Professor of Geometry at Oxford from 1649. It should be mentioned that in Thomas Harriot England had a brilliant scientist whose work both in algebra and the calculus preceded some of the methods discussed in this chapter. But only his *Artis analyticae praxis* (' Practice of the analytical art ' : *1631a*), which contains his less important work, was published (posthumously) at this time ; thus his unpublished results will not be considered.

The period provides several good examples of the independent and almost simultaneous discovery of methods with striking resemblance, which often gave rise to disputes about priority and charges of plagiarism. Today, we are able to establish that as a rule these charges were unfounded ; but at the time this was not possible, since it was not common to publish one's treatises. For this there were two principal reasons. First, after 1640 publishers were reluctant to print mathematical literature, which was not very profitable ; and second, mathematicians were reticent about publishing their new methods, wanting to release the results only. Many treatises had to wait a very long time for their publication : several were left unprinted until the end of the 19th and the beginning of the 20th centuries, and some remain unpublished to this day.

Not until the last third of the 17th century did scientific periodicals come into existence ; before that time mathematicians communicated by letter. Here the Frenchman Marin Mersenne played an important part, for he kept in touch with many European scientists by correspondence and meetings which he held at his convent in Paris. To the mathematicians he sent the problems which he could not solve himself, and took care that the results and manuscripts he received were circulated among those interested in them.

1.3. *Geometrical curves and associated problems*

In the 17th century the calculus was closely bound up with the investigation of curves, since there was as yet no explicit concept of the variable or of functional relationships between variables. The first curves to be dealt with were those inherited from the Greeks : the conic sections, Hippias's quadratrix, the Archimedean spiral, the conchoid of Nicomedes, and the cissoid of Diocles. (For the definition and the history of these and the following curves see, for example, Loria *1902a*.)

As the century went on, these curves were augmented by, among others, the cycloid, the higher parabolas and hyperbolas ($y^m = kx^n$ and $ky^m x^n = 1$ respectively, m and n being natural numbers and k a constant), the spiral of Galileo, and the conchoid to a circle, also termed ‘ the limaçon of [Etienne] Pascal ’, which is in turn a variant of the curves called ‘ the ovals of Descartes ’.

Next to the conic sections the cycloid, the curve traced by a point on the circumference of a circle which rolls along a horizontal line, was the curve most often investigated. Its early history is connected with a problem called ‘ Aristotle's wheel ’ (see Drabkin *1950a*). When solving this problem Roberval generalised the motion which generates the curve, and considered the curtate and the prolate cycloid (which are traced by points on a radius and respectively outside and inside the circle) as well as the ordinary cycloid. In 1658 Blaise Pascal arranged a competition designed to find the area of a section of the cycloid, its centre of gravity, the volumes of solids obtained by revolving the section about certain axes, and the centres of gravity of these volumes (Pascal *1658a* and *1658b*).

In *La géométrie* (*1637a*) Descartes introduced his oval as a curve involved in the solution of various optical problems. One of these problems was to determine the form of a lens which makes all the rays that come from a single point or that are parallel converge at another unique point, after having passed through the lens (Descartes *1637a*, 362 ; *1925a*, 135).

Similarly, Galileo's spiral was the attempted solution of a physical problem concerning the path of a body which moves uniformly around a centre and at the same time descends towards the centre with constant acceleration. The recognition of the shape of another of Galileo's curves, namely, the catenary, caused the mathematicians many difficulties. This curve has the form of a chain suspended from two points (see section 2.8).

The three last-mentioned curves are examples of an interplay between physics and mathematics. Before discussing this topic further we shall answer the question : what kind of problems concerning curves did the mathematicians solve in the period before 1660 ?

Pascal's competition of 1658 relates to certain typical problems which were solved. Other problems consisted in finding tangents, surface areas and extreme values ; furthermore, some inverse tangent problems (that is, to find a curve which has tangents with a specific property) were considered. Finally, about the middle of the century, the rectification of arcs became a question of interest. Although there are earlier examples of rectifications, Christopher Wren's rectification of the cycloidal arc in the late 1650s was the first widely known one. He sent the result to Pascal outside the competition (see Wren *1659a*, or Wallis *Works*, vol. 1, 532–541).

Even though the solutions to these problems could be applied both to physics and to astronomy, their inspiration owed more to Greek mathematics than to physics and astronomy. The Greeks had worked on all the types of problem mentioned above ; one may therefore consider work on them as a continuation of the tradition of the Greek mathematicians. This does not mean that there was no correlation between mathematics and physics. This continued to happen, if for no other reason than that in this period important physicists were often also important mathematicians. It is nevertheless difficult to point unambiguously to a concrete physical problem which inspired the mathematicians to take up the above-mentioned problems. In the late 1650s, however, a new mathematical problem cropped up which sprang from physics, namely the study of evolutes, which was started by Huygens in connection with his work on the pendulum clock.

1.4. *Algebra and geometry*

When the Greeks came to realise the existence of incommensurable magnitudes, which meant that the rational numbers are not sufficient for purposes of measurement, they made geometry the foundation of that part of mathematics which was not number theory, the straight line being a substitute for a continuous field of numbers. This attitude resulted in the geometric algebra on which Euclid, Archimedes and Apollonius based their calculations.

In the course of time the theory of equations became separated from geometry, and a good deal of symbolism was gradually developed for this discipline. Viète contributed much to the introduction of symbols with his work *In artem analyticen isagoge* (' Introduction to the analytic art ' : *1591a*), in which he emphasised the advantage of using symbols to indicate not only unknown but also known quantities (Viète *1591a*, ch. V, 5 ; *Works*, 8, or *1973a*, 52). In this way he could deal with equations in general.

Viète also connected algebra and geometry by determining the equations which correspond to various geometrical constructions. He only employed this technique when the geometrical problems were determinate and led to determinate equations in one unknown quantity. The next step was to use an indeterminate equation in two unknown quantities when solving problems concerning geometric loci. Fermat and Descartes took this step almost simultaneously.

Fermat's treatise *Ad locos planos et solidos isagoge* (' Introduction to plane and solid loci ' : *1637a*) contains a pedagogic introduction to analytic geometry and some of its applications. However, the treatise did not have any great influence, for the simple reason that Descartes's *La géométrie* was published before it was generally known. *La géométrie* treats many subjects with supreme skill, but it starts with an introduction to analytic geometry that was not easy for the uninitiated to follow. Notwithstanding this fact, the work had a tremendous influence, especially after van Schooten had published it in Latin translation and with commentaries in 1659. Its success was mainly due to Descartes's notation, which bore the hallmark of genius. It will not surprise the modern reader, as it is the beginning of the notation still in use; but for the time it was revolutionary. There is no doubt that the notation and the thoughts embodied in *La géométrie* had a positive—if only indirect—influence on the development of the calculus.

1.5. *Descartes's method of determining the normal, and Hudde's rule*

In *La géométrie* Descartes described his technique of determining the normal to an algebraic curve at any point. He attached great importance to the method, as can be seen from the following introductory remarks (*1637a*, 341; *1925a*, 95):

> This is my reason for believing that I shall have given here a sufficient introduction to the study of curves when I have given a general method of drawing a straight line making right angles with a curve at an arbitrarily chosen point upon it. And I dare say that this is not only the most useful and most general problem in geometry that I know, but even that I ever desired to know.

Let the algebraic curve *ACE* be given and let it be required to draw the normal to the curve at *C* (see figure 1.5.1). Descartes supposed the line *CP* to be the solution of the problem. Let $CM = x$, $AM = y$,

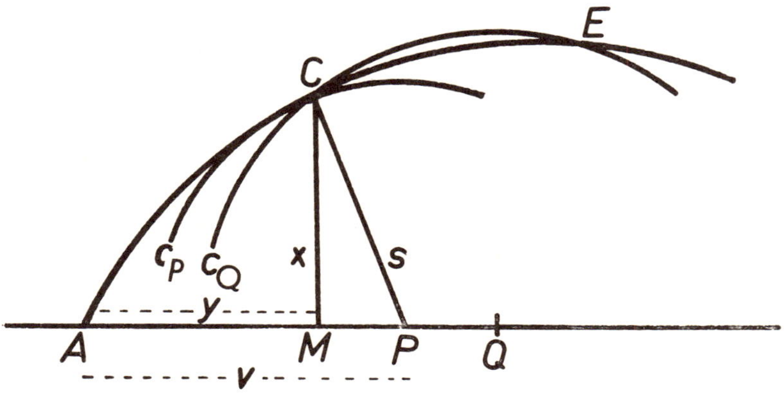

Figure 1.5.1.

$AP = v$ and $CP = s$. Although he always used a particular example, for the sake of convenience we shall suppose the curve to have the following equation :

$$x = f(y). \tag{1.5.1}$$

We shall also modernise his notation to some extent.

Besides the curve, Descartes considered the circle c_P with centre at P and passing through C; that is, the circle with the equation

$$x^2 + (v - y)^2 = s^2. \tag{1.5.2}$$

This circle will touch the curve CE at C without cutting it, whereas the circle c_Q

$$x^2 + (v_Q - y)^2 = s_Q{}^2 \tag{1.5.3}$$

with centre at a point Q different from P and passing through C will cut the curve not only at C but also in another point. Let this point be E. This means that the equation obtained by eliminating x from (1.5.1) and (1.5.3),

$$(f(y))^2 + (v_Q - y)^2 - s_Q{}^2 = 0, \tag{1.5.4}$$

has two distinct roots ;[1] but 'the more C and E approach each other, the smaller the difference of the two roots, and at last, when the points coincide, the roots are exactly equal, that is to say when the circle through C touches the curve at the point C without cutting it' (Descartes *1637a*, 346–347 ; *1925a*, 103–104).

Thus the analysis has brought Descartes to the conclusion that CP will be a normal to the curve at C when P (that is, v) is so determined that the equation

$$(f(y))^2 + (v - y)^2 - s^2 = 0 \tag{1.5.5}$$

has two roots equal to y_0 (or the corresponding equation with y eliminated has one pair of equal roots). With modern conceptions it is not difficult to realise that this requirement gives the correct expression,

$$v - y_0 = f'(y_0) \cdot f(y_0), \tag{1.5.6}$$

for the sub-normal MP.

Descartes illustrated his method by finding, among other things, the normal to the ellipse (*1637a*, 347 ; *1925a*, 104). Putting its equation in the form

$$x^2 = ry - \frac{r}{q} y^2, \tag{1.5.7}$$

he found the equation corresponding to (1.5.5) to be

[1] Descartes only considered curves for which $(f(y))^2$ is a polynomial in y or y^2 a polynomial in x.

$$y^2+\left(\frac{rq-2vq}{q-r}\right)y+\frac{qv^2-qs^2}{q-r}=0. \tag{1.5.8}$$

This equation has two roots equal to y_0 when

$$\frac{rq-2vq}{q-r}=-2y_0 \quad \text{and} \quad \frac{qv^2-qs^2}{q-r}=y_0{}^2; \tag{1.5.9}$$

because the point C is given, the value y_0 is known, and from (1.5.9) the sub-normal $v-y_0$ can be determined:

$$v-y_0=\frac{r}{2}-\frac{r}{q}y_0. \tag{1.5.10}$$

Although an indication, not to say a full account, of what happens when the two points C and E coincide would involve limit-considerations,[1] Descartes, by taking the double contact of the circle with the curve as a characteristic of the normal, has avoided the use of infinitesimals and obtained an algebraic method. His correspondence indicates that in solving some of his problems he did employ methods which involved the use of infinitesimals. However, he did not consider them precise enough to be published.

In principle, Descartes's method is applicable to any algebraic curve. But when the equation of the curve is not a simple algebraic equation, the method becomes tedious because of the laborious calculations which it is necessary to carry out in order to determine v by comparing the coefficients.

The Dutch mathematician (later Burgomaster of Amsterdam) Johann Hudde invented a rule for determining double roots. He described his method in a letter to Frans van Schooten, who published it in his 1659 Latin edition of Descartes's *La géométrie* (Hudde *1659a*, 507):

> If in an equation two roots are equal, and if the equation is multiplied by any arithmetical progression in such a way that the first term of the equation is multiplied by the first term of the progression and so on, I say that the product will be an equation in which the given root is found again.

[1] If we let the coordinates of E be $(y_0+\Delta y, f(y_0+\Delta y))$, then the requirement that C and E be on the same circle with centre at Q on the axis gives us the condition:

$$AQ=y_0+\frac{\Delta y}{2}+\left(\frac{f(y_0+\Delta y)-f(y_0)}{\Delta y}\right)\left(\frac{f(y_0+\Delta y)+f(y_0)}{2}\right).$$

(To obtain this result, let F be the mid-point of CE and note that $QF\perp CE$.) P and v are then determined by the coincidence of the points C and E, that is:

$$v=AP=\lim_{\Delta y\to 0} AQ=f'(y_0)f(y_0)+y_0.$$

For this rule Hudde gave a proof which in modern notation may be rendered as follows. Let $x = x_0$ be a double root in the polynomial $p(x)$, that is,

$$\begin{aligned} p(x) &= (x - x_0)^2 \sum_{i=0}^{n} \alpha_i x^i \\ &= \sum_{i=0}^{n} \alpha_i (x^{i+2} - 2x_0 x^{i+1} + x_0^2 x^i), \end{aligned} \tag{1.5.11}$$

and let $a, a+d, \ldots, a+(n+2)d$ be an arbitrary arithmetical progression. We then multiply the constant term $\alpha_0 x_0^2$ in $p(x)$ by a, the term of the first degree by $a+d$, and so on. Let the result of this procedure be denoted by $(p(x), a, d)$; that is,

$$\begin{aligned} (p(x), a, d) = \sum_{i=0}^{n} \alpha_i \{ &(a+(i+2)d)x^{i+2} - 2(a+(i+1)d)x_0 x^{i+1} \\ &+ (a+id)x_0^2 x^i \}. \end{aligned} \tag{1.5.12}$$

(Note that

$$(p(x), a, d) = ap(x) + dxp'(x), \tag{1.5.13}$$

where $p'(x)$ is the derivative of $p(x)$ and 'dx' means $d \times x$.) If we put $x_0 = x$, the expression in curled brackets in (1.5.12) vanishes. We therefore have $(p(x_0), a, d) = 0$.

This necessary condition for a polynomial to have one pair of equal roots made Descartes's method easier to apply, because one might so arrange the arithmetical progression that a difficult term might be multiplied by 0. We see that in his studies in autumn 1664 Newton found the sub-normal to a curve by using a combination of Descartes's method and Hudde's rule (Newton *Papers*, vol. 1, 217 ff.).

Hudde applied his rule to the determination of extreme values, acting on the assumption that if α is a value which makes $p(x)$ extreme, then the equation $p(x) = p(\alpha)$ has two equal roots (see Haas *1956a*, 250–255). He also extended his procedure to a rule for determining sub-tangents (*1659b*). He did not prove this rule, but it is interesting because it is one of the first general rules. Let the equation of the curve be $p(x, y) = 0$, where p is a polynomial in x and y; Hudde's rule then states that the sub-tangent t to a point (x, y) is given by

$$t = \frac{-x(p(x, y), a, d)_y}{(p(x, y), a, d)_x}. \tag{1.5.14}$$

The subscripts mean that in the numerator $p(x, y)$ is to be considered as a polynomial in y and in the denominator as a polynomial in x. From (1.5.13) we have

$$t = \frac{-x(ap(x, y) + dyp_y'(x, y))}{ap(x, y) + dxp_x'(x, y)} \tag{1.5.15}$$

(where the prime indicates differentiation with respect to the subscript variable), or, since $p(x, y) = 0$,

$$t = \frac{-yp_y'(x, y)}{p_x'(x, y)}. \tag{1.5.16}$$

Hudde's method was not forgotten after the introduction of the differential calculus ; for example, l'Hôpital commented on it in his *1696a*, ch. 10, para. 192 (see also section 2.5 below).

1.6. *Roberval's method of tangents*

In the late 1630s Gilles Personne de Roberval and Evangelista Torricelli independently found a method of tangents which used arguments from kinematics. In 1644, in his *Opera geometrica*, Torricelli published an application of his method to the parabola (Torricelli *1644a*, 119–121 ; *Works*, vol. 2, 122–124). In the same year Mersenne, in his *Cogitata physico mathematica* (' Physico-mathematical thoughts '), mentioned Roberval's method and applied it also to the parabola (Mersenne *1644a*, 115–116 ; see Jacoli *1875a*). One of Roberval's pupils, François du Verdus, wrote a treatise on Roberval's method. It was eventually published in 1693 (Roberval *Observations*) and became quite well-known, so the kinematic method came to bear Roberval's name.

The method rests on two basic ideas. The first is to consider a curve as the path of a moving point which is simultaneously impressed by two motions. The second is to consider the tangent at a given point as the direction of motion at that very point. If the two generating motions are independent, then the direction of the resultant motion is found by the parallelogram law for compounding motions. However, Roberval also applied his method to curves like the quadratrix and the cissoid, where the generating motions which he considered were dependent. He ingeniously compensated for the dependence when compounding the motions, as we shall see.

Roberval succeeded in determining the correct tangents to all the curves which were generally considered at his time. For the conic sections, however, the tangents were not determined correctly, because he took the generating motions to be the motions away from the foci or from the focus and the directrix, and wrongly used the parallelogram rule in compounding these motions (see Pedersen *1968a*, 165 ff.).

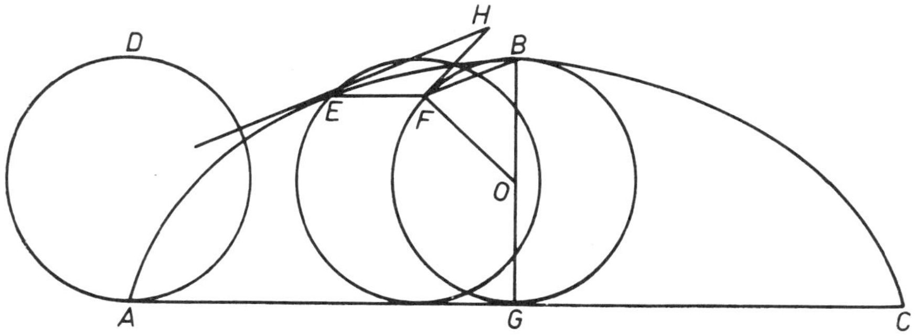

Figure 1.6.1.

To illustrate the method, we shall first see how Roberval determined the tangents to the cycloids (Roberval *Works*$_2$*a*, 58–63). Let ABC be a cycloid generated by the circle AD; that is, ABC is the path of the point A when the circle makes one turn on the line AC (compare figure 1.6.1, where the ordinary cycloid is drawn). The motion of A is then compounded of a uniform motion with direction AC or EF, and a uniform rotation about the centre of the generating circle, the direction of this at a point E being the tangent to the generating circle at E or the line FH. The ratio between the speeds of these motions is equal to the ratio between AC and the perimeter ADA, so if the point H is determined by

$$EF : FH = AC : \text{perimeter } ADA, \qquad (1.6.1)$$

then EH will be the tangent to the cycloid at E. For the ordinary cycloid, the ratio on the right hand side is equal to unity, and Roberval proved geometrically that EH is parallel to FB.

Thus the method is easily applied to the cycloid; but to see how general it is, let us also consider Roberval's determination of the tangent to the quadratrix. In figure 1.6.2 we let the two sides AD and CD of a square $ABCD$ move simultaneously, AD being rotated uniformly about A and CD being paralleledly displaced in such a way that AD and CD coincide with AB at the same time. The point of intersection between the two lines will then describe a quadratrix DFH. Let F—the point of intersection between IN and AD_1—be one of the points of the quadratrix and let us see how he determines the tangent at F. (Actually he considers a point on DFH's prolongation, but the principle is the same.)

Roberval starts by letting the line FK represent the velocity of the line IN. From the definition of the quadratrix follows that F describes the line FK in the same time as D_1 describes the arc D_1B, whence arc D_1B represents the speed of D_1's circular motion. As

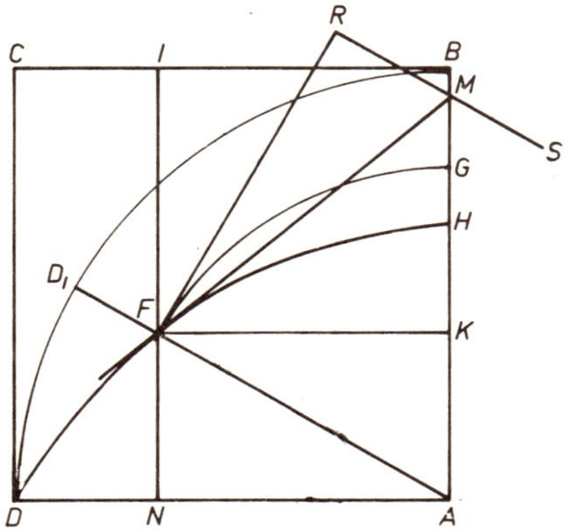

Figure 1.6.2.

$$\text{(the speed of the circular motion of } F) : \text{(the speed of the circular motion of } D_1) = AF : AD_1 = \text{arc } FG : \text{arc } D_1B, \qquad (1.6.2)$$

the arc FG represents the speed of F's circular motion ; and further, as the direction of this latter motion is perpendicular to AF, the circular motion of F will be represented by the line-segment FR on the perpendicular with length equal to arc FG. To obtain F's direction of movement he then draws the line RS through R parallel to AF and seeks the point of intersection, M, between RS and AB (which is the line through K parallel to IF) and connects F and M. FM will then be the tangent.

Roberval used this general approach in other cases too. His argument for it is not quite clear, but it has a great deal in common with the following. F's motion can be considered in two ways :

(1) F's motion on the quadratrix is compounded of the motion F has by taking part in AF's motion (with the instantaneous velocity FR) and the motion F has on AF because it has to be the point of intersection; the direction of the last motion is AF or RS. By compounding these two motions we see that the line of direction of the movement of F starts at F and ends on the line RS.

(2) Similarly, it is realised, by compounding the motion F has when it takes part in the motion of IF with its motion on IF, that its direction of motion is a line starting at F and ending on AB.

As both the conclusion of (1) and (2) must be fulfilled, the above construction follows.

By taking the instantaneous direction of motion as known, Roberval and Torricelli had avoided the use of infinitesimals in their method. Their method had the further advantage of being applicable to curves which are not referred to a Cartesian coordinate system. The method, however, was not general as long as the velocities could not be generally determined.

It is interesting to note that Newton's method of tangents from 1666 is inspired by the same ideas as Roberval's. For algebraic curves Newton only had to use the method once to obtain the sub-tangent expressed by a formula ; but for transcendental curves like the quadratrix he found the tangent in almost the same manner as had Roberval (Newton *Papers*, vol. 1, 416–418).

1.7. *Fermat's method of maxima and minima*

About 1636 there was circulated among the French mathematicians a memoir of Fermat entitled *Methodus ad disquirendam maximam et minimam* (' Method of investigating maxima and minima ' : *Methodus*). It was remarkable, for it gave the first known general method of determining extreme values. It contained another striking feature, namely, the idea of giving an increment to a magnitude, which we might interpret as the independent variable.

The memoir opens with the sentence : ' The entire theory of determining maxima and minima is based on two positions expressed in symbols and this single rule '. The rule is the following :

I. Let A be a term related to the problem ;
II. The maximum or minimum quantity is expressed in terms containing powers of A ;
III. A is replaced by $A+E$, and the maximum or minimum is then expressed in terms involving powers of A and E ;
IV. The two expressions of the maximum or minimum are made ' adequal ', which means something like ' as nearly equal as possible ' ;[1]
V. Common terms are removed ;
VI. All terms are divided by a power of E, so that at least one term does not contain E ;
VII. The terms which still contain E are ignored ;
VIII. The rest are made equal.

The solution of the last equation will give the value of A which makes the expression take an extreme value. Fermat illustrated his method by finding the point E on the line-segment AC which makes the rectangle $AE \, . \, EC$ a maximum. Let $AC=b$ and let us replace Fermat's A by x (so that $AE=x$), and his E by e ; we then have to

[1] Fermat used the word ' adaequo '. Mahoney has translated this as ' set adequal ' (*1973a*, 162). The idea of adequality derives from Diophantus (*ibid.*, 163–165).

maximize the expression $x(b-x)$. In accordance with the method, we have

$$(x+e)(b-(x+e)) \approx x(b-x), \tag{1.7.1}$$

where $\approx$ signifies the adequality. Removing common terms, we have

$$be \approx 2xe+e^2, \tag{1.7.2}$$

and dividing by e,

$$b \approx 2x+e. \tag{1.7.3}$$

Finally we ignore the term e and obtain $b=2x$.

It is tempting to reproduce Fermat's method by letting $A=x$, $E=\Delta x$, and the quantity $=f(x)$; the rule then tells us

$$\text{IV, V} \qquad f(x+\Delta x)-f(x) \approx 0, \tag{1.7.4}$$

$$\text{VI} \qquad \frac{f(x+\Delta x)-f(x)}{\Delta x} \approx 0, \tag{1.7.5}$$

$$\text{VII, VIII} \qquad \left(\frac{f(x+\Delta x)-f(x)}{\Delta x}\right)_{\Delta x=0} = 0. \tag{1.7.6}$$

For differentiable functions this might be interpreted in modern terms as if the x which makes $f(x)$ a local extreme value is determined by the equation

$$f'(x) = \lim_{\Delta x \to 0} \left\{\frac{f(x+\Delta x)-f(x)}{\Delta x}\right\} = 0. \tag{1.7.7}$$

However, this would be to read too much into the method. Primarily, Fermat did not think of a quantity as a function. Secondly, he did not say anything about E being an infinitesimal, or even a small magnitude, and the method does not involve any concept of limits ; it is purely algebraic. Thirdly, the statement in VI makes no sense in this interpretation, as we always have to divide by E to the first degree. Nevertheless, his examples show us that on occasion he divided by higher powers of E than one. The reason for this is that, if the quantity contained a square root, he squared the adequality before applying the last steps of the rule. Note that he did not emphasise that his method gave only a necessary condition.

Few results in the history of science have been so closely examined as Fermat's method of maxima and minima. He wrote about a dozen short memoirs where he explained and applied his method. Historians have been puzzled by his very short descriptions, and disagree about the dating of the memoirs and about the order of his ideas. To me it seems probable that he developed his ideas in the way that he intimated in his manuscript ' Syncriseos et anastrophes ' (*Syncriseos* ; see Mahoney *1973a*, 145–165).

Fermat says here that he got the idea of a process for determining extreme values by studying Viète's theory of equations and combining it with the expression ' μοναχός ' used by Pappus to characterise a minimal ratio (see Pappus *Collections*, book VII, theorem 61). Fermat takes ' μοναχὸς ' to mean ' singular ' in the sense of ' unique ' (see his *Works*, vol. 1, 142, 147), and gives an illustrative example of what he meant. The line-segment of the length B has to be divided by a point so that the product of the segments is maximum. The required point is the midpoint which makes the maximum equal to $B^2/4$. If $Z < B^2/4$, then the equation

$$X(B-X)=Z \tag{1.7.8}$$

will have two roots. Let them be A and E. Following Viète, Fermat obtains

$$A(B-A)=E(B-E) \tag{1.7.9}$$

or

$$BA-BE=A^2-E^2. \tag{1.7.10}$$

By dividing by $A-E$, it is seen that $B=A+E$. The closer that Z approaches $B^2/4$, the smaller will be the difference between A and E; at last, when $Z=B^2/4$, A will be equal to E, and $B=2A$, which is the unique solution leading to the maximum product. In other words, to find the maximum you have to equate the two roots.

As it can be complicated to divide by the binomial $A-E$, Fermat chose to let the two roots be A and $A+E$; then he divided by E, and finally equated the two roots by putting $E=0$. After these considerations he repeated his procedure from *Methodus* sketched in I–VIII at the beginning of this section. In this procedure he did not put $E=0$, but ignored the terms still containing E. However, the process is the same, and it became common practice to put E, or a corresponding magnitude, equal to 0 when his method was applied.

Until it was realised that the important process is

$$\lim_{\Delta x\to 0}\left\{\frac{f(x+\Delta x)-f(x)}{\Delta x}\right\}, \tag{1.7.11}$$

the procedure that involved dividing by E and putting $E=0$ was a thorn in the mathematicians' side. They were severely criticised for it, and they admitted that it was unsatisfactory.

Huygens who knew, applied and simplified Fermat's method, tried in vain to justify it logically (manuscript from 1652 printed in Huygens *Works*, vol. 12, 61). Instead he found another method, and one of which he could give a proof (*ibid.*, 62 ff.). This method combined Fermat's idea of an extreme value as unique with Descartes's idea of a double-root which he used in his method of normals. Briefly and in modern terms: Let $p(x)$ be a polynomial and let $p(x_0)$ be a maximum; when $a < p(x_0)$, the equation $p(x) = a$ has two roots which will be equal when $a = p(x_0)$. By a comparison of coefficients, x_0 may then be determined from the relation

$$p(x) - p(x_0) = (x^2 - 2xx_0 + x_0^2)p_1(x), \qquad (1.7.12)$$

where $p_1(x)$ is again a polynomial. As the applicability of this method is very limited, and as it is intricate to use, Huygens admitted that Fermat's method was easier to operate, and he himself accepted it.

Among others, Pierre Brûlart requested Fermat to give a proof of his method. In his answer *1643a* Fermat took another line, considering the coefficients of the powers of E in the development of $f(A \pm E)$. Although he could not prove it rigorously, he made it seem plausible that a maximum or minimum can be determined from the equation obtained by putting the coefficient of E equal to 0. Further, he showed that he understood that the coefficient of E^2 must be smaller than 0 for a maximum and greater for a minimum.

To Fermat it was more important to see that a method worked in practice than to give an exact proof of it. The method of maxima and minima had proved its value, for it gave the correct results when applied to a series of problems. Among these was the determination of the points of inflection of a curve in the manuscript ‘ Doctrinam tangentium ’ (Fermat *Works*, vol. 1, 166–167).

Fermat, however, did not stop at that; he extended the use of the procedure III–VIII from *Methodus* to other fields. This enabled him to determine tangents to curves (as will be seen in the next section), centres of gravity (*1638a*), and the sine law of refraction (*1662a*).

1.8. *Fermat's method of tangents*

In *Methodus*, Fermat made a determination of the tangent to the parabola, and presented this as an application of his method of maxima and minima. Before discussing the method we shall consider the example (Fermat *Works*, vol. 1, 134–136). Let the parabola DB with axis DC be given as in figure 1.8.1. Fermat wants to find the tangent at B; suppose it to be BE, and let the sub-tangent be EC. He takes an arbitrary point O on BE and draws IO parallel to the ordinate BC. Let P be a point of intersection of IO with the parabola.

From the inequality $IO > IP$, and from the property of the parabola

$$DC : DI = CB^2 : IP^2, \qquad (1.8.1)$$

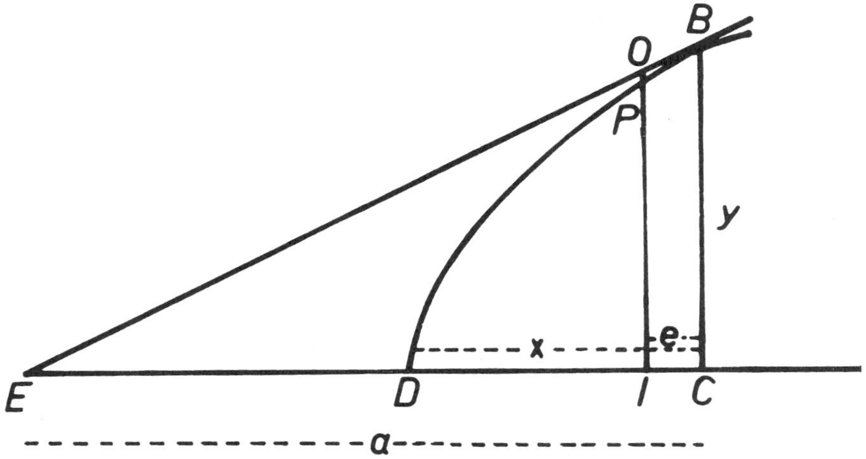

Figure 1.8.1.

it follows that

$$DC : DI > CB^2 : IO^2. \tag{1.8.2}$$

Since the triangles EIO and ECB are similar, we have

$$CB^2 : IO^2 = EC^2 : EI^2. \tag{1.8.3}$$

Thus

$$DC : DI > EC^2 : EI^2. \tag{1.8.4}$$

Let $DC = x$ (x is known since the point B is given), $EC = a$ (the unknown quantity) and $IC = e$. Then (1.8.4) becomes

$$x : (x-e) > a^2 : (a-e)^2, \tag{1.8.5}$$

or

$$xa^2 + xe^2 - 2xae > xa^2 - a^2e. \tag{1.8.6}$$

Fermat replaces this inequality by the adequality

$$xa^2 + xe^2 - 2xae \approx xa^2 - a^2e. \tag{1.8.7}$$

By using the procedure of the method of maxima and minima he obtains $a = 2x$, and thereby determines the tangent.

In a letter to Mersenne of January 1638 Descartes objected to this determination, maintaining that it did not solve the problem of an extreme value (see Fermat *Works*, vol. 2, 126–132, or Descartes *Works*, vol. 1, 486–493). He also accused Fermat of not having used the specific property of the curve, so that the determination would give the same result for all curves. The last objection is clearly wrong, and may be ascribed to the hostile attitude which Descartes took to Fermat after

Fermat had criticised his *Dioptrique* (*1637a*). The first objection, however, is worth examining.

The inequality $IO > IP$ holds for curves concave with respect to the axis, and the inequality $IO < IP$ for convex curves. For curves without points of inflection it is possible from these inequalities to find a magnitude depending on $a-e$ and $x-e$ which has an extreme value for $x-e=x$ (see Itard *1947a*, 597, and Mahoney *1973a*, 167). As $x(=DC)$ is known, a may be determined from the requirement for an extreme value. Neither in *Methodus* nor in Fermat's later writings, however, is there any indication that this was the way he related his method of tangents to his method of maxima and minima. In the memoir *1638b* of June 1638, Fermat, after having explained his method, wanted to show that there was a relation between the method of maxima and minima and that of tangents. However, by solving a problem of extrema he did not find the tangent to the curve, but rather the normal. This gave an algorithm quite different from the one used in *Methodus* and explained in the memoir. He is therefore not likely to have used this relation when he established his method of tangents. (By the way, the problem of extreme values which Fermat solved was suggested by Descartes in his first attack on Fermat's method.) So Descartes was right after all in raising the objection that the method of tangents was not a direct application of the method of maxima and minima.

When, in the memoir just mentioned, Fermat explained his method of tangents to Descartes, he clearly showed that he used only the procedure drawn from the method of maxima and minima. Descartes thereafter accepted the method. In modern notation Fermat's explanation can be reproduced in the following way. Let B be the point (x, y) on the curve $f(x, y)=0$ and let $DI=x-e$ (see figure 1.8.1). From the similar triangles EOI and EBC we obtain

$$IO=\frac{y(a-e)}{a}. \qquad (1.8.8)$$

Since IO is almost equal to PI, Fermat writes

$$f\left(x-e, \frac{y(a-e)}{a}\right)\approx 0. \qquad (1.8.9)$$

This is the *adequality* to which he applied his procedure from the method of maxima and minima. It is not difficult to see that it will lead to an expression for a corresponding to

$$a=-\frac{yf_y'}{f_x'}. \qquad (1.8.10)$$

If we have the parabola $\alpha x = y^2$, we obtain from (1.8.9)

$$\alpha(x-e) - \frac{y^2(a-e)^2}{a^2} \approx 0, \qquad (1.8.11)$$

or

$$y^2(a-e)^2 \approx a^2\alpha(x-e)\,; \qquad (1.8.12)$$

and since $y^2 = \alpha x$, then

$$x(a-e)^2 \approx a^2(x-e), \qquad (1.8.13)$$

which is (1.8.7).

As the method requires a development of

$$f\left(x-e, \frac{y(a-e)}{a}\right),$$

it was in its original presentation only applicable to algebraic curves (because in Fermat's time only algebraic functions were developed). However, in 'Doctrinam tangentium' Fermat extended its field of application to include some transcendental curves. He introduced two principles (Fermat *Works*, vol. 1, 162), stating that it was allowed

> (1) ... to replace the ordinates to the curves by the ordinates to the tangents [already] found ...
>
> (2) ... to replace the arc lengths of the curves by the corresponding portions of tangents already found ... .

These two principles enabled him to determine the tangent to the cycloid (*ibid.*, 163). Let HCG be a cycloid with vertex C and generating circle CMF (figure 1.8.2), and RB be the tangent at an arbitrary point R. For the sake of convenience we reproduce his analysis with use of some modern symbols. Let $CD = x$, $RD = f(x)$, $MD = g(x)$, and the magnitude to be investigated $DB = a$. The specific property of the cycloid is the following:

$$f(x) = RM + MD = \text{arc } CM + g(x). \qquad (1.8.14)$$

Let $DE = e$, and draw NE parallel to RD intersecting RB at N and the circle at O; as usual in the method of tangents, we have that

$$NE = \frac{f(x)(a-e)}{a} \approx f(x-e), \qquad (1.8.15)$$

where

$$f(x-e) = \text{arc } CO + g(x-e) = \text{arc } CM - \text{arc } OM + g(x-e). \qquad (1.8.16)$$

Let MA be the tangent to the circle at M intersecting NE at V, and let $MA = d$ and $AD = b$.

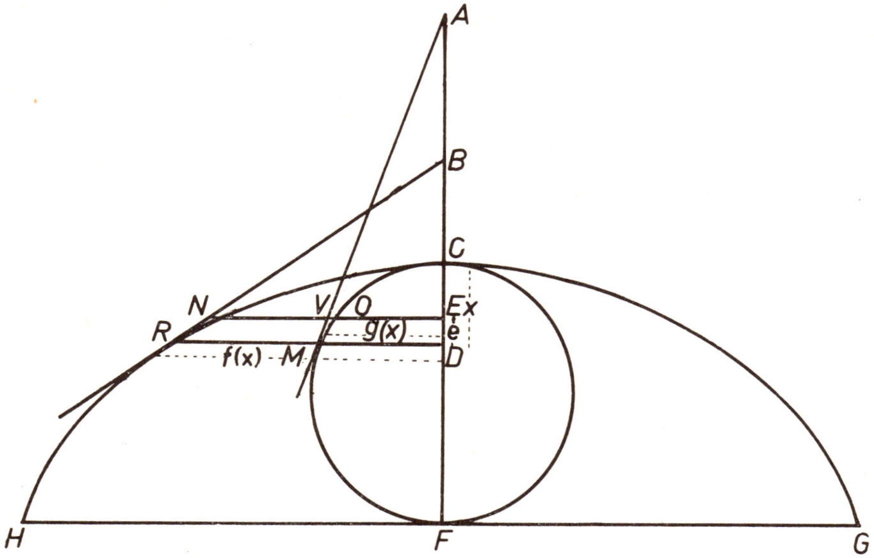

Figure 1.8.2.

From the first principle Fermat obtains

$$g(x-e) \approx EV = \frac{g(x)(b-e)}{b}, \tag{1.8.17}$$

and from the second

$$\text{arc } OM \approx MV = \frac{de}{b}. \tag{1.8.18}$$

Thus

$$f(x-e) \approx \text{arc } CM - \frac{de}{b} + \frac{g(x)(b-e)}{b}, \tag{1.8.19}$$

which together with (1.8.14) and (1.8.15) gives

$$\frac{(\text{arc } CM + g(x))(a-e)}{a} \approx \text{arc } CM - \frac{de}{b} + \frac{g(x)(b-e)}{b}. \tag{1.8.20}$$

Hence, by the standard procedure,

$$\frac{\text{arc } CM + g(x)}{a} = \frac{d + g(x)}{b}, \tag{1.8.21}$$

or

$$\frac{f(x)}{a} = \frac{d + g(x)}{b}. \tag{1.8.22}$$

Geometrically it is seen that

$$\frac{d+g(x)}{b}=\frac{g(x)}{x}, \tag{1.8.23}$$

so that the tangent at R is parallel to MC.

1.9. *The method of exhaustion*

The method of geometrical integration which was considered in the first part of the 17th century to be ideal was the exhaustion method, which had been invented by Eudoxus and improved by Archimedes. The name is unfortunate because the idea of the method is to avoid the infinite, and the method therefore does not lead to an exhaustion of the figure to be determined, as will be seen from the following outline of the idea behind it (see Dijksterhuis *1956a*, 130–132).

The method aims at showing that an area, a surface or a volume to be investigated, X, is equal to a known magnitude of the same kind K (for example, X may be the surface of a sphere and K four great circles on the sphere). A monotone ascending sequence I_n and a monotone descending sequence C_n of, respectively, inscribed and circumscribed figures to X are constructed. Thus we have the result :

$$\text{for all } n,\ I_n < X < C_n. \tag{1.9.1}$$

It is then shown either that for any magnitude $\epsilon > 0$ there exists a number N such that

$$C_N - I_N < \epsilon\,; \tag{1.9.2}$$

or that for any two magnitudes of the same kind μ and ν where $\mu > \nu > 0$, there exists a number N such that

$$C_N : I_N < \mu : \nu, \tag{1.9.3}$$

and further that

$$\text{for all } n, \quad I_n < K < C_n. \tag{1.9.4}$$

From (1.9.1), (1.9.2) or (1.9.3), and (1.9.4), it follows by a *reductio ad absurdum* that $K = X$.

This last demonstration always proceeds in the same manner, independent as it is of the magnitudes in question. Nevertheless, whenever applying the method, the Greek mathematicians wrote out the argument down to the last detail. The reason may be that they did not have a notation which made it easy for them to deal with the general case. Furthermore, it is rather complicated to establish the basic inequalities of the proof, especially (1.9.4), and the method can be used only if K

is known in advance. This means that it needs to be supplemented by another method, if results are to be produced.

Among the mathematicians of the early 17th century there was a desire to find such a method of obtaining results which, in contrast to the method of exhaustion, would be direct. It would be as well if the new method, apart from giving results, could be used to prove the relations achieved. Such a direct method might have been obtained had it been realised that

$$\lim_{n\to\infty} C_n = \lim_{n\to\infty} I_n, \tag{1.9.5}$$

and had X been put equal to that limit; however, this was not within the style of expression and power of abstraction of 17th-century mathematicians.

The path which they followed was that of an intuitive understanding of the geometric magnitudes. They imagined an area to be filled up, for example, by an infinite number of parallel lines. When, in 1906, Heiberg found Archimedes's *The method*, it was discovered that Archimedes too had adopted this point of view in his search for results. However, he did not regard it as sufficiently rigorous to be applied in proofs. Kepler, too, had used techniques involving such intuitive considerations, and it was the purpose of the first systematic exposition of the method of indivisibles to legitimise such techniques. This exposition, *Geometria indivisibilibus continuorum nova quadam ratione promota* (' Geometry by indivisibles of the continua advanced by a new method ' : *1635a*, hereafter referred to as *Geometria*), by Cavalieri, appeared in 1635, when he was a professor of mathematics at the University of Bologna. The ideas that it contained were developed in 1627, as can be seen in a letter from Cavalieri to Galileo (Galileo *Works*, vol. 13, 381).

The mathematicians differed on the importance to attach to a proof by the method of indivisibles. Most of those who thought about the matter regarded the method of indivisibles as heuristic, and thought that an exhaustion proof was still necessary. The exhaustion method was therefore modified and extended during the 17th century (see Whiteside *1961a*, 333–348). In many cases, however, mathematicians confined themselves to the remark that the results achieved by the method of indivisibles could be easily demonstrated by an exhaustion proof.

1.10. *Cavalieri's method of indivisibles*

Geometria, and Cavalieri's later work *Exercitationes geometricae sex* (' Six geometrical exercises ' : *1647a*), became well-known among mathematicians. The works inspired many of them to find their own methods, whereas others like Fermat and Roberval found their integration methods independently of Cavalieri.

Cavalieri presented two methods of indivisibles in his *Geometria*, and called them the 'collective' and the 'distributive' methods respectively. The first six of the seven books of *Geometria* embody the collective method, and a summary of it is given in *Exercitationes*, Book I. The framework of this section cannot possibly allow for a full account of the wide spectrum of concepts and ideas which Cavalieri introduced and developed in these six books, but the following outline gives a rough idea of his approach.

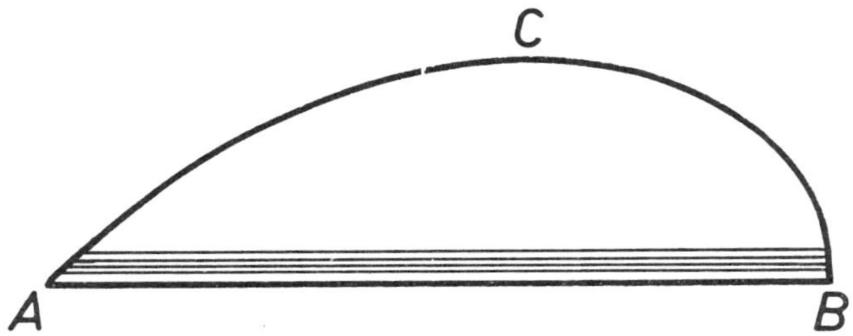

Figure 1.10.1.

Let there be given a plane figure $F=ABC$ limited by the curve ABC, and the straight line AB, called the 'regula' (figure 1.10.1). Cavalieri imagined that a straight line starting along AB is uniformly displaced parallel to AB, and considered the bunch of parallel line-segments which made up the section between F and the line during the motion. He named these line-segments 'all the lines of the given figure' ('omnes lineae propositae figurae'), and sometimes referred to them as 'the indivisibles of the given figure'; let us denote them by $\mathcal{O}_F(l)$.

Expressed in modern terms, Cavalieri constructed a mapping

$$F \to \mathcal{O}_F(l) \qquad (1.10.1)$$

from the set of plane figures into a set consisting of bunches of parallel line-segments. He then extended Eudoxus's theory of magnitudes (see book V of Euclid's *Elements*) to include his new magnitudes $\{\mathcal{O}_F(l)\}$. Thereafter he established—although not in a mathematically satisfactory manner—the fundamental relation

$$F_1 : F_2 = \mathcal{O}_{F_1}(l) : \mathcal{O}_{F_2}(l) \qquad (1.10.2)$$

between two plane figures (Cavalieri *1635a*, Book II, Theorem 3). By letting the regula be a plane he obtained in a similar way the relation

$$S_1 : S_2 = \mathcal{O}_{S_1}(p) : \mathcal{O}_{S_2}(p), \qquad (1.10.3)$$

where S_i is a solid and $\mathcal{O}_{S_i}(p)$ all the planes belonging to it, $i = 1, 2$.

Cavalieri's aim was to find the ratio on the left hand side of (1.10.2) by calculating the ratio on the right hand side. In doing so he was greatly helped by a postulate which leads to 'Cavalieri's theorem' (described below), a skilful use of previous results, theorems about similar figures, and the concept of powers of line-segments.

The postulate (*1635a*, Corollarium to Theorem 4 of Book II) states that if in two figures F_1 and F_2 with the same altitude every pair of corresponding line-segments (that is, line-segments at equal distances from the common regula) has the same ratio, then $\mathcal{O}_{F_1}(l)$ and $\mathcal{O}_{F_2}(l)$ have this ratio too. In modern notation and using figure 1.10.2,

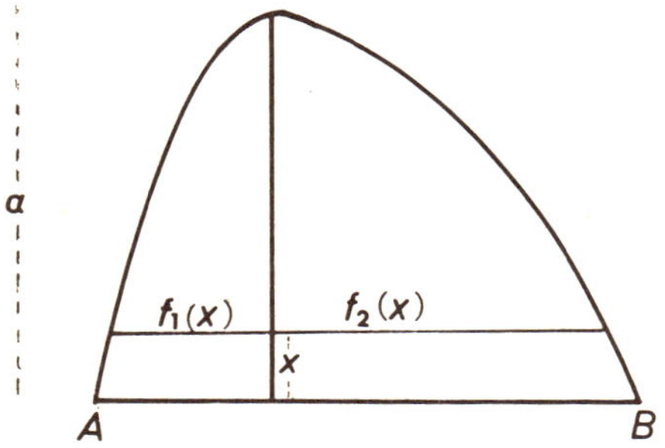

Figure 1.10.2.

if $f_1(x) : f_2(x) = b : c$ for all x $0 < x < a$,

$$\text{then } \mathcal{O}_{F_1}(l) : \mathcal{O}_{F_2}(l) = b : c. \qquad (1.10.4)$$

This, together with (1.10.2), immediately gives 'Cavalieri's theorem':

If $f_1(x) : f_2(x) = b : c$ for all x $0 < x < a$,

$$\text{then } F_1 : F_2 = b : c \qquad (1.10.5)$$

(*1635a*, Book II, Theorem 4).

Cavalieri's skilful employment of his previous results may be illustrated by a simple example. It is easily realised from figure 1.10.3 that

$$\mathcal{O}_{ACF}(l) = \mathcal{O}_{CDF}(l) \quad \text{and} \quad \mathcal{O}_{ACDF}(l) = \mathcal{O}_{ACF}(l) + \mathcal{O}_{CDF}(l). \qquad (1.10.6)$$

From these relations follows the theorem that the parallelogram *ACDF* is the double of each of the triangles *ACF* and *CDF*. However, Cavalieri was capable of interpreting them in a more general way.

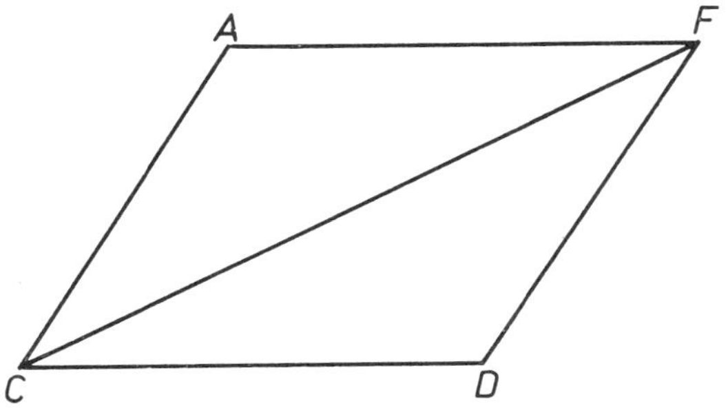

Figure 1.10.3.

By setting $AC = CD$ and using concepts which we cannot go into here he obtained a result which he could use every time he needed a proportion corresponding to

$$\int_0^a x\,dx : \int_0^a a\,dx = 1 : 2 \tag{1.10.7}$$

(*1635a*, Corollarium II to Theorem 19 of Book II: compare figure 1.10.4).

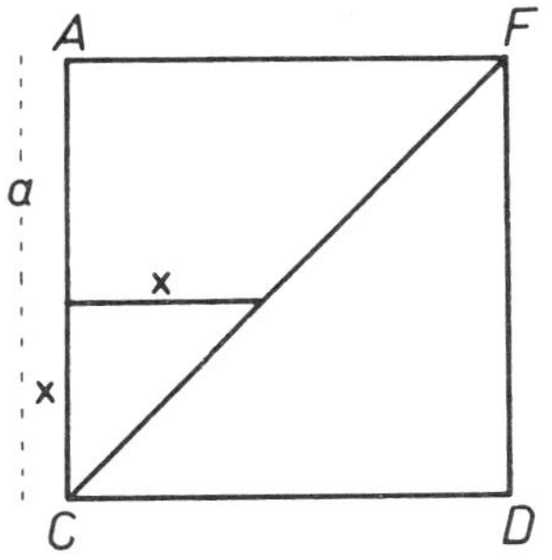

Figure 1.10.4.

Cavalieri found an alternative to integrating x^2 by introducing the squares of line-segments. If, instead of considering the line-segments of $\mathcal{O}_F(l)$, we take their squares situated in parallel planes, we obtain what he called ' all the quadrates of the given figure ' (' omnia quadrata propositae figurae '); this aggregate will be denoted by $\mathcal{O}_F(\square l)$.

Let us illustrate the use of this concept by an example. For each l in the parallelogram $ACGE$ in figure 1.10.5 we have

$$\square R_lT_l + \square T_lV_l = 2\square R_lS_l + 2\square T_lS_l, \tag{1.10.8}$$

where $\square R_lT_l$ means the square on the side R_lT_l. From this relation Cavalieri concluded that

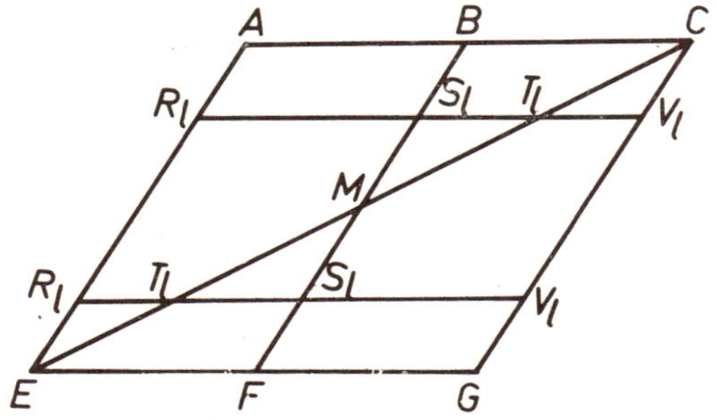

Figure 1.10.5.

$$\mathcal{O}_{AEC}(\square l)+\mathcal{O}_{CEG}(\square l)= 2\mathcal{O}_{ABFE}(\square l)+2(\mathcal{O}_{MEF}(\square l)+\mathcal{O}_{CBM}(\square l)). \quad (1.10.9)$$

Since the triangles AEC and CEG are congruent, we have

$$\mathcal{O}_{AEC}(\square l)=\mathcal{O}_{CEG}(\square l), \quad (1.10.10)$$

and similarly

$$\mathcal{O}_{MEF}(\square l)=\mathcal{O}_{CBM}(\square l). \quad (1.10.11)$$

He further proved that, since the triangles CEG and MEF are similar, the following relation holds :

$$\mathcal{O}_{CEG}(\square l):\mathcal{O}_{MEF}(\square l)=EG^3:EF^3=8:1. \quad (1.10.12)$$

In the same way he found that

$$\mathcal{O}_{ACGE}(\square l):\mathcal{O}_{ABFE}(\square l)=EG^2:EF^2=4:1. \quad (1.10.13)$$

From (1.10.9)–(1.10.13) it follows that

$$\mathcal{O}_{ACGE}(\square l)=3\mathcal{O}_{CEG}(\square l) \quad (1.10.14)$$

(*1635a*, Book II, Theorem 24). This result has as an immediate consequence that a cylinder is three times the inscribed cone. Cavalieri applied the relation (1.10.14) to a series of problems concerning conics, interpreting it by analogy with (1.10.6) as a relation which was an alternative to

$$\int_0^a x^2\,dx=\tfrac{1}{3}a^3. \quad (1.10.15)$$

The first six books of *Geometria* are in their general style a copy of the Greek classical mathematical works, built up of definitions and postulates from which the theorems are carefully deduced, all verbally. Although Cavalieri ingeniously used his concepts to obtain many results, this made the reading of the book rather tedious. Perhaps he felt this himself ; at least, he wrote to Galileo in 1634 that he composed the seventh book of *Geometria* to help those who found the concept of ' all the lines ' too difficult (Galileo *Works*, vol. 16, 113). In this last book, concerning the distributive method, he turned to a more intuitive treatment of the indivisibles.

As we saw in the relation (1.10.2), by the collective method Cavalieri found the ratio between two figures by comparing the aggregates of indivisibles. In the distributive method, two figures with the same altitude were compared by comparing *corresponding* indivisibles. The basic relation in this theory was Cavalieri's theorem (1.10.5), for which he gave a new proof without using the concepts from the collective theory.

A part of the criticism to which Cavalieri's methods were exposed was levelled against the nature of his indivisibles and the problem of the structure of the continuum. Some mathematicians took him to mean that a plane figure was made up of indivisibles and that these were line-segments. This was against the Aristotelian view of a continuum as divisible into parts of the same kind as the original magnitude, the parts again being infinitely divisible. To avoid his seeming error of dimensionality, they tried tentatively to conceive a plane figure as composed of rectangles with infinitesimal breadth. But the distinction was of theoretical interest only, for it remained usual to consider the ratio between two areas, so that an eventually missing Δx was cancelled by the relation

$$\frac{A}{B}=\frac{\sum a_n\Delta x}{\sum b_n\Delta x}=\frac{\sum a_n}{\sum b_n}, \tag{1.10.16}$$

where a_n and b_n are the altitudes in the rectangles of which the areas A and B are composed.

The conception of an area as a kind of a sum $\sum a_n\Delta x$ did not solve the problem, because it was still uncertain what was meant by an infinitesimal magnitude and by an infinite sum. Despite the lack of rigour in their foundations, the methods were useful insofar as they provided the mathematicians with new results.

1.11. *Wallis's method of arithmetic integration*

To determine the area under the spiral of Galileo, Fermat used an arithmetic quadrature which he described in a letter to Mersenne in 1638 (Mersenne *Correspondence*, vol. 7, 377–380). In his *Traité des indivisibles*[1] Roberval squared many figures on the basis of arithmetical

[1] *Traité* uses a method of infinitesimals which Roberval worked out about 1630. The date of the composition of the *Traité* is, however, unknown. It was first printed in 1693 (Roberval $Works_1$) and reprinted in 1730 ($Works_2$).

considerations. Pascal observed in his treatise *Potestatum numericarum summa* (' sum of numerical powers ') that his results concerning the sums $\sum_{i=0}^{m}(A+id)^n$ (where A, d and n are natural numbers) could be applied to the quadratures of curves (Pascal *Works*$_1$, vol. 3, 364 ; *Works*$_2$, vol. 2, 1272). Using proofs by complete induction he also established the rules for determining the binomial coefficients (*1654a*).

But most of the results based on a method of arithmetic integration were achieved by John Wallis. His treatise on the subject, *Arithmetica infinitorum* (' The arithmetic of infinites ' : *1655a*), is not burdened with proofs, for he relied boldly and confidently on his really astounding intuition as to the correlation between the sums of different series. He called his favourite method in the treatise ' modus inductionis ' : later it was termed ' incomplete induction '. One might also call it ' conclusion by analogy '.

Wallis started the treatise by establishing by this method that

$$\frac{\sum_{i=0}^{l} i}{(l+1)l}=\tfrac{1}{2},\quad \frac{\sum_{i=0}^{l} i^2}{(l+1)l^2}=\tfrac{1}{3}+\frac{1}{6l},\quad \frac{\sum_{i=0}^{l} i^3}{(l+1)l^3}=\tfrac{1}{4}+\frac{1}{4l},\tag{1.11.1}$$

and similarly that

$$\frac{\sum_{i=0}^{l} i^n}{(l+1)l^n}=\frac{1}{n+1}+\frac{a_1}{l}+\dots\frac{a_{n-1}}{l^{n-1}},\tag{1.11.2}$$

where the a_i's are rational numbers and $n=4$, 5, 6 (Wallis *1655a*. Propositions I, XIX and XXXIX ; *Works*, vol. 1, 365, 373 and 382), From this he concluded that

$$\lim_{l\to\infty}\left\{\frac{\sum_{i=0}^{l} i^n}{(l+1)l^n}\right\}=\frac{1}{n+1}\tag{1.11.3}$$

(*ibid*., 384). This relation enabled him to square the curves $y=x^n$ in figure 1.11.1 to obtain

$$\frac{\sum_{x=0}^{a} y}{\sum_{x=0}^{a} b}=\lim_{l\to\infty}\left\{\frac{\left(\frac{a\,.\,0}{l}\right)^n+\left(\frac{a\,.\,1}{l}\right)^n+\dots\left(\frac{a\,.\,l}{l}\right)^n}{a^n+a^n+\dots a^n}\right\}$$

$$=\lim_{l\to\infty}\left\{\frac{\sum_{i=0}^{l} i^n}{(l+1)l^n}\right\}=\frac{1}{n+1},\tag{1.11.4}$$

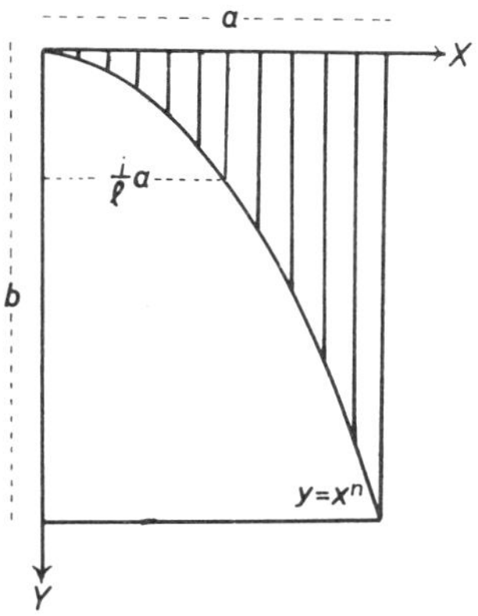

Figure 1.11.1.

a result which corresponds to

$$\frac{\int_0^a x^n\,dx}{a^{n+1}} = \frac{1}{n+1}. \tag{1.11.5}$$

This result was not new, and indeed it had been found by many of Wallis's predecessors ; but he did not stop there. He extended the range of n in (1.11.3) to include at least all rational numbers except -1. The foundation of his extension is an observation which he made in connection with the formula (1.11.3), namely : If the numbers l^{n_1}, $l^{n_2}, \ldots, l^{n_r}$ are in geometric progression (where $n_1, n_2, \ldots, n_r$ are non-negative whole numbers), then

$$\lim_{l\to\infty}\left\{\frac{\sum_{i=0}^{l+1} l^{n_j}}{\sum_{i=0}^{l} i^{n_j}}\right\}, \quad j=1, 2, \ldots r, \tag{1.11.6}$$

will be in arithmetic progression (*ibid.*, 387). Further, from the fact that for $0 \leqslant p \leqslant q$ $(q=1, 2, 3, \ldots)$

$l^0, l^{1/q}, l^{2/q}, \ldots l^{p/q}, \ldots l^1$ are in geometric progression,

$1, 1+\frac{1}{q}, 1+\frac{2}{q}, \ldots 1+\frac{p}{q}, \ldots 2$ are in arithmetic progression,

and the first and last members of the latter sequence are the reciprocals of the values of the right hand side of (1.11.3) for $n=0$ and $n=1$ respectively, he concluded that

$$\lim_{l\to\infty}\left\{\frac{\sum_{i=0}^{l+1} i^{p/q}}{\sum_{i=0}^{l} l^{p/q}}\right\}=\frac{1}{1+\frac{p}{q}} \tag{1.11.7}$$

(*ibid.*, 390). He did not doubt that the relation (1.11.7) held good for all $p/q \geqq 0$; he even said that it was valid for an irrational exponent, such as $\sqrt{3}$ (*ibid.*, 395), and as he extended the concept of power to include negative powers he considered (1.11.7) to be valid for them too—except -1 (*ibid.*, 408). By means of (1.11.7), he was now able to determine, when p/q was a rational number different from -1, the ratios between the areas under the curves $y=x^{p/q}$ and the circumscribed rectangles. He could also determine the ratios between the volumes obtained by a revolution of these areas about an axis and the circumscribed cylinders.

After that, Wallis proceeded to study polynomials in x; he applied the formula (1.11.7) to binomial expansions of $(x^p(D^n-x^n))^m$ when p, n and m are small natural numbers and D is a constant, and by analogy deduced that

$$\frac{\int_0^D [x^p(D^n-x^n)]^m\,dx}{D^{m(n+p)+1}}=\frac{n\,.\,2n\,.\,\ldots\,mn}{(mp+1)(mp+n+1)(mp+2n+1)\ldots(mp+mn+1)} \tag{1.11.8}$$

(*ibid.*, 419–420 and 425–430), a result which he put into various tables. (For clarity I render the last of his sums as integrals.) He further extended (1.11.8) to include the case where p and n are positive rational numbers (*ibid.*, 433).

One of Wallis's purposes was to square the circle; he stressed that from (1.11.8) and its extension we know for $m=0, 1, 2, 3\ldots$ the 'sums'

$$\frac{\int_0^R (R^2-x^2)^m\,dx}{R^{2m+1}} \quad \text{and} \quad \frac{\int_0^D (xD-x^2)^m\,dx}{D^{2m+1}}, \tag{1.11.9}$$

and for $m=1, 2, 3, \ldots$ the 'sum'

$$\frac{\int_0^R (R^{1/m}-x^{1/m})^m\,dx}{R^2}, \tag{1.11.10}$$

where R is the radius and D the diameter of the circle. He wished to find the values of these 'sums' for $m = \frac{1}{2}$, and he introduced the symbol '$\square$' to signify the reciprocal of (1.11.10):

$$\square = \frac{R^2}{\int_0^R (R^2 - x^2)^{1/2}\, dx} \left(= \frac{4}{\pi} \right). \tag{1.11.11}$$

By a principle of interpolation which we cannot go into here, he succeeded in establishing the formula

$$\frac{R^{n-1}}{\int_0^R (R^2 - x^2)^{(n/2)-1}\, dx} = a_n \quad \text{for} \quad n = 1, 2, 3, \ldots, \tag{1.11.12}$$

where

$$\left. \begin{aligned} & a_1 = \frac{\square}{2}, \quad a_2 = 1, \quad a_3 = \square, \\ & a_{n+2} = \frac{3 \,.\, 5 \,.\, 7 \ldots (n+1)}{2 \,.\, 4 \,.\, 6 \ldots \quad n}, \quad n = 2, 4, 6, \ldots \\ & a_{n+2} = \frac{4 \,.\, 6 \,.\, 8 \ldots (n+1)}{3 \,.\, 5 \,.\, 7 \ldots \quad n} \square, \quad n = 3, 5, 7, \ldots \end{aligned} \right\} \tag{1.11.13}$$

(see Prag *1929a*, 389–392, and Whiteside *1961a*, 237–241). From the fact that

$$\frac{a_{n+2}}{a_n} = \frac{n+1}{n} \quad \text{for} \quad n = 1, 2, 3 \ldots, \tag{1.11.14}$$

he concluded that the ratio a_{n+1}/a_n is continuously decreasing,[1] so that

$$\frac{n+2}{n+1} = \frac{a_{n+3}}{a_{n+1}} = \frac{a_{n+3}}{a_{n+2}} \,.\, \frac{a_{n+2}}{a_{n+1}} < \left(\frac{a_{n+2}}{a_{n+1}} \right)^2 < \frac{a_{n+2}}{a_{n+1}} \,.\, \frac{a_{n+1}}{a_n} = \frac{a_{n+2}}{a_n} = \frac{n+1}{n}, \tag{1.11.15}$$

and hence

$$\sqrt{\left(\frac{n+2}{n+1} \right)} < \frac{a_{n+2}}{a_{n+1}} < \sqrt{\left(\frac{n+1}{n} \right)}. \tag{1.11.16}$$

From the formulae (1.11.13) he obtained for odd n the inequalities

[1] Wallis was lucky that his sequence behaved in this way, for a sequence defined by

$$a_1 = k, \; a_2 = 1, \; a_{2n+1} = \frac{2n}{2n-1} a_{2n-1}, \quad \text{and} \quad a_{2n+2} = \frac{2n+1}{2n} a_{2n}$$

will not generally have a_{n+1}/a_n continuously decreasing.

$$\frac{3.3.5.5.7.7\ldots(n-2).n.n}{2.4.4.6.6.8\ldots(n-1)(n-1)(n+1)}\sqrt{\left(\frac{n+2}{n+1}\right)} < \square$$
$$< \frac{3.3.5.5.7.7\ldots(n-2).n.n}{2.4.4.6.6.8\ldots(n-1)(n-1)(n+1)}\sqrt{\left(\frac{n+1}{n}\right)}. \qquad (1.11.17)$$

In the limit as $n \to \infty$, these give a result now called ' Wallis's product ' :

$$\frac{4}{\pi} = \square = \frac{3.3.5.5.7.7.9.9\ldots}{2.4.4.6.6.8.8.10\ldots} \qquad (1.11.18)$$

(Wallis *Works*, vol. 1, 469).

1.12. *Other methods of integration*

Most of the methods of integration in use before the time of Newton and Leibniz made use of an equidistant sub-division of intervals and compared the area or volume to be found with a known one, as we have seen with Cavalieri and Wallis. However, Fermat had a method which allowed him to make an absolute calculation of an area, employing a sub-division which meant that the areas of the infinitesimal rectangles to be summed were in a geometric progression with quotient less than unity. We may illustrate this by means of an example from his treatise on quadrature *De aequationum*, which he wrote about 1658 using ideas he had already had in the 1640s (see Mahoney *1973a*, 243 f.).

Fermat considered the hyperbolas

$$yx^n = k, \quad k \text{ is a constant}, \quad n = 2, 3, 4, \ldots . \qquad (1.12.1)$$

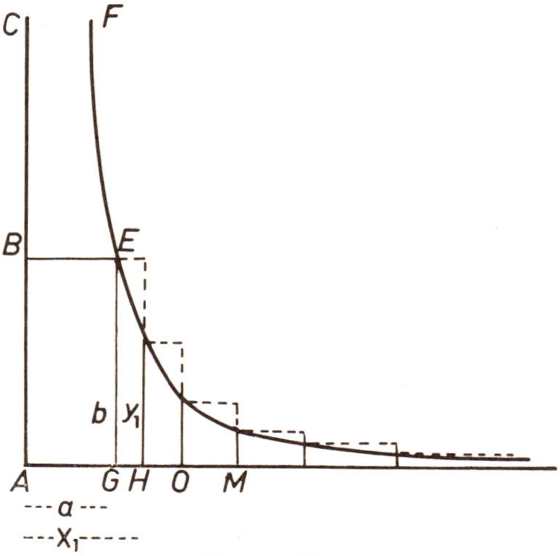

Figure 1.12.1.

For convenience I reproduce his arguments in modern terms. He divided the x-axis to the right of the point G (see figure 1.12.1) in intervals GH, HO, OM, ... of lengths x_1-a, x_2-x_1, x_3-x_2, ... $(a=AG)$, so that

$$\frac{a}{x_1}=\frac{x_1}{x_2}=\frac{x_2}{x_3}=\dots \qquad (1.12.2)$$

and hence

$$\frac{a}{x_1}=\frac{x_1-a}{x_2-x_1}=\frac{x_2-x_1}{x_3-x_2}=\dots. \qquad (1.12.3)$$

He then considered the circumscribed rectangles

$$\left.\begin{aligned} &R_1=b(x_1-a), \text{ where } b=GE, \\ &R_r=y_{r-1}(x_r-x_{r-1}), \quad r=2, 3, \dots. \end{aligned}\right\} \qquad (1.12.4)$$

From (1.12.1)–(1.12.4) it follows that

$$\frac{R_1}{R_2}=\frac{b(x_1-a)}{y_1(x_2-x_1)}=\frac{x_1{}^{n-1}}{a^{n-1}}=\frac{x_{n-1}}{a}, \qquad (1.12.5)$$

$$\frac{R_r}{R_{r+1}}=\frac{y_{r-1}(x_r-x_{r-1})}{y_r(x_{r+1}-x_r)}=\frac{x_r{}^n a}{x_{r-1}{}^n\, x_1}=\frac{x_1{}^n a}{a^n x_1}=\frac{x_{n-1}}{a}, \qquad (1.12.6)$$

which means that the circumscribed rectangles are in a geometric progression with quotient a/x_{n-1}.

To determine the sum S of a geometric progression with first term α and quotient u/v $(u<v)$, Fermat used the following relation :

$$\frac{v-u}{u}=\frac{\alpha}{S-\alpha} \qquad (1.12.7)$$

(this is equivalent to $S=\alpha/(1-u/v)$). Hence, if S denotes the sum of the rectangles R_r, we have :

$$\frac{x_{n-1}-a}{a}=\frac{b(x_1-a)}{S-b(x_1-a)} \qquad (1.12.8)$$

or

$$\frac{x_{n-1}-a}{x_1-a}=\frac{ba}{S-b(x_1-a)}. \qquad (1.12.9)$$

He then imagined the intervals x_1-a, x_2-x_1, ... to be sufficiently small and almost equal, and he concluded that the left hand side of (1.12.9) by adequality is equal to $n-1$. Further, as the intervals are small, he concluded that $S-b(x_1-a)$ in the relation (1.12.9) can be set equal to the area σ defined by the hyperbola and the lines GH and GE. Hence

$$n-1=\frac{ba}{\sigma}=\frac{AG\,.\,GE}{\sigma}, \tag{1.12.10}$$

and the quadrature is achieved.[1] We could have obtained (1.12.10) by taking the limit of both sides of (1.12.9) for x_1 approaching a, but he did not use limits. He observed that his method could not be applied when $n=1$ as the rectangles R_r will then be equal.

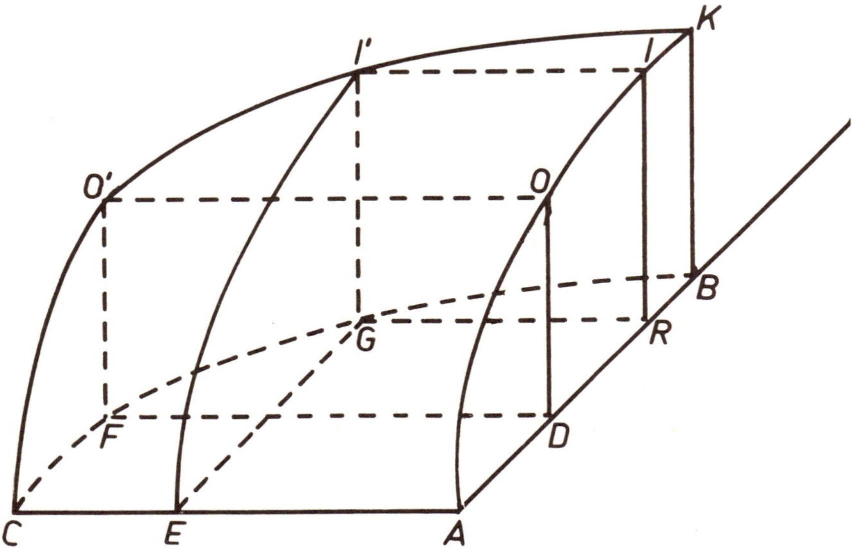

Figure 1.12.2.

[1] Fermat called his method ' logarithmic ' (*Works*, Vol. 1, 265). In his time the word ' logarithmic ' was used to characterise a connection between a geometric and an arithmetic progression ; hence ' logarithmic ' was also used at that time where today we would say ' exponential '. Let us indicate in modern terms how his expression and proof can be interpreted. If we let $a=\exp(t_0)$ and $x_r=\exp(t_0+r\Delta t)$, $r=1, 2, 3, \ldots$, then we have a sub-division which is equivalent to (1.12.2). An easy calculation shows that

$$R_r=k\exp[-(n-1)(t_0+(r-1)\Delta t)](\exp[\Delta t]-1). \tag{1}$$

Hence

$$S=\sum_{r=1}^{\infty}R_r=k\exp[-(n-1)t_0](\exp[\Delta t]-1) : (1-\exp[-(n-1)\Delta t]), \tag{2}$$

and

$$\lim_{\Delta t\to 0} S=(k\exp[-(n-1)t_0]) : (n-1)=(a\,.\,b) : (n-1). \tag{3}$$

An ingenious use of geometrical considerations and arguments from statics led mathematicians to many transformations corresponding to transformations of integrals, which could be applied to find connections between various problems solved by quadratures and cubatures. In his *1658c* Pascal systematically drew up schedules in which appear the sums necessary to determine areas and volumes as well as their centres of gravity. He found a fundamental theorem for these connections by conceiving the volume $KCAB$ (see figure 1.12.2) both as composed of the rectangles $FDOO' = FD \,.\, DO$ and as composed of the areas $EGI' = ARI$ (Pascal *1658c*, 'Lemme générel' in the section 'Traité des trilignes rectangles'; *Works*$_1$, vol. 9, 3–5). That is,

$$\sum_{AB} FD \,.\, DO = \sum_{AC} EGI'. \tag{1.12.11}$$

If we put $AB=a$, $AC=b$, $AD=x$, $FD=y=f(x)$ and $DO=z=g(x)$ (both being monotone functions), the relation corresponds to

$$\int_0^a f(x)g(x)\,dx = \int_0^b \left(\int_0^{f^{-1}(y)} g(t)\,dt \right) dy, \tag{1.12.12}$$

which can be obtained by an integration by parts. Since $f(a)=0$ we have:

$$\int_0^a f(x)g(x)\,dx = -\int_0^a \left(\int_0^x g(t)\,dt \right) f'(x)\,dx$$

$$= \int_0^b \left(\int_0^{f^{-1}(y)} g(t)\,dt \right) dy. \tag{1.12.13}$$

When $g(x)=x$, we obtain

$$\int_0^a xy\,dx = \int_0^b \frac{x^2}{2}\,dy. \tag{1.12.14}$$

Roberval found the summation form of (1.12.14) in his *Traité* in a way similar to that of Pascal (Roberval *Works*$_2a$, 271), and it was used by Fermat too (*Works*, vol. 1, 272). Among other things, it could be applied to the determination of the centre of gravity of the area $\int_0^a y\,dx$. Let the x-coordinate of this point be ξ; in modern notation the argument is the following (see figure 1.12.3). If we consider a lever AC and let the area $\int_0^a y\,dx$ operate on the arm ξ on the one side, and at the other let all the rectangles $y\Delta x$ of the area $\int_0^a y\,dx$ or BDC operate each on the arm x, then there will be equilibrium. Hence we have

$$\xi \int_0^a y\,dx = \int_0^a xy\,dx. \tag{1.12.15}$$

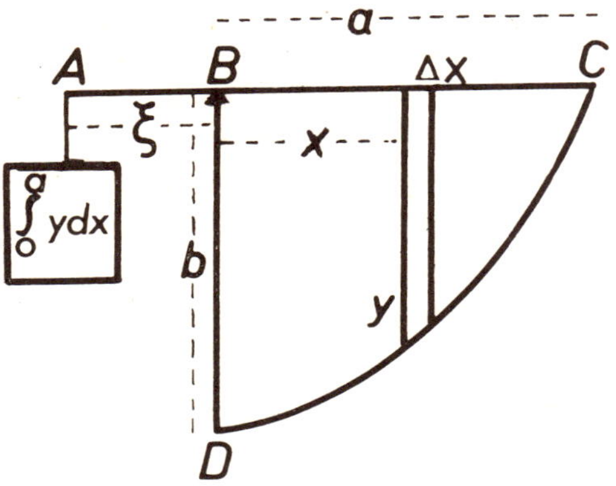

Figure 1.12.3.

Therefore, by (1.12.14),

$$\xi = \frac{\int_0^b \frac{x^2}{2}\, dy}{\int_0^a y\, dx}, \tag{1.12.16}$$

which gives the x-coordinate of the centre of gravity. The y-coordinate can be found in a similar way.

(1.12.16) is equivalent to the relation

$$\pi \int_0^b x^2\, dy = 2\pi\xi \int_0^a y\, dx, \tag{1.12.17}$$

which states that the volume obtained by revolving the area *BCD* about the axis *BD* (compare figure 1.12.3) is equal to the product of the area and the distance traversed by the centre of gravity. This is a special case of the theorem now known as ' Pappus-Guldin's theorem ', formulated by Paul Guldin in *Centrobaryca* (*1635–1641a*, vol. 2, 147) in the following way : ' the product of a rotating quantity and the path of rotation [that is, the circumference of the circle traversed by the centre of gravity], is equal to the quantity generated by the rotation '. The theorem is also found in Book VII of Pappus's *Collections*, but it may be a later addition (see, for example, Ver Eecke *1932a*).

1.13. *Concluding remarks*

The examples given in sections 1.5–1.8 and 1.10–1.12 illustrate the remark in the introductory section 1.1 about the special character of the infinitesimal methods in the period 1630–1660. In the case of the methods of quadrature we saw that they were all naturally founded on the conception of an area as an infinitesimal sum. However, mathematicians differed in their ways of approaching the problems raised by that concept. And not only were the methods of the various mathematicians based on different ideas ; some of them also developed different methods, each one adapted to solve special problems of quadrature.

Some of the methods of solving tangent or normal problems led to fixed rules—of which the most general one was Hudde's rule for determining the sub-tangent to an algebraic curve—while others only suggested a procedure. The ideas behind the methods differed widely. Descartes used an argument about the number of points of intersection between a circle and the curve ; Fermat employed similar triangles and the concept of *adequality* ; while Roberval's method was founded on an intuitive conception of instantaneous velocity and the law of parallelogram of velocities. The characteristic triangle (with sides Δx, Δy and Δs) did not explicitly play a part in the deduction of the tangent methods. Nevertheless, it was applied by (for example) Pascal in connection with a transformation of a sum (see section 2.3) ; but not until Leibniz was the importance of this triangle fully recognised.

Thus the period did not in itself bring any perception of basic concepts which were applicable to the determination of tangents as well as to quadratures. An important reason why mathematicians failed to see the general perspectives inherent in their various methods was probably the fact that to a great extent they expressed themselves in ordinary language without any special notation and so found it difficult to formulate the connections between the problem they dealt with. As an illustration we may consider one of the results achieved by the different quadrature methods outlined in the preceding sections. This result can be expressed in modern terms as

$$\int_0^a x^n \, dx = \frac{a^{n+1}}{n+1}, \tag{1.13.1}$$

where n is a natural number different from -1. The mathematicians of that period, however, could not express their result so simply ; they had to refer to areas under special parabolas. Their terminology did not prevent them from seeing connections such as that between the rectification of the parabola and the quadrature of the hyperbola, or the relation of certain inverse tangent problems to quadratures ; but it may have barred their way to a deeper insight into the meaning of these connections.

These remarks are not to be taken in the negative sense at all. It is not the task of a historian of mathematics to evaluate the work of earlier mathematicians by present mathematical standards, nor to emphasise the inadequacy of their concepts as compared to modern ones. On the contrary, a historian of mathematics ought to enter into the mode of thought of the period under consideration in order to bring out the development of the mathematical ideas in its historical context. Briefly, it may be said that the mathematicians in the period preceding the invention of the calculus blazed the trail for its invention. They did so by employing heuristic methods, by making the geometry analytical, and by seeking methods for solving problems of quadratures and tangents.[1]

[1] I am grateful to Dr. John North of Oxford University for correcting some of my linguistic mistakes, and to Dr. D. T. Whiteside of Cambridge University for his valuable comments on the manuscript.

Chapter 2

Newton, Leibniz and the Leibnizian Tradition

H. J. M. Bos

2.1. *Introduction and biographical summary*

The starting-point of this chapter is the ' invention ', or rather ' inventions ', of the calculus. Both Newton (in 1664–1666) and Leibniz (in 1675) created, independently of each other, an infinitesimal calculus. Their inventions were very different in concepts and style, but each contains so much of what we now recognise as essential to the calculus that the expression ' invention of the calculus ' is justified in both cases. I go on to consider the subsequent development of the calculus till about 1780. In this development the Leibnizian type of calculus with differentials and integrals proved more successful than the Newtonian fluxional calculus ; therefore I concentrate on the former.

Many great and lesser mathematicians were involved in the development of the calculus in the period covered by this chapter. I shall restrict myself to those who played the prime roles in the story : Isaac Newton, Lucasian professor of mathematics at Cambridge and later Master of the Mint in London ; Gottfried Wilhelm Leibniz, historian and scientist at the ducal court of Hanover ; Jakob Bernoulli, professor of mathematics at Basle ; his brother Johann Bernoulli, younger by thirteen years, who after a professorate at Groningen succeeded Jakob in Basle in 1705 ; Guillaume François Marquis de l'Hôpital, a French nobleman living by private means, and an able mathematician eagerly interested in the new developments in infinitesimal methods ; and finally Leonhard Euler, who studied with Johann Bernoulli and then entered a career in the typically 18th-century scientific institutions, the academies. He was professor at the St. Petersburg (now Leningrad) Academy from 1730 to 1741 and from 1766 till his death ; in the intervening years he served the Berlin Academy as professor.

Many of the great ideas that were to make Isaac Newton famous in mathematics and natural science came to him in the years 1664–1666.

At that time he was a graduate student at Trinity College, Cambridge, but for some time during those two years he lived in Lincolnshire, staying away from Cambridge for fear of the Plague (compare Whiteside *1966a*). His ideas on gravity, which he was to work out later and present to the world in his famous *Principia* (*1687a*), date from that period, as well as his theory of colours, published in the treatise *Opticks* in 1704, the binomial series theorem and his fluxional calculus, which we shall discuss in more detail in section 2.2.

As with gravity and colours, publication of these mathematical ideas in print was long delayed. Newton did compose several accounts of his findings in infinitesimal calculus. In October 1666 he summarised the discoveries of the fruitful two years in a tract on fluxions (*1666a*); in 1669 he wrote a treatise on infinite series, the *De analysi* (*1669a*), which circulated in manuscript form among members of the Royal Society; from 1671 dates a treatise on the method of fluxions and infinite series (*1671a*); and in about 1693 he composed a treatise on the quadrature of curves (*1693a*). However, the 1666 tract and the treatise on the method of fluxions were not published in his lifetime, the *De analysi* was published only in 1711, and the treatise on quadratures of curves in 1704. Meanwhile the *Principia* of 1687 had brought for the first time to the general public indications of his methods in infinitesimal calculus, but these were not enough to show the scope and power of his mathematical discoveries.

About the turn of the century a fair amount was published about Leibniz's calculus (as we shall see in sections 2.5–2.8 below), and sufficient information about Newton's calculus was available to show that both men had found new methods in essentially the same mathematical field. This caused a nasty quarrel over priority, in which feelings of personal and national pride combined with insufficient insight in the mathematics involved (at least in the case of the lesser participants in the debate) to create a distasteful muddle of misunderstandings and insinuations which has only been cleared up through patient historical research in the present century. The net result of the historical research is that Leibniz found his calculus later than Newton and independently of him, and that he published it earlier.

In 1669 Newton had succeeded Isaac Barrow as Lucasian professor, but in the 1690s he grew dissatisfied with his position at Cambridge. He visited London often, to attend meetings of the Royal Society, of which he was a fellow from 1672, and to be present at sessions of Parliament as a member for the Cambridge University constituency. He moved finally to London in 1696 when he was offered the office of Warden of the Mint. In 1703 he became president of the Royal Society, a post which he held till his death. His position as the most eminent British scientist was further emphasised by a knighthood in 1705.

By the 1710s so much on the fluxional calculus was in print that the method was taken up and applied by others. However, this further development of the Newtonian type of calculus remained restricted to Great Britain, and it did not achieve much. Reasons of the lack of success lie in the isolation from the Continental developments in analysis because of the priority dispute, in the lack of mathematicians in Britain of sufficient stature to really develop Newton's calculus, and in an over-stressed loyalty to Newton's conception of the calculus and to his notations, which were less versatile than Leibniz's.

On the Continent Leibniz's inventions gave rise to a much more intense development, to whose origins in the 1670s we now turn.

Before Leibniz entered the service of the house of Hanover in 1676 he had spent four years in Paris on a diplomatic mission, which left him ample time to pursue his interest in mathematics, the sciences, history, philosophy and many other things. He met many French philosophers and made two visits to London to the Royal Society. The Paris years were his formative period. When he arrived in 1672 his knowledge of mathematics was slight, despite the fact that he had published a small tract on combinatorics. He was trained in law at the university of his home town of Leipzig. In Paris Christiaan Huygens, who lived there at that time, recognised Leibniz's mathematical abilities and guided his first studies in the higher mathematics. Leibniz's 'growth to mathematical maturity' (see Hofmann *1949a*) was indeed impressive; it led to his discovery of the calculus in 1675, the elaboration of that calculus in the following years and its publication in 1684–1686. He contributed to other branches of mathematics as well, for instance to algebra (solvability of equations, determinants) and to nearly all other fields of human learning, including religion, politics, history, physics, mechanics, technology, mathematics, geology, linguistics and natural history. Many of his results were not immediately published and became known only gradually, through correspondence (from his comparative intellectual isolation in Hanover Leibniz corresponded with over a thousand scholars), through publication of short articles in journals (he was one of the founders of the first scientific journal in Germany, the *Acta eruditorum*), and later through the publication of his manuscripts, most of which he kept and which are now stored at the Leibniz archive in Hanover.

Leibniz's publication of his calculus in two articles in the *Acta* of 1684 and 1686 did not provoke great commotion in mathematical circles. The articles were rather short, and they were marred by misprints and in places deliberately obscure, so that it is in fact surprising that in the following decade they were understood at all.

Jakob and Johann Bernoulli studied the articles from 1687, and by 1690 they showed, in articles published in the *Acta*, that they had mastered the Leibnizian symbolism and its use. They both started a correspondence with Leibniz ; the contact between Johann and Leibniz was especially intensive and productive. After 1690 a stream of articles in the *Acta* and in other journals, written by the Bernoullis and Leibniz and later joined by l'Hôpital and others, showed the learned world that the new calculus was something to be reckoned with.

However, for people of lesser mathematical calibre than the Bernoullis, it would have been very difficult actually to learn the calculus from these articles. What was wanted was a proper textbook of the calculus. Such a textbook came, though only of the differential calculus, in 1696 with l'Hôpital's *Analyse des infiniment petits pour l'intelligence des lignes courbes* (' Analysis of infinitely small quantities for the understanding of curved lines ' : *1696a*).

The Marquis de l'Hôpital was introduced to the calculus by Johann Bernoulli, who, after finishing his medical studies in 1690, had travelled to Paris, where he impressed learned circles by a method to determine, by means of differentials, the curvature of arbitrary curves—a problem which by the methods of Cartesian analytic geometry was well nigh unsolvable. l'Hôpital was most impressed and asked Bernoulli to give him, for a good fee, lectures on the new method. Bernoulli accepted and the lectures were given, in Paris and at the country chateau of the Marquis. They were written out and both men kept copies. After about a year Bernoulli left Paris but agreed to continue instructing l'Hôpital by letter. In fact the agreement was that Bernoulli, for a handsome monthly salary, would answer all l'Hôpital's questions concerning mathematics, would send him all his mathematical discoveries and would give no one else access to these findings (see Bernoulli *Correspondence*, 144) ; a most curious and hardly honourable agreement which put Bernoulli's originality strictly in l'Hôpital's service. From the start Bernoulli did not quite keep to the letter of the contract, and l'Hôpital soon realised that he could not bind a brilliant mathematician in this way. But when in 1696 l'Hôpital published his textbook, and Bernoulli saw that most of its content was taken from his lectures with not more than a passing reference to the Marquis's indebtedness to Bernoulli, he could only be angry in silence, being bound by the contract.

Later, after l'Hôpital's death, Johann Bernoulli did try to get his part in the *Analyse* acknowledged, but by that time his credibility in priority questions had become very low because of open quarrels on such matters with his brother. Jakob Bernoulli was a rather introverted personality, but he was sensitive to praise from members of the mathematical community and he resented being overshadowed by his brilliant younger brother. Johann, on the other hand, liked his own success too much to spare his brother's feelings. So there appeared insinuating remarks in articles, and later a quarrel exploded and went on quite openly. Johann Bernoulli's claim to much of the content of the *Analyse* was found to be justified only when in 1921 the manuscript of his Paris lectures on the differential calculus was found (see Johann Bernoulli *1924a*).

However strained their mutual relations, through the writings of these men the Leibnizian calculus became known and proved its power. By the first decade of the 18th century other mathematicians devoted themselves to the new calculus, such as Jakob Hermann, Pierre Varignon, Niklaus Bernoulli (a nephew) and Daniel Bernoulli (son of Johann). The family Bernoulli continued to yield famous mathematicians throughout the 18th century.

In these early days the new calculus consisted mainly of rather loosely connected methods, and problems solved by these methods. The man who reshaped the Leibnizian calculus into a soundly organised body of mathematical knowledge was Leonhard Euler. Euler was the central figure of continental mathematics in the middle years of the 18th century. He published an enormous number of books and articles on mathematics, mechanics, optics, astronomy, navigation, hydrodynamics, technical matters such as artillery and shipbuilding, and very many other topics. He maintained this impressive productivity despite losing the sight of one eye in 1735 and becoming completely blind in 1766. His position at the academies involved him in many other tasks besides scientific research, such as advice on the performance of new inventions as fire-engines and pumps, and on technological enterprises like canal-building and the construction of water-works in the park of the royal palace *Sans Souci* of Prussia's Frederick the Great.

Euler's greatest influence on the calculus and on analysis in general was through his great textbooks, in which he gave analysis a definitive form, which it was to keep until well into the 19th century. These textbooks, written in Latin, were : *Introductio ad analysin infinitorum* (' Introduction to the analysis of infinites ' : *1748a*), *Institutiones calculi differentialis* (' Textbooks on the differential calculus ' : *1755b*), and *Institutiones calculi integralis* (' Textbooks on the integral calculus ' : *1768–1770a*).

These were the men who created the calculus and shaped the Leibnizian tradition in analysis. In sections 2.3–2.8 I shall describe the mathematics involved, but first I shall devote the next section to an overview of the Newtonian calculus.

2.2. *Newton's fluxional calculus*

As was mentioned above, Newton's main mathematical discoveries in the infinitesimal calculus date from 1664 to 1666. (For a detailed account of his achievements in this period, see Newton *Papers*, vol. 1, 145–154, and *Works*$_2$, vol. 1, viii–xiii.) Autodidactically he quickly acquired adequate knowledge of existing theories in the field, benefitting especially from reading Descartes's *La géométrie* in van Schooten's edition with commentaries, and from the works of Wallis. Starting from these studies he developed in these fruitful two years his *fluxional calculus*.

In Newton's discoveries, complex, deep and many-sided as they are, a number of central themes may be distinguished. These are: series expansions, algorithms, the inverse relationship of differentiation and integration, the conception of variables as moving in time, and the doctrine of prime and ultimate ratios. Although these themes are interconnectedly present in almost all of his studies in the infinitesimal calculus, I shall deal with them separately.

Newton valued *power-series expansions* very highly, because they provide a means to reduce the analytical formulae of curves to a form in which all terms simply consist of a constant times a power of the variable. Thus transcendental curves (admitting no algebraic equation), as well as algebraic curves with complicated equations, can be represented by much simpler equations (be it with an infinite number of terms). Newton saw that this has two great advantages. Firstly, series expansion makes it possible to apply rules and algorithms which are defined for simple equations only, to a much wider range of curves. In particular, the relation

$$\int x^n \, dx = \frac{1}{n+1} x^{n+1}, \qquad (2.2.1)$$

which was known in various forms by the 1660s (see sections 1.10 and 1.11) can be used, in combination with power-series expansions, to provide series expressions for the quadratures of almost all curves. Secondly, series expansion provides a ready means for the approximation and simplification of formulae through the discarding of higher-order terms—a feature which he used with virtuosity in his applications of his mathematical methods to physical problems.

Newton's most famous series expansion is the 'binomial theorem', which he found in the winter of 1664–1665 and which states that the well-known binomial expansion for integer powers n,

$$(a+x)^n = a^n + \frac{n}{1} a^{n-1}x + \frac{n(n-1)}{1\,.\,2} a^{n-2}x^2 + \ldots + x^n, \qquad (2.2.2)$$

can be generalised for fractional powers $\alpha = p/q$, in which case the right hand side of

$$(a+x)^{\alpha} = a^{\alpha} + \frac{\alpha}{1} a^{\alpha-1}x + \frac{\alpha(\alpha-1)}{1\,.\,2} a^{\alpha-2}x^2 + \ldots \qquad (2.2.3)$$

is an infinite series. He found the theorem in connection with the problem of squaring the circle $y=(1-x^2)^{1/2}$. He compared the formulae $(1-x^2)^0$, $(1-x^2)^{1/2}$, $(1-x^2)^{2/2}$, $(1-x^2)^{3/2}$, $(1-x^2)^{4/2}$, ... The first, third, fifth, . . . formulae involve no root, and therefore the quadratures of the corresponding curves are easily found :

$$\left.\begin{aligned} &\text{quadrature of } y=(1-x^2)^0 \text{ is } x, \\ &\text{quadrature of } y=(1-x^2)^{2/2} \text{ is } x-\tfrac{1}{3}x^3, \\ &\text{quadrature of } y=(1-x^2)^{4/2} \text{ is } x-\tfrac{2}{3}x^3+\tfrac{1}{5}x^5. \end{aligned}\right\} \qquad (2.2.4)$$

On examining the coefficients in these expansions, Newton noted that the denominators are the odd numbers 1, 3, 5, 7, . . . and that the numerators are, in the successive expansions, {1}, {1, 1}, {1, 2, 1}, {1, 3, 3, 1}, . . ., that is, the numbers in the ' Pascal triangle ', which he knew could be expressed for successive integral values of n as

$$\left\{1,\ n,\ \frac{n(n-1)}{1\,.\,2},\ \frac{n(n-1)(n-2)}{1\,.\,2\,.\,3},\ \ldots\right\}.$$

He then guessed that, by analogy, the same expressions would apply for *fractional* values of n. When $n=\frac{1}{2}$ this yields :

quadrature of $y=(1-x^2)^{1/2}$ is

$$x - \frac{\frac{1}{2}x^3}{3} - \frac{\frac{1}{8}x^5}{5} - \frac{\frac{1}{16}x^7}{7} - \frac{\frac{5}{128}x^9}{9} - \ldots\,. \qquad (2.2.5)$$

He then saw that this procedure of guessing, or ' interpolating ', expansions such as (2.2.5) from the scheme of the series (2.2.4) could be applied to the equations of the curves as well as to their quadratures, and in this way he found that

$$(1-x^2)^{1/2} = 1 - \tfrac{1}{2}x^2 - \tfrac{1}{8}x^5 - \tfrac{1}{16}x^6 - \tfrac{5}{128}x^8 - \ldots\,. \qquad (2.2.6)$$

Not satisfied with the reliability of the interpolation procedure, he checked (2.2.6) in two ways. He showed that the product of the right hand side of (2.2.6) with itself yields $1-x^2$ (that is, all further coefficients in the product series are zero), and he saw that a common method of root extraction known as the ' galley method ', applied formally to $1-x^2$, yields the same series. In the same way as with root extraction, he used the algorithm of long division to obtain series expansions, for instance,

$$\frac{1}{1+x}=1-x+x^2-x^3+x^4-\ldots, \quad (2.2.7)$$

which provided the quadrature of the hyperbola $y=1/(1+x)$. He also obtained (2.2.7) by assuming that the binomial expansion applied when $n=-1$.

In the *De analysi* (*1669a*), in which these methods of series expansions are explained and used, Newton also provides a general rule to compute, for a given polynomial equation

$$\sum a_{ij}x^i y^j=0 \quad (2.2.8)$$

between x and y, the first coefficients of the pertaining series

$$y=\sum b_i x^i \quad (2.2.9)$$

(*Papers*, vol. 2, 222–247).

Both in the way that Newton found the binomial theorem and in the application of series expansions in general, the relation, which we now write as

$$\int x^n\, dx=\frac{1}{n+1}\, x^{n+1}, \quad (2.2.10)$$

plays an important role. He mentioned this 'quadrature of simple curves' at the outset of his *De analysi*: 'RULE 1. If $ax^{m/n}=y$, then will $(na/(m+n))x^{(m+n)/n}$ equal the area ABD' (*ibid.*, 206–207; see figure 2.2.1). Later in that treatise he gave a general procedure (of which rule 1 is a direct consequence) for finding the relation between the quadrature of a curve (as AD in figure 2.2.1) and its ordinate. The procedure makes it clear that Newton recognised the *inverse relationship of integration and differentiation* (although, of course, he did not use these terms). He explains his method by means of an example, from which, however, the generality of the procedure is quite clear. He

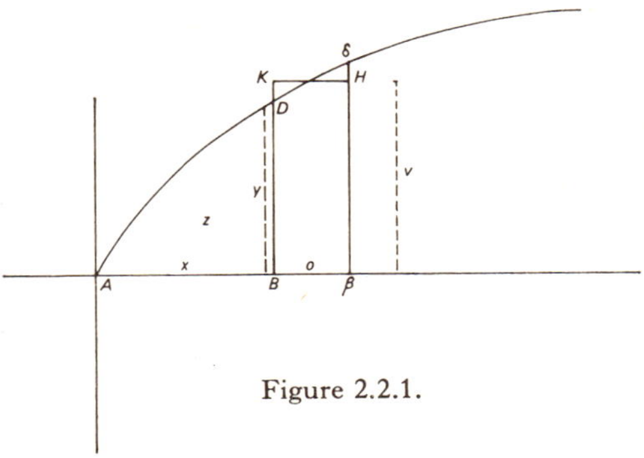

Figure 2.2.1.

proceeds as follows (*ibid.*, 242–245). In figure 2.2.1 let area $ABD=z$, $BD=y$ and $AB=x$; let further $B\beta=o$ and let $BK=v$ be chosen such that area $BD\delta\beta=$ area $BKHB=ov$. Consider, as example, the curve for which

$$z=\tfrac{2}{3}x^{3/2}, \tag{2.2.11}$$

that is (removing roots to get a polynomial equation),

$$z^2=\tfrac{4}{9}x^3; \tag{2.2.12}$$

then also

$$(z+ov)^2=\tfrac{4}{9}(x+o)^3, \tag{2.2.13}$$

from which

$$z^2+2zov+o^2v^2=\tfrac{4}{9}(x^3+3x^2o+3xo^2+o^3). \tag{2.2.14}$$

Now by removing the terms without o, which are equal on both sides from (2.2.12), and dividing the remainder by o, we obtain

$$2zv+ov^2=\tfrac{4}{9}(3x^2+3xo+o^2). \tag{2.2.15}$$

Now Newton takes $B\beta$ ' infinitely small ', in which case, as the figure suggests, $v=y$ and the terms with o vanish :

$$2zy=\tfrac{4}{3}x^2. \tag{2.2.16}$$

Inserting the value of z from (2.2.11), he obtains

$$y=x^{1/2}. \tag{2.2.17}$$

Clearly the procedure is applicable to all polynomial relations between x and z. It consists in essence of calculating the derivative (in this case the y) for any algebraic function z of x.

Newton saw clearly that the problem of quadratures was to be approached in this inverse way: by calculating y for all manner of algebraic z, he could find all manner of curves (y, x) which are quadrable. Indeed, he calculated many such quadrable curves, writing them together in extensive lists, which are thus nothing less than the first tables of integrals (compare *Papers*, vol. 1, 404–411).

The essential element in the foregoing procedure is the substitution of ' small ' corresponding increments o and ov for x and z in the equation. In studies on the determination of maxima and minima, tangents and curvature, Newton had extensively made use of this method, and he had worked out various ***algorithms*** for these problems, by which he could calculate the slope of the tangent or the curvature in any point of an algebraic curve. (In modern terms, he had developed algorithms to determine the derivative of any algebraic function.) Later he reformulated these algorithms and their proofs in terms of fluents and fluxions, and we shall come back to them after discussing these concepts.[1]

[1] Compare, for instance, Newton *1671a*, in *Papers*, vol. 3, 72–73.

The terms ' fluents ' and ' fluxions ' indicate Newton's conception of variable quantities in analytical geometry : he saw these as ' flowing quantities ', that is, *quantities that change with respect to time.* Thus, when considering the curve of figure 2.2.1, he would conceive the point D as moving along the curve, while correspondingly the ordinate y, the abscissa x, the quadrature z or any other variable quantity connected with the curve would increase or decrease, or in general change or ' flow '. He called these flowing quantities ' fluents ' (as opposed to the constant quantities occurring in the figure or in the problem at hand), and he called their rate of change with respect to time their ' fluxion '. In his earlier researches he indicated fluxions by separate letters ; in *1671a* he introduced the dot-notation, where the fluxions of the fluents x, y, z are $\dot{x}$, $\dot{y}$, $\dot{z}$ respectively.

It should be remarked that the way in which the fluents vary with time is arbitrary. Newton often makes, for simplicity, an additional assumption about the movement of the variables, supposing that one of the variables, say x, moves uniformly, so that $\dot{x}=1$. Such assumptions can be made because the values of the fluxions themselves are not of interest but rather their ratio, such as $\dot{y}/\dot{x}$, which gives the slope of the tangent. By this conception of quantities moving in time Newton thought himself able to solve the foundational difficulties inherent in considering ' small ' corresponding increments of variables, which are so small that we may discard them, and yet are not equal to zero, as we want to divide through by them. In his approach to this problem, his theory of *prime and ultimate ratios*, which we shall discuss in section 2.10, his conception of flowing quantities is essential ; through this conception he comes very near to a use of limits as foundation of the calculus.

We now return to the *algorithms* mentioned above. The corresponding increments of variables, can be expressed in terms of fluxions : let o now be an infinitesimal element of time, then the corresponding increments of the fluents x, y, z, . . . are $\dot{x}o$, $\dot{y}o$, $\dot{z}o$, . . . respectively. The ratio of $\dot{y}$ to $\dot{x}$ can now be determined in a way which is evident in the following example, which Newton gives himself in *1671a* (*Papers*, vol. 3, 79–81). Let a curve be given with equation

$$x^3-ax^2+axy-y^3=0. \tag{2.2.18}$$

Substituting $x+\dot{x}o$ and $y+\dot{y}o$ for x and y respectively yields

$$\begin{aligned}(x^3+3\dot{x}ox^2+3\dot{x}^2o^2x+\dot{x}^3o^3)-(ax^2+2a\dot{x}ox+a\dot{x}^2o^2)\\ +(axy+a\dot{x}oy+a\dot{y}ox+a\dot{x}\dot{y}o^2)\\ -(y^3+3\dot{y}oy^2+3\dot{y}^2o^2y+\dot{y}^3o^3)=0.\end{aligned} \tag{2.2.19}$$

Deleting $x^3-ax^2+axy-y^3$ as equal to zero from (2.2.18), dividing through by o and discarding the terms in which o is left, yields

$$3\dot{x}x^2-2a\dot{x}x+a\dot{x}y+a\dot{y}x-3\dot{y}y^2=0, \qquad (2.2.20)$$

from which the ratio of $\dot{y}$ and $\dot{x}$ is easily obtained :

$$\frac{\dot{y}}{\dot{x}}=\frac{3x^2-2ax+ay}{3y^2-ax}. \qquad (2.2.21)$$

We note that the numerator and the denominator in the result are (apart from a sign) the partial derivatives f_x and f_y of $f(x, y)=x^3-ax^2+axy-y^3$, the left hand side of the equation of the curve. Thus

$$\frac{\dot{y}}{\dot{x}}=-\frac{f_x}{f_y}. \qquad (2.2.22)$$

Indeed, this relation is implicit in the algorithms which, as we mentioned before, Newton worked out for problems of tangents, maxima and minima, and curvature. He even at one time introduced special notations in this connection (see *Papers*, vol. 1, 289–294), writing $\mathscr{X}$ for the left hand side of the equation of the curve (with the right hand side zero). He then wrote $\cdot\mathscr{X}$ and $\mathscr{X}\cdot$ for what we would write as xf_x and yf_y respectively (the so-called 'homogeneous partial derivatives'), using further symbols for homogeneous higher-order partial derivatives occurring in connection with curvature. However, the connection of Newton's $\cdot\mathscr{X}$ and $\mathscr{X}\cdot$ with modern partial derivatives should not be considered without some qualifications ; he defined them formally as modifications of the formula $\mathscr{X}$, and he did not explicitly view $\mathscr{X}$ as a function of two variables which assumes also other values than the zero in the equation.

With these algorithms, and further finesses which we cannot go into here, Newton was able to solve what he formulated as one of the two fundamental problems in infinitesimal calculus : given the fluents and their relations, to find the fluxions.

The second problem is the converse of the first : given the relation of the fluxions, to find the relation of the fluents. Transposed in modern terminology, this means : given a differential equation, to find its solution. This of course is a much harder problem than the first. Newton did more about the problem than formulate it ; his integral tables, already mentioned, form a means toward its solution, and he also studied various individual differential equations (or rather, fluxional equations).

As we have seen in the previous section, Newton's calculus was not to have the influence which Leibniz's achieved. Therefore, within the space and organisation of this chapter, we must leave it at this short summary of the fluxional calculus and some more remarks on its foundations in section 2.10, turning now our attention to the more successful rival, the Leibnizian calculus.

2.3. *The principal ideas in Leibniz's discovery*

One of the most precious documents of the Leibniz archive at Hanover is a set of mathematical manuscripts dated 25, 26 and 29 October, and 1 and 11 November, 1675.[1] On these sheets Leibniz wrote down his thoughts, more or less as they came to him, during a study of that most important problem of 17th-century mathematics : to find methods for the quadrature of curves. In the course of these studies he came to introduce the symbols '$\int$' and 'd', to explore the operational rules which they obey in formulas, and to apply them in translating many geometrical arguments about the quadrature of curves into symbols and formulas. In short, these manuscripts contain the record of Leibniz's 'invention' of the calculus. We will discuss them in more detail below, but first we will mention three principal ideas which guided him in those fateful studies in 1675.

The first principal idea was a philosophical one, namely Leibniz's idea of a *characteristica generalis*, a general symbolic language, through which all processes of reason and argument could be written down in symbols and formulas ; the symbols would obey certain rules of combination which would guarantee the correctness of the arguments. This idea guided him in much of his philosophical thinking ; it also explains his great interest in notation and symbols in mathematics and in general his endeavour to translate mathematical statements and methods into formulas and algorithms. Thus, in studying the geometry of curves, he was interested in methods rather than in results, and especially in ways to transform these methods into algorithms performable with formulas. In short, he was looking for a *calculus* for infinitesimal-geometrical problems.

The second principal idea concerned difference sequences. In studying sequences $a_1, a_2, a_3, \ldots$, and the pertaining difference sequences $b_1 = a_1 - a_2$, $b_2 = a_2 - a_3$, $b_3 = a_3 - a_4, \ldots$, Leibniz had noted that

$$b_1 + b_2 + \ldots + b_n = a_1 - a_{n+1}. \tag{2.3.1}$$

This means that difference sequences are easily summed, an insight which he put to good use in solving a problem which Huygens suggested

[1] They are discussed in Hofmann *1949a*, and an English translation is given in Child *1920a*.

to him in 1672: to sum the series $\frac{1}{1}+\frac{1}{3}+\frac{1}{6}+\frac{1}{10}+\frac{1}{15}+\ldots$, the denominators being the so-called 'triangular numbers' $r(r+1)/2$. He found that the terms can be written as differences,

$$\frac{2}{r(r+1)}=\frac{2}{r}-\frac{2}{r+1}, \tag{2.3.2}$$

and hence

$$\sum_{r=1}^{n}\frac{2}{r(r+1)}=2-\frac{2}{n+1}. \tag{2.3.3}$$

In particular, the series, when summed to infinity has sum 2. This result motivated him to study a whole scheme of related sum and difference sequences, which he put together in his so-called 'harmonic triangle' (figure 2.3.1), in which the oblique rows are successive difference sequences, so that their sums can be easily read off from the scheme (Leibniz *Writings*, vol. 5, 405: compare Hofmann *1949a*, 12; *1974a*, 20).

$$\begin{array}{ccccccccccccc}
 & & & & & & 1 & & & & & & \\
 & & & & & \frac{1}{2} & & \frac{1}{2} & & & & & \\
 & & & & \frac{1}{3} & & \frac{1}{6} & & \frac{1}{3} & & & & \\
 & & & \frac{1}{4} & & \frac{1}{12} & & \frac{1}{12} & & \frac{1}{4} & & & \\
 & & \frac{1}{5} & & \frac{1}{20} & & \frac{1}{30} & & \frac{1}{20} & & \frac{1}{5} & & \\
 & \frac{1}{6} & & \frac{1}{30} & & \frac{1}{60} & & \frac{1}{60} & & \frac{1}{30} & & \frac{1}{6} & \\
\frac{1}{7} & & \frac{1}{42} & & \frac{1}{105} & & \frac{1}{140} & & \frac{1}{105} & & \frac{1}{42} & & \frac{1}{7} \\
\cdot & & \cdot & & \cdot & & \cdot & & \cdot & & \cdot & & \cdot
\end{array}$$

Figure 2.3.1.

Leibniz's 'harmonic triangle'. The numbers in the n-th row are

$$\left[(n-1)\binom{n}{k}\right]^{-1}.$$

Summations can be read off from the scheme as, for example:

$$\frac{1}{3}+\frac{1}{12}+\frac{1}{30}+\frac{1}{60}+\frac{1}{105}+\ldots=\frac{1}{2}.$$

These results were not exactly new, but they did make Leibniz aware that the forming of difference sequences and of sum sequences are mutually inverse operations. This principal idea became more significant when he transposed it to geometry. The curve in figure 2.3.2 defines a sequence of equidistant ordinates y. If their distance is 1, the sum of the y's is an approximation of the quadrature of the curve, and the difference of two successive y's yields approximately the slope of the pertaining tangent. Moreover, the smaller the unit 1 is chosen, the better the approximation. Leibniz concluded that if the unit could

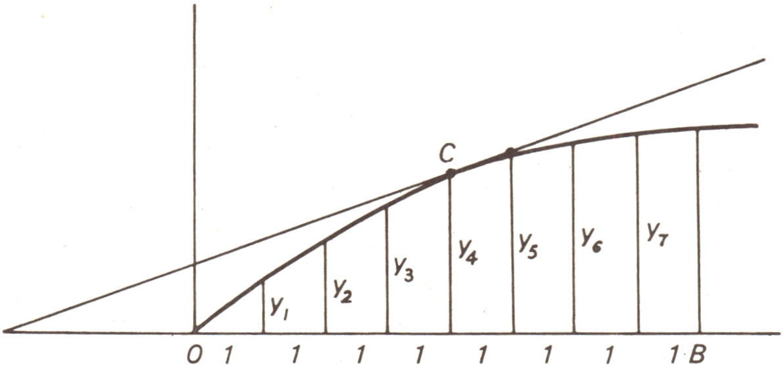

Figure 2.3.2.

be chosen *infinitely small*, the approximations would become exact : in that case the quadrature would be equal to the sum of the ordinates, and the slope of the tangent would be equal to the difference of the ordinates. In this way, he concluded from the reciprocity of summing and taking differences that the determination of quadratures and tangents are also mutually inverse operations.

Thus Leibniz's second principal idea, however vague as it was in about 1673, suggested already an infinitesimal calculus of sums and differences of ordinates by which quadratures and tangents could be determined, and in which these determinations would occur as inverse processes. The idea also made plausible that, just as in sequences the determination of differences is always possible but the determination of sums is not, so in the case of curves the tangents are always easily to be found, but not so the quadratures.

The third principal idea was the use of the ' characteristic triangle ' in transformations of quadratures. In studying the work of Pascal, Leibniz noted the importance of the small triangle *cc′d* along the curve in figure 2.3.3, for it was (approximately) similar to the triangles formed

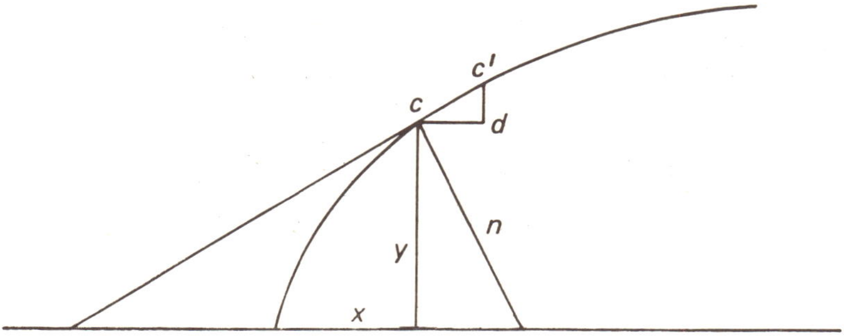

Figure 2.3.3.

by ordinate, tangent and sub-tangent, or ordinate, normal and sub-normal. The configuration occurs in many 17th-century mathematical works ; Pascal's use of it concerned the circle. Leibniz saw its general use in finding relations between quadratures of curves and other quantities like moments and centres of gravity. For instance, the similarity of the triangles yields $cc' \times y = cd \times n$; hence

$$\sum cc' \times y = \sum cd \times n. \qquad (2.3.4)$$

The left hand side can be interpreted as the total moment of the curve arc with respect to the x-axis (the moment of a particle with respect to an axis is its weight multiplied by its vertical distance to the axis), whereas the right hand side can be interpreted as the area formed by plotting the normals along the x-axis.

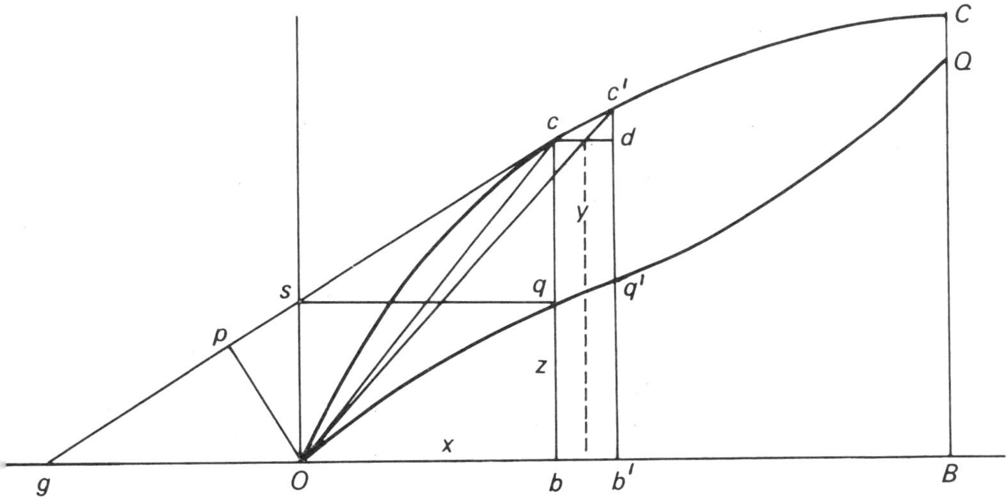

Figure 2.3.4.

As an example of Leibniz's use of the characteristic triangle, here is his derivation of a special transformation of quadratures which he called 'the transmutation' and which, for good reasons, he valued highly (compare Hofmann *1949a*, 32–35 (*1974a*, 54–60), and Leibniz *Writings*, vol. 5, 401–402). In figure 2.3.4 let the curve $Occ'C$ be given, with characteristic triangle cdc' at c. Its quadrature $\mathcal{Q} = \widehat{OCB}$, the sum of the strips $bcc'b'$, can also be considered as the sum of the triangles Occ' supplemented by the triangle OBC :

$$\mathcal{Q} = \sum \Delta Occ' + \Delta OBC. \qquad (2.3.5)$$

Now

$$\Delta Occ' = \tfrac{1}{2}cc' \times Op$$
$$= \tfrac{1}{2}cd \times Os$$

(since the characteristic triangle cdc' is similar to ΔOsp)

$$= \tfrac{1}{2}bqq'b'. \qquad (2.3.6)$$

Now for each c on $Occ'C$ we can find the corresponding q by drawing the tangent, determining s and taking $bq = Os$. Thus we form a new curve $Oqq'Q$, and we have from (2.3.5):

$$\mathcal{Q} = \tfrac{1}{2}\,(\text{quadrature } Oqq'Q) + \Delta OCB. \qquad (2.3.7)$$

This is Leibniz's transmutation rule which, through the use of the characteristic triangle, yields a transformation of the quadrature of a curve into the quadrature of another curve, related to the original curve through a process of taking tangents. It can be used in those cases where the quadrature of the new curve is already known, or bears a known relation to the original quadrature. Leibniz found this for instance to be the case with the general parabolas and hyperbolas (see section 1.3), for which the rule gives the quadratures very easily. He also applied his transmutation rule to the quadrature of the circle, in which investigation he found his famous arithmetical series for π:

$$\frac{\pi}{4} = 1 - \tfrac{1}{3} + \tfrac{1}{5} - \tfrac{1}{7} + \tfrac{1}{9} - \tfrac{1}{11} + \ldots. \qquad (2.3.8)$$

The success of the transmutation rule also convinced him that the analytical calculus for problems of quadratures which he was looking for would have to cover transformations such as this one by appropriate symbols and rules.

The transmutation rule as Leibniz discovered it in 1673 belongs to the style of geometrical treatment of problems of quadrature which was common in the second half of the 17th century. Similar rules and methods can be found in the works of Huygens, Barrow, Gregory and others. Barrow's *Lectiones geometricae* (*1670a*), for instance, contain a great number of transformation rules for quadratures which, if translated from his purely geometrical presentation into the symbolism and notation of the calculus, appear as various standard alogrithms of the differential and integral calculus. This has even been used (by J. M. Child in his *1920a*) as an argument to give to Barrow, rather than Newton or Leibniz, the title of inventor of the calculus. However, this view can be sustained only when one disregards completely the effect of the translation of Barrow's geometrical text into analytical formulas. It is the very possibility of the analytical expression of methods, and hence the understanding of their logical coherence and generality, which was the great advantage of Newton's and Leibniz's discoveries.

It is appropriate to illustrate this advantage by an example. To do this, I shall give a translation, with comments, of Leibniz's transmutation rule into analytical formulas.

The ordinate z of the curve $Oqq'Q$ is, by construction,

$$z = y - x\frac{dy}{dx} \tag{2.3.9}$$

(note the use of the characteristic triangle). The transmutation rule states that, for $OB = x_0$,

$$\int_0^{x_0} y\, dx = \tfrac{1}{2}\int_0^{x_0} z\, dx + \tfrac{1}{2}x_0 y_0. \tag{2.3.10}$$

Inserting z from (2.3.9), we find

$$\int_0^{x_0} y\, dx = \tfrac{1}{2}\int_0^{x_0}\left(y - x\frac{dy}{dx}\right) dx + \tfrac{1}{2}x_0 y_0$$

$$= \tfrac{1}{2}\int_0^{x_0} y\, dx - \tfrac{1}{2}\int_0^{x_0} x\frac{dy}{dx}\, dx + \tfrac{1}{2}x_0 y_0.$$

Hence

$$\int_0^{x_0} y\, dx + \int_0^{x_0} x\frac{dy}{dx}\, dx = x_0 y_0, \tag{2.3.11}$$

so that we recognise the rule as an instance of 'integration by parts'.

Apart from the indication of the limits of integration $(0, x_0)$ along the $\int$-sign, the symbolism used above was found by Leibniz in 1675. The advantages of that symbolism over the geometrical deduction and statement of the rule are evident: the geometrical construction of the curve $Oqq'Q$ is described by a simple formula (2.3.9), and the formalism carries the proof of the rule with it, as it were. (2.3.11) follows immediately from the rule

$$d(x\,y) = x\, dy + y\, dx. \tag{2.3.12}$$

These advantages, manipulative ease and transparency through the rules of the symbolism, formed the main factors in the success of Leibniz's method over its geometrical predecessors.

But we have anticipated in our story. So we return to October 1675, when the transmutation rule was already found but not yet the new symbolism.

2.4. *Leibniz's creation of the calculus*

In the manuscripts of 25 October–11 November 1675 we have a close record of studies of Leibniz on the problem of quadratures. We find him attacking the problem from several angles, one of these being the use of the Cavalierian symbolism ' omn.' in finding, analytically (that is, by manipulation of formulas) all sorts of relations between quadratures. ' Omn. ' is the abbreviation of ' omnes lineae ', ' all lines ' ; in section 1.10 it was represented by the symbol ' $\mathcal{O}$ '.

A characteristic example of Leibniz's investigations here is the following. In a diagram such as figure 2.4.1 he conceived a sequence of

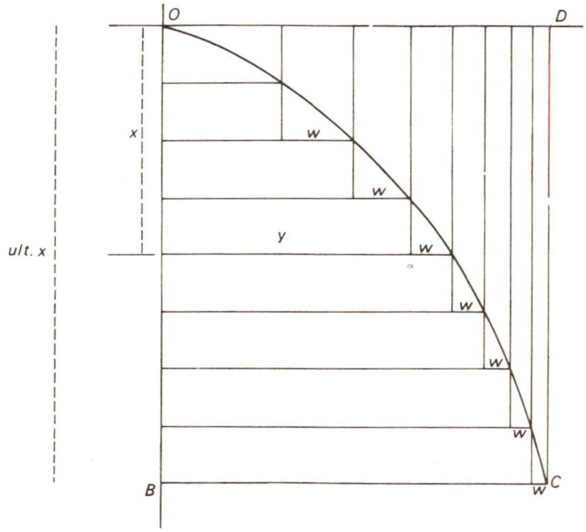

Figure 2.4.1.

ordinates y of the curve $\overset{\frown}{OC}$; the distance between successive ordinates is the (infinitely small) unit. The differences of the successive ordinates are called w. OBC is then equal to the sum of the ordinates y. The rectangles like $w \times x$ are interpreted as the moments of the differences w with respect to the axis OD (moment = weight × distance to axis). Hence the area OCD represents the total moment of the differences w. $\overset{\frown}{OCB}$ is the complement of $\overset{\frown}{OCD}$ within the rectangle $ODCB$, so that Leibniz finds that ' The moments of the differences about a straight line perpendicular to the axis are equal to the complement of the sum of the terms ' (Child *1920a*, 20). The ' terms ' are the y. Now w is the difference sequence of the sequence of ordinates y ; hence, conversely, y is the sum-sequence of the w's, so that we may eliminate y

and consider only the sequence *w* and its sum-sequences, which yields : ' and the moments of the terms are equal to the complement of the sum of the sums ' (*ibid.*). Here the ' terms ' are the *w*. Leibniz writes this result in a formula using the symbol ' Omn. ' for what he calls ' a sum '. We give the formula as he gave it, and we add an explanation under the accolades ; $\sqcap$ is his symbol for equality, ' ult. x ' stands for *ultimus* x, the last of the x, that is, OB, and he uses overlining and commas where we would use brackets (*ibid.*) :

$$\underbrace{\text{omn. }\overline{xw}}_{\substack{\text{moments of}\\ \text{the terms } w}} \sqcap \underbrace{\underbrace{\text{ult. } x, \overline{\text{omn. } w},}_{\text{total}} - \underbrace{\overline{\text{omn. omn. } w}}_{\substack{\text{sum of the sums}\\ \text{of the terms}}}}_{\substack{\text{complement of the sum of the sums}\\ \text{of the terms}}} \qquad (2.4.1)$$

(Compare the form of (2.4.1) with that of (2.3.11).) Immediately he sees the possibility to obtain from this formula, by various substitutions, other relations between quadratures. For instance, substitution of $xw = a$, $w = a/x$ yields

$$\text{omn. } a \sqcap \text{ult. } x, \text{ omn. } \frac{a}{x} - \text{omn. omn. } \frac{a}{x}, \qquad (2.4.2)$$

which he interprets as an expression of the ' sum of the logarithms in terms of the quadrature of the hyperbola ' (*ibid*,. 71). Indeed, omn. a/x is the quadrature of the hyperbola $y = a/x$, and this quadrature is a logarithm, so that omn. omn. a/x is the sum of the logarithms.

We see in these studies an endeavour to deal analytically with problems of quadrature through appropriate symbols and notations, as well as a clear recognition and use of the reciprocity relation between difference and sum sequences. In a manuscript of some days later, these insights are pushed to a further consequence. Leibniz starts here from the formula (2.4.1), now written as

$$\text{omn. } xl \sqcap x \text{ omn. } l - \text{omn. omn. } l. \qquad (2.4.3)$$

He stresses the conception of the sequence of ordinates with infinitely small distance : ' . . . l is taken to be a term of the progression, and x is the number which expresses the position or order of the l corresponding to it ; or x is the ordinal number and l is the ordered thing ' (*ibid.*, 80). He now notes a rule concerning the dimensions in formulas like (2.4.3), namely that omn., prefixed to a line, such as l, yields an area (the quadrature) ; omn., prefixed to an area, like xl, yields a solid, and so on. Such a law of dimensional homogeneity was well-known from the Cartesian analysis of curves, in which the formulas must consist of terms all of the same dimension. (In (2.4.3) all terms are of three dimensions, in $x^2+y^2=a^2$ all terms are of two dimensions ; an expression like a^2+a is, if dimensionally interpreted, unacceptable, for it would express the sum of an area and a line.)

This consideration of dimensional homogeneity seems to have suggested to Leibniz to use a single letter instead of the symbol ' omn. ', for he goes on to write : ' It will be useful to write $\int$ for omn, so that $\int l$ stands for omn. l or the sum of all l's ' (*ibid.*). Thus the $\int$-sign is introduced. ' $\int$ ' is one of the forms of the letter ' s ' as used in script (or italics print) in Leibniz's time : it is the first letter of the word *summa*, sum. He immediately writes (2.4.3) in the new formalism :

$$\int xl = x\int l - \int\int l\,; \qquad (2.4.4)$$

he notes that

$$\int x = x^2/2 \quad \text{and} \quad \int x^2 = x^3/3, \qquad (2.4.5)$$

and he stresses that these rules apply for ' series in which the differences of the terms bear to the terms themselves a ratio that is less than any assigned quantity ' (*ibid.*), that is, series whose differences are infinitely small.

Some lines further on we also find the introduction of the symbol ' d ' for differentiating. It occurs in a brilliant argument which may be rendered as follows : The problem of quadratures is a problem of summing sequences, for which we have introduced the symbol ' $\int$ ' and for which we want to elaborate a *calculus*, a set of useful algorithms. Now summing sequences, that is, finding a general expression for $\int y$ for given y, is usually not possible, but it *is* always possible to find an expression for the differences of a given sequence. This finding of differences is the reciprocal calculus of the calculus of sums, and therefore we may hope to acquire insight in the calculus of sums by working out the reciprocal calculus of differences. To quote Leibniz's own words (*ibid.*, 82) :

> Given l, and its relation to x, to find $\int l$. This is to be obtained from the contrary calculus, that is to say, suppose that $\int l = ya$. Let $l = ya/d$; then just as $\int$ will increase, so d will diminish the dimensions. But $\int$ means a sum, and d a difference. From the given y, we can always find y/d or l, that is, the difference of the y's.

Thus the ' d '-symbol (or rather the symbol ' $1/d$ ') is introduced Because Leibniz interprets $\int$ dimensionally, he has to write the ' d ' in the denominator : l is a line, $\int l$ is an area, say ya (note the role of ' a ' to make it an area), the differences must again be lines, so we must write ' ya/d '. In fact he soon becomes aware that this is a notational disadvantage which is not outweighed by the advantage of dimensional interpretability of $\int$ and d, so he soon writes '$d(ya)$' instead of 'ya/d' and henceforth re-interprets 'd' and '$\int$' as dimensionless symbols. Nevertheless, the consideration of dimension did guide the decisive steps of choosing the new symbolism.

In the remainder of the manuscript Leibniz explores his new symbolism, translates old results into it and investigates the operational rules for $\int$ and d. In these investigations he keeps for some time to the idea that $d(uv)$ must be equal to $du\,dv$, but finally he finds the correct rule

$$d(uv) = u\,dv + v\,du. \tag{2.4.6}$$

Another problem is that he still for a long time writes $\int x$, $\int x^2$, . . . for what he is later to write consistently as $\int x\,dx$, $\int x^2\,dx$, ...

A lot of this straightening out of the calculus was still to be done after 11 November 1675 ; it took Leibniz roughly two years to complete it. Nevertheless, the manuscripts which we discussed contain the essential features of the new, the Leibnizian, calculus : the concepts of the differential and the sum, the symbols d and $\int$, their inverse relation and most of the rules for their use in formulas.

Let us summarise shortly the main features of these Leibnizian concepts (compare Bos *1974a*, 12–35). The *differential* of a variable y is the infinitely small difference of two successive values of y. That is, Leibniz conceives corresponding sequences of variables such as y and x in figure 2.4.2. The successive terms of these sequences lie infinitely close. dy is the infinitely small difference of two successive ordinates y, dx is the infinitely small difference of two successive abscissae x, which, in this case, is equal to the infinitely small distance of two successive y's. A sum (later termed 'integral' by the Bernoullis) like $\int y\,dx$ is the sum of the infinitely small rectangles $y \times dx$. Hence the quadrature of the curve is equal to $\int y\,dx$.

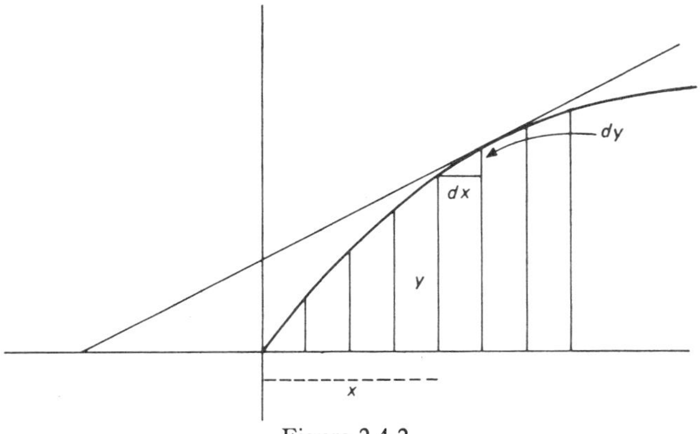

Figure 2.4.2.

Leibniz was rather reluctant to present his new calculus to the general mathematical public. When he eventually decided to do so, he faced the problem that his calculus involved infinitely small quantities, which were not rigorously defined and hence not quite acceptable in mathematics. He therefore made the radical but rather unfortunate decision to present a quite different concept of the differential which was not infinitely small but which satisfied the same rules. Thus in his first publication of the calculus, the article 'A new method for

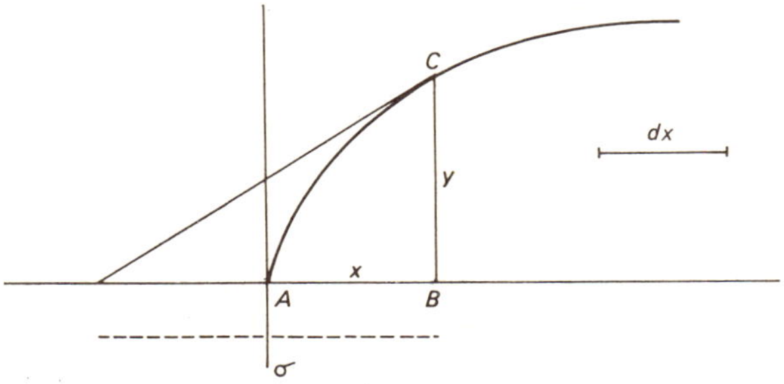

Figure 2.4.3.

maxima and minima as well as tangents' (*1684a*) in the issue for October 1684 of the *Acta*, he introduced a fixed finite line-segment (see figure 2.4.3) called dx, and he defined the dy at C as the line-segment satisfying the proportionality

$$y : \sigma = dy : dx, \tag{2.4.7}$$

σ being the length of the sub-tangent, or

$$dy = \frac{y}{\sigma}\, dx. \tag{2.4.8}$$

So defined, dy is also a finite line-segment. Leibniz presented the rules of the calculus for these differentials, and indicated some applications. In an article published two years later (*1686a*) he gave some indications about the meaning and use of the $\int$-symbol. This way of publication of his new methods was not very favourable for a quick and fruitful reception in the mathematical community. Nevertheless, the calculus was accepted, as we shall see in the following sections.

2.5. *l'Hôpital's textbook version of the differential calculus*

Leibniz's publications did not offer an easy access to the art of his new calculus, and neither did the early articles of the Bernoullis. Still, a good introduction appeared surprisingly quickly, at least to the differential calculus, namely l'Hôpital's *Analyse* (*1696a*).

As a good textbook should, the *Analyse* starts with definitions, of variables and their differentials, and with postulates about these differentials. The definition of a differential is as follows: 'The infinitely small part whereby a variable quantity is continually increased or decreased, is called the differential of that quantity' (ch. 1). For further explanation l'Hôpital refers to a diagram (figure 2.5.1), in which,

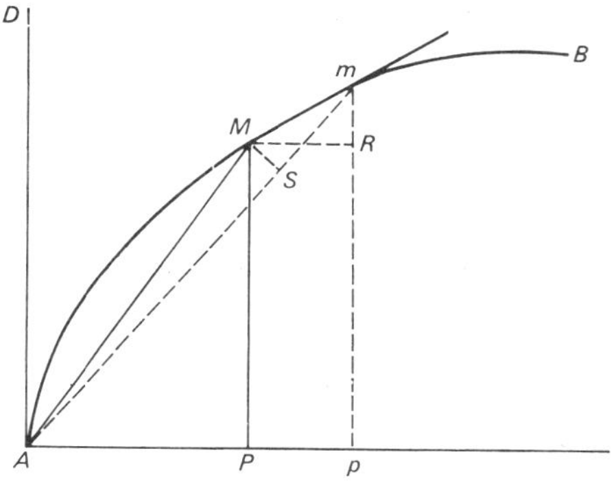

Figure 2.5.1.

with respect to a curve AMB, the following variables are indicated: abscissa $AP=x$, ordinate $PM=y$, chord $AM=z$, arc $\widehat{AM}=s$ and quadrature $\widehat{AMP}=\mathcal{Q}$. A second ordinate pm 'infinitely close' to PM is drawn, and the differentials of the variables are seen to be: $dx=Pp$, $dy=mR$, $dz=Sm$, $ds=Mm$ (the chord Mm and the arc Mm are taken to coincide) and $d\mathcal{Q}=MPpm$. l'Hôpital explains that the 'd' is a special symbol, used only to denote the differential of the variable written after it. The small lines Pp, mR, . . . in the figure have to be considered as 'infinitely small'. He does not enter into the question whether such quantities exist, but he specifies, in the two postulates, how they behave (*ibid.*):

> Postulate I. Grant that two quantities, whose difference is an infinitely small quantity, may be used indifferently for each other: or (which is the same thing) that a quantity, which is increased or decreased only by an infinitely smaller quantity, may be considered as remaining the same.

This means that AP may be considered equal to Ap (or $x=x+dx$), MP equal to mp ($y=y+dy$), and so on.

The second postulate claims that a curve may be considered as the assemblage of an infinite number of infinitely small straight lines, or equivalently as a polygon with an infinite number of sides. The first postulate enables l'Hôpital to derive the rules of the calculus, for instance :

$$\left.\begin{aligned} d(xy) &= (x+dx)(y+dy)-xy \\ &= x\,dy+y\,dx+dx\,dy \\ &= x\,dy+y\,dx \end{aligned}\right\} \quad (2.5.1)$$

' because $dx\,dy$ is a quantity infinitely small, in respect of the other terms $y\,dx$ and $x\,dy$: for if, for example, you divide $y\,dx$ and $dx\,dy$ by dx, we shall have the quotients y. and dy, the latter of which is infinitely less than the former ' (*ibid.*, ch. 1, para. 5). l'Hôpital's concept of differential differs somewhat from Leibniz's. Leibniz's differentials are infinitely small differences between successive values of a variable. l'Hôpital does not conceive variables as ranging over a sequence of infinitely close values, but rather as continually increasing or decreasing ; the differentials are the infinitely small parts by which they are increased or decreased.

In the further chapters l'Hôpital explains various uses of differentiation in the geometry of curves : determination of tangents, extreme values and radii of curvature, the study of caustics, envelopes and various kinds of singularities in curves. For the determination of tangents he remarks that postulate 2 implies that the infinitesimal part Mm of the curve in figure 2.5.2, when prolonged, gives the tangent.

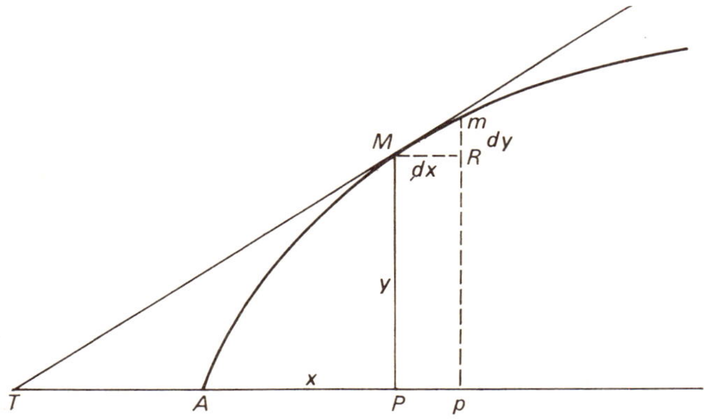

Figure 2.5.2.

Therefore $Rm : RM$, or $dy : dx$, is equal to $y : PT$, so that $PT = y(dx/dy)$, and the tangent can be constructed once we have determined $y\,dx/dy$ (*ibid.*, ch. 2, para. 9) :

> Now by means of the difference of the given equation you can obtain a value of dx in terms which all contain dy, and if you multiply by y and divide by dy you will obtain an expression for the sub-tangent PT entirely in terms of known quantities and free from differences, which will enable you to draw the required tangent MT.

To explain this, consider for example the curve $ay^2 = x^3$. The 'difference of the equation' is derived by taking differentials left and right:

$$2ay\, dy = 3x^2\, dx. \tag{2.5.2}$$

dx can now be expressed in terms of dy:

$$dx = \frac{2ay}{3x^2}\, dy. \tag{2.5.3}$$

Hence

$$PT = \frac{y\, dx}{dy} = y\, \frac{2ay}{3x^2} = \frac{2ay^2}{3x^2}, \tag{2.5.4}$$

which provides the construction of the tangent.

The 'difference of the equation' is a true differential equation, namely an equation between differentials. l'Hôpital considers expressions like 'dy/dx' actually as quotients of differentials, not as single symbols for derivatives.

2.6. *Johann Bernoulli's lectures on integration*

In 1742, more than fifty years after they were written down, Johann Bernoulli published his lectures to l'Hôpital on 'the method of integrals' in his collected works (Bernoulli *1691a*), stating in a footnote that he omitted his lectures on differential calculus as their contents were now accessible to everyone in l'Hôpital's *Analyse*. His lectures may be considered as a good summary of the views on integrals and their use in solving problems which were current around 1700.

Bernoulli starts with defining the integral as the inverse of the differential: the integrals of differentials are those quantities from which these differentials originate by differentiation. This conception of the integral—the term, in fact, was introduced by the Bernoulli brothers—differs from Leibniz's, who considered it as a sum of infinitely small quantities. Thus, in Leibniz's view, $\int y\, dx = \mathcal{Q}$ means that the sum of the infinitely small rectangles $y \times dx$ equals $\mathcal{Q}$; for Bernoulli it means that $d\mathcal{Q} = y\, dx$.

Bernoulli states that the integral of $ax^p\, dx$ is $(a/(p+1))x^{p+1}$, and he gives various methods usable in finding integrals; among them is the method of substitution, explained by several examples, such as the following (*1691a*, lecture 1):

> Suppose that one is required to find the integral of
>
> $$(ax+xx)\,dx\sqrt{(a+x)}.$$
>
> Substituting $\sqrt{(a+x)}=y$ we shall obtain $x=yy-a$, and thus $dx=2y\,dy$, and the whole quantity
>
> $$(ax+xx)\,dx\sqrt{(a+x)}=2y^6\,dy-2ay^4\,dy.$$
>
> It is now easy and straightforward to integrate this expression; its integral is $\frac{2}{7}y^7-\frac{2}{5}ay^5$ and, after substituting the value of y, we find the integral to be $\frac{2}{7}(x+a)^3\sqrt{(x+a)}-\frac{2}{5}a(x+a)^2\sqrt{(x+a)}$.

The principal use of the integral calculus, Bernoulli goes on to explain, is in the squaring of areas. For this the area has to be considered as divided up into infinitely small parts (strips, triangles, or quadrangles in general as in figure 2.6.1). These parts are the differentials of the areas; one has to find an expression for them 'by means of determined letters and only one kind of indeterminate' (*ibid.*, lecture 2), that is, an expression $f(u)\,du$ for some variable u. The required area is then equal to the integral $\int f(u)\,du$.

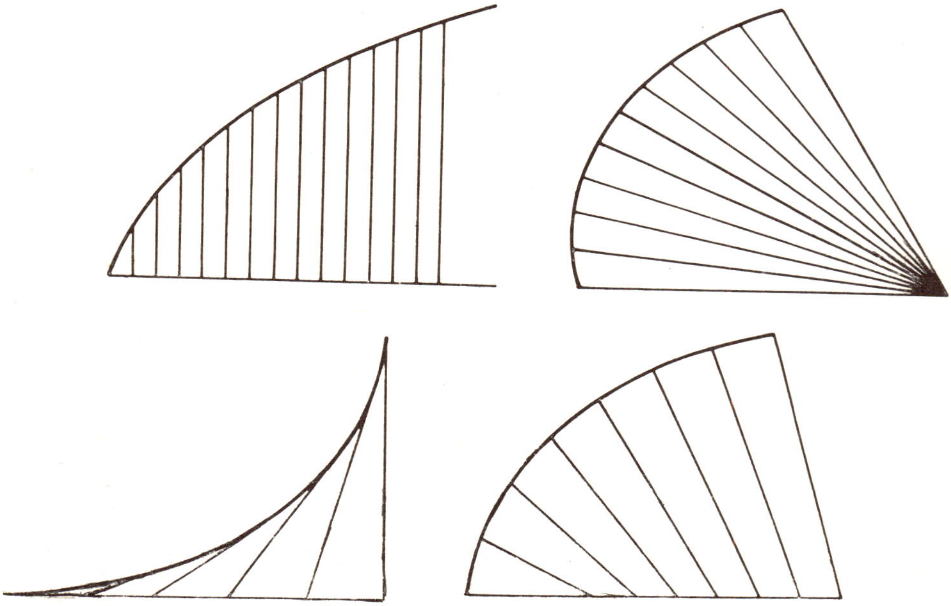

Figure 2.6.1.

The further use of the method of integrals is in the so-called ' inverse method of tangents ' (*ibid.*, lecture 8). The method, or rather the type of problem which Bernoulli has in mind here, originated in the 17th century ; it concerns the determination of a curve from a given property of its tangents. He teaches that the given property of the tangents has to be expressed as an equation involving differentials, that is, a differential equation. From this differential equation the equation of the curve itself has to be found by means of the method of integrals. His first example is (*ibid.*, lecture 8 ; see figure 2.6.2) :

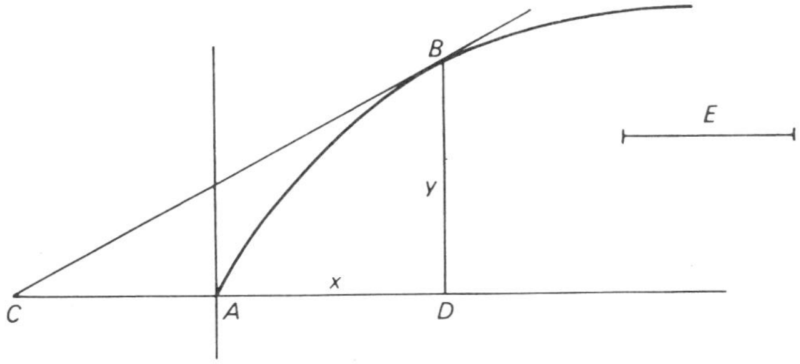

Figure 2.6.2.

It is asked what kind of curve AB it is whose ordinate BD is always the middle proportional between a given line E and the subtangent CD (that is, $E : BD = BD : CD$). Let $E = a$, $AD = x$, $DB = y$, then $CD = yy : a$. Now $dy : dx = y : CD = yy/a$ (that is, $CD = yy/a$) ; therefore we get the equation $y\,dx = yy\,dy : a$ or $a\,dx = y\,dy$; and after taking integrals on both sides, we get $ax = \frac{1}{2}yy$ or $2ax = yy$; which shows that the required curve AB is the parabola with parameter $= 2a$.

In the further lectures Bernoulli solves many instances of inverse tangent problems. He devotes considerable attention to the question how to translate the geometrical or often mechanical data of the problem into a treatable differential equation. The problems treated in his lectures concern, among other things, the rectification (computation of the arc-length) of curves, cycloids, logarithmic spirals, caustics (linear foci occurring when light-rays reflect or refract on curved surfaces), the catenary (see section 2.8 below), and the form of sails blown by the wind.

2.7. *Euler's shaping of analysis*

In the (about) 50 years after the first articles on the calculus appeared, the Leibnizian calculus developed from a loose collection of methods for problems about curves into a coherent mathematical discipline: Analysis. Though many mathematicians, such as Jean le Rond d'Alembert, Alexis Clairaut, the younger generation of Bernoullis, and others, contributed to this development, it was in a large measure the work of one man: Leonhard Euler. Not only did Euler contribute many new discoveries and methods to analysis, but he also unified and codified the field by his three great textbooks mentioned already in section 2.1.

Shaping analysis into a coherent branch of mathematics meant first of all making clear what the subject was about. In the period of Leibniz, the elder Bernoullis and l'Hôpital, the calculus consisted of analytical methods for the solution of problems about curves; the principal objects were *variable geometrical quantities* as they occurred in such problems. However, as the problems became more complex and the manipulations with the formulas more intricate, the geometrical origin of the variables became more remote and the calculus changed into a discipline merely concerning formulas. Euler accentuated this transition by affirming explicitly that analysis is a branch of mathematics which deals with *analytical expressions*, and especially with *functions*, which he defined (following Johann Bernoulli) as follows: ' a function of a variable quantity is an analytical expression composed in whatever way of that variable and of numbers and constant quantities ' (*1748a*, vol. 1, para. 4). Expressions like x^n, $(b+x)^2ax$ (with constants a and b) were functions of x. Algebraic expressions in general, and also infinite series, were considered as functions. The constants and the variable quantities could have imaginary or complex values.

Euler undertook the inventorisation and classification of that wide realm of functions in the first part of his *Introduction to the analysis of infinites* (*1748a*). The *Introduction* is meant as a survey of concepts and methods in analysis and analytical geometry preliminary to the study of the differential and integral calculus. He made of this survey a masterly exercise in introducing as much as possible of analysis without using differentiation or integration. In particular, he introduced the elementary transcendental functions, the logarithm, the exponential function, the trigonometric functions and their inverses without recourse to integral calculus—which was no mean feat, as the logarithm was traditionally linked to the quadrature of the hyperbola and the trigonometric functions to the arc-length of the circle.

Euler had to use some sort of infinitesimal process in the *Introduction*, namely, the expansion of functions in power-series (through long division, binomial expansion or other methods) and the substitution of infinitely large or infinitely small numbers in the formulas. A characteristic example of this approach is the deduction of the series expansion for a^z (*1748a*, vol. 1, paras. 114–116), where he proceeds as follows. Let $a>1$, and let ω be an 'infinitely small number, or a fraction so small that it is just not equal to zero'. Then

$$a^{\omega}=1+\psi \tag{2.7.1}$$

for some infinitely small number ψ. Now put

$$\psi=k\omega \tag{2.7.2}$$

in which k depends only on a; then

$$a^{\omega}=1+k\omega \tag{2.7.3}$$

and

$$\omega=\log(1+k\omega) \tag{2.7.4}$$

if the logarithm is taken to the base a.

Euler shows that for $a=10$ the value of k can be found (approximately) from the common table of logarithms. He now writes

$$a^{i\omega}=(1+k\omega)^i \tag{2.7.5}$$

for any (real) number i, so that by the binomial expansion

$$a^{i\omega}=1+\frac{i}{1}k\omega+\frac{i(i-1)}{1\,.\,2}k^2\omega^2+\frac{i(i-1)(i-2)}{1\,.\,2\,.\,3}k^3\omega^3+\ldots. \tag{2.7.6}$$

If z is any finite positive number, then $i=z/\omega$ is infinitely large, and by substituting $\omega=z/i$ in (2.7.6) we obtain

$$a^z=a^{i\omega}=1+\frac{1}{1}kz+\frac{1(i-1)}{1\,.\,2i}k^2z^2+\frac{1(i-1)(i-2)}{1\,.\,2i\,.\,3i}k^3z^3+\ldots. \tag{2.7.7}$$

But if i is infinitely large, $(i-1)/i=1$, $(i-2)/i=1$, and so on, and we arrive at

$$a^z=1+\frac{kz}{1}+\frac{k^2z^2}{1\,.\,2}+\frac{k^3z^3}{1\,.\,2\,.\,3}+\frac{k^4z^4}{1\,.\,2\,.\,3\,.\,4}+\ldots. \tag{2.7.8}$$

The *natural logarithms* arise if a is chosen such that $k=1$. Euler gives that value of a up to 23 decimals, introduces the now familiar notation e for that number and writes (*ibid.*, para. 123):

$$e^z=1+\frac{z}{1}+\frac{z^2}{1\,.\,2}+\frac{z^3}{1\,.\,2\,.\,3}+\ldots. \tag{2.7.9}$$

In the next chapter Euler deals with trigonometric functions. He writes down the various sum-formulas and adds: 'Because $(\sin.z)^2+(\cos.z)^2=1$, we have, by factorising, $(\cos.z+\sqrt{-1}.\sin.z)(\cos.z-\sqrt{-1}.\sin.z)=1$, which factors, although imaginary, nevertheless are of immense use in comparing and multiplying arcs' (*ibid.*, para. 132).

He further finds that

$$(\cos y \pm \sqrt{-1} \sin y)(\cos z \pm \sqrt{-1} \sin z) = \cos (y+z) \pm \sqrt{-1} \sin (y+z), \qquad (2.7.10)$$

and hence

$$(\cos z \pm \sqrt{-1} \sin z)^n = \cos nz \pm \sqrt{-1} \sin nz, \qquad (2.7.11)$$

a relation usually called ' de Moivre's formula ' as it occurs already in the work of Abraham de Moivre (see Schneider *1968a*, 237–247).

By expanding (2.7.11) Euler obtains expressions for cos nz and sin nz. Now taking z to be infinitely small (so that $\sin z = z$ and $\cos z = 1$), $nz = v$ finite and hence n infinitely large, he arrives, by methods similar to those above, at

$$\cos v = 1 - \frac{v^2}{1\,.\,2} + \frac{v^4}{1\,.\,2\,.\,3\,.\,4} - \frac{v^6}{1\,.\,2\,.\,3\,.\,4\,.\,5\,.\,6} + \ldots, \qquad (2.7.12)$$

$$\sin v = v - \frac{v^3}{1\,.\,2\,.\,3} + \frac{v^5}{1\,.\,2\,.\,3\,.\,4\,.\,5} - \frac{v^7}{1\,.\,2\,.\,3\,.\,4\,.\,5\,.\,6\,.\,7} + \ldots \qquad (2.7.13)$$

(*ibid.*, para. 134). Some paragraphs later (art. 138) we find, derived by similar methods, the identities :

$$\exp(\pm v\sqrt{-1}) = \cos v + \sqrt{-1} \sin v, \qquad (2.7.14)$$

$$\cos v = \tfrac{1}{2}(\exp[v\sqrt{-1}] + \exp[-v\sqrt{-1}]), \qquad (2.7.15)$$

$$\sin v = \frac{1}{2\sqrt{-1}}(\exp[v\sqrt{-1}] - \exp[-v\sqrt{-1}]). \qquad (2.7.16)$$

Euler's *Textbooks on the differential calculus* (*1755b*) starts with two chapters on the calculus of finite differences and then introduces the differential calculus as a calculus of infinitely small differences, thus returning to a conception more akin to Leibniz's than to l'Hôpital's : ' The analysis of infinites . . . will be nothing else than a special case of the method of differences expounded in the first chapter, which occurs, when the differences, which previously were supposed finite, are taken infinitely small ' (*1755b*, para. 114). He considers infinitely small quantities as being in fact equal to zero, but capable of having finite ratios ; according to him, the equality $0\,.\,n = 0$ implies that $0/0$ may in cases be equal to n. The differential calculus investigates the values of such ratios of zeros. Euler proceeds to discuss the differentiation of functions of one or several variables, higher-order differentiation and differential equations. He also obtains the equality

$$\frac{\partial^2 V}{\partial x \partial y} = \frac{\partial^2 V}{\partial y \partial x} \qquad (2.7.17)$$

for a function V of x and y (though not using this notation, and without obtaining a fully rigorous proof ; *1755b*, paras. 288 ff.).

In his discussion of higher-order differentiation Euler gives a prominent role to the *differential coefficients*, p, q, r, . . . defined, for a function $y = f(x)$, as follows :

$$dy = p\,dx \tag{2.7.18}$$

(where p is the coefficient with which to multiply the constant dx in order to obtain dy, so that p is again a function of x) ; and similarly,

$$dp = q\,dx \text{ (so that } ddy = q(dx)^2\text{)}, \tag{2.7.19}$$

$$dq = r\,dx \text{ (so that } dddy = r(dx)^3\text{)}, \ldots . \tag{2.7.20}$$

These differential coefficients are, though differently defined, equal to the first- and higher-order *derivatives* of the function f. In his textbook on the integral calculus he treats higher-order differential equations in terms of these differential coefficients, thus, in some measure, paving the way for the replacement of the differential by the derivative as fundamental concept of the calculus.

The three-volume *Textbooks on the integral calculus* (*1768–1770a*) give a magisterial close to the trilogy of textbooks. Here Euler gives a nearly complete discussion of the integration of functions in terms of algebraic and elementary transcendental functions, he discusses various definite integrals (including those now called the beta and gamma functions), and he gives a host of methods for the solution of ordinary and partial differential equations.

Apart from determining, through these textbooks, the scope and style of analysis for at least the next fifty years, Euler contributed to the infinitesimal calculus in many other ways. Two of these contributions are worth special emphasis. Firstly, he gave a thorough treatment of the calculus of variations, whose beginnings lie in the studies by the Bernoullis of the brachistochrone and of isoperimetric problems (see section 2.8 below). Secondly, he applied analysis, and indeed worked out many new analytical methods, in the context of studies in mechanics, celestial mechanics, hydrodynamics and many other branches of natural sciences, thus transforming these subjects into strongly mathematised form. In the next section I shall describe one example of each of these ways.

2.8. *Two famous problems : the catenary and the brachistochrone*

In writing the history of the calculus, it is customary to devote much attention to the fundamental concepts and methods. This tends to obscure the fact that most mathematicians spend most of their time not in contemplating these concepts and methods, but in using them to solve problems. Indeed, in the 18th century the term ' mathematics ' comprised much more than the calculus and analysis, for it ranged from arithmetic, algebra and analysis through astronomy, optics, mechanics and hydrodynamics to such technological subjects as artillery, ship-building and navigation. In this section I discuss two famous problems whose solution was made possible by the new methods of the differential and integral calculus ; in the next section I shall say something about what more was made possible through these methods.

The catenary problem

The catenary is the form of a hanging fully flexible rope or chain (the name comes from *catena*, which means ' chain '), suspended on two points (see figure 2.8.1). The interest in this curve originated with

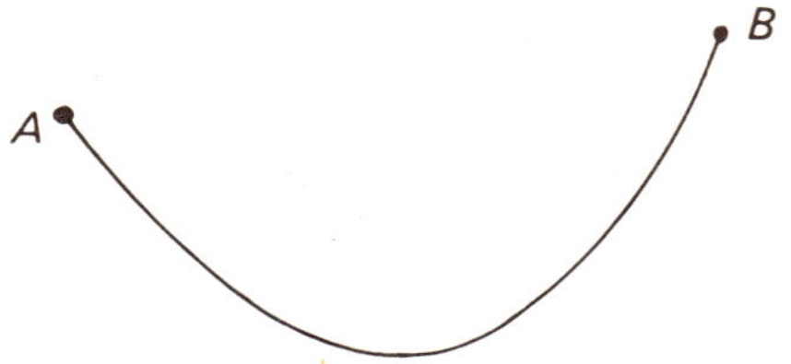

Figure 2.8.1.

Galileo, who thought that it was a parabola. Young Christiaan Huygens proved in 1646 that this cannot be the case. What the actual form was remained an open question till 1691, when Leibniz, Johann Bernoulli and the then much older Huygens sent solutions of the problem to the *Acta* (Jakob Bernoulli, *1690a*, Johann Bernoulli *1691b*, Huygens *1691a* and Leibniz *1691a*), in which the previous year Jakob Bernoulli had challenged mathematicians to solve it. As published, the solutions did not reveal the methods, but through later publications of manuscripts these methods have become known. Huygens applied with great virtuosity the by then classical methods of 17th-century infinitesimal mathematics, and he needed all his ingenuity to reach a satisfactory solution. Leibniz and Bernoulli, applying the new calculus, found the solutions in a much more direct way. In fact, the catenary was a test-case between the old and the new style in the study of curves, and only because the champion of the old style was a giant like Huygens, the test-case can formally be considered as ending in a draw.

A short summary of Johann Bernoulli's solution (he recapitulated it in his *1691a*, lectures 12 and 36), may provide an insight in how the

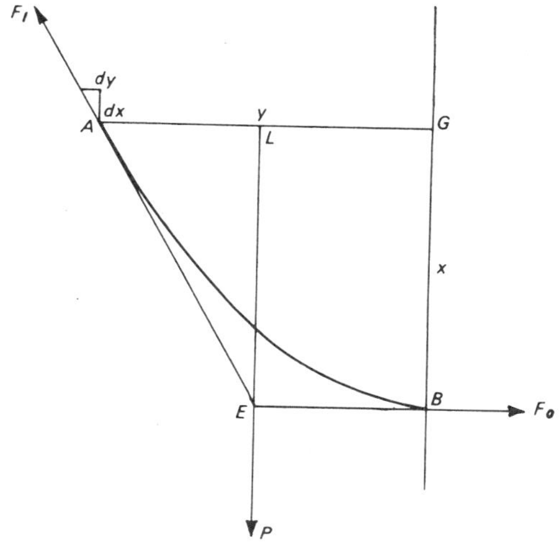

Figure 2.8.2.

new method was applied. In figure 2.8.2 let AB be part of the catenary. Using arguments from mechanics, he inferred that the forces F_0 and F_1, applicable in B and A to keep the part AB of the chain in position, are the same (in direction and quantity) as the forces required to keep the weight P of the chain AB in position, suspended as a mass at E on weightless cords AE and BE, which are tangent to the curve as in the figure. Moreover, the force F_0 at B does not depend on the choice of the position of A along the chain. P may be put equal to the length s of the chain from B to A; $F_0=a$, a constant; and from composition of forces we have

$$P : F_0 = s : a = dx : dy. \tag{2.8.1}$$

Hence

$$\frac{dy}{dx} = \frac{a}{s}. \tag{2.8.2}$$

This is the differential equation of the curve, though in a rather intractable form as x and y occur implicitly in the arc-length s. Through skilful manipulation Bernoulli arrives at the equivalent differential equation

$$dy = \frac{a\,dx}{\sqrt{(x^2-a^2)}}. \tag{2.8.3}$$

I shall not follow his argument here in detail, but the equivalence can be seen by going backwards and calculating ds from (2.8.3):

$$ds = \sqrt{(dy^2+dx^2)} = \sqrt{\left(\frac{a^2}{x^2-a^2}+1\right)}\,dx = \frac{x\,dx}{\sqrt{(x^2-a^2)}}. \tag{2.8.4}$$

Hence by integration

$$s = \int \frac{x\,dx}{\surd(x^2-a^2)} = \surd(x^2-a^2) = a\frac{dx}{dy}. \tag{2.8.5}$$

Through a substitution $x \to x+a$ Bernoulli reduces (2.8.3) to

$$dy = \frac{a\,dx}{\surd(x^2+2ax)}. \tag{2.8.6}$$

This substitution is needed to move the origin to B. In the differential equation (2.8.6) the variables are separated, so that the solution is

$$y = \int \frac{a\,dx}{\surd(x^2+2ax)}, \tag{2.8.7}$$

and the question is left to find out what the right hand side means. At that time, in the early 1690s, Bernoulli had not yet the analytical form of the logarithmic function at his disposal to express the integral as we would (namely, as $a \log(a+x+\surd[x^2+2ax])$). Instead he gave geometrical interpretations of the integral, namely, as quadratures of curves. He noted that the integral represents the area under the curve

$$z = \frac{a^2}{\surd(x^2+2ax)}. \tag{2.8.8}$$

But he also interpreted (through transformations which again we shall not present in detail) the integral as an area under a certain hyperbola and even as an arc-length of a parabola. By these last two interpretations, or ' constructions ' as this procedure of interpreting integrals was called, he proved that the form of the catenary ' depended on the quadrature of the hyperbola ' (we would say : involves only the transcendental function the logarithm) and with this proof the problem was, to the standards of the end of the 17th century, adequately solved.

The brachistochrone problem

If a body moves under influence of gravity, without friction or air resistance along a path γ (see figure 2.8.3), then it will take a certain time, say T_γ, to move to B starting from rest in A. T_γ depends on the form of γ. The *brachistochrone* (literally : shortest time) is the curve γ_0 from A to B for which T_γ is minimal. It can easily be seen that the fall along a straight line from A to B does not take the minimal time, so there is a problem : to determine the brachistochrone.

The problem was publicly proposed by Johann Bernoulli in the *Acta* of June 1696 (Bernoulli *1696a*) and later in a separate pamphlet. Several solutions reached the *Acta* and were published in May 1697 (Johann

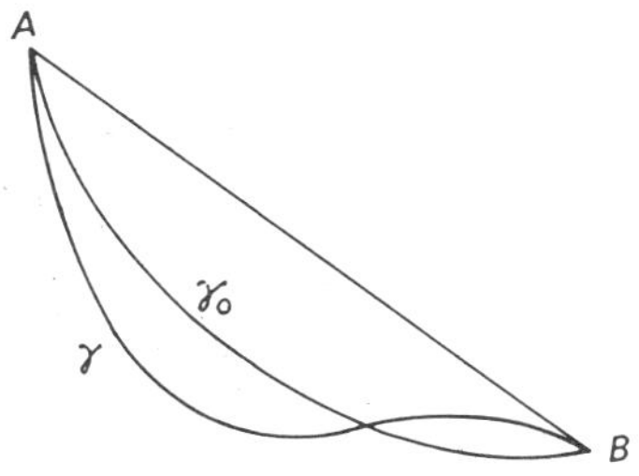

Figure 2.8.3.

Bernoulli *1697a*, l'Hôpital *1697a*, Leibniz *1697a* and Newton *1697a*; see Hofmann *1956a*, 35–36). Bernoulli's own solution used an analogy argument: he saw that the problem could be reduced to the problem of the refraction of a light-ray through a medium in which the density, and hence the refraction index, is a function of the height only. Leibniz and Jakob Bernoulli first considered the position of two consecutive straight line-segments (see figure 2.8.4) such that T_γ from P to Q is minimal. This is an extreme value problem depending on one variable and therefore solvable. Extending this to three consecutive straight segments and considering these as infinitely small, they arrived at a differential equation for the curve, which they solved. They found, as did Johann Bernoulli, that the brachistochrone is a *cycloid* (compare section 1.8) through A and B with vertical tangent at A. Newton had also reached this conclusion.

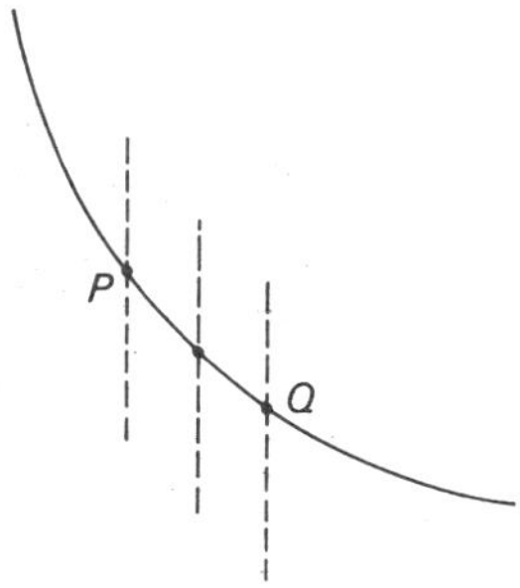

Figure 2.8.4.

The problem of the brachistochrone is very significant in the history of mathematics, as it is an instance of a problem belonging to the *calculus of variations*. It is an extreme value problem, but one in which the quantity (T_γ), whose extreme value is sought, does not depend on one or a finite number of independent variables but on the *form of a curve*.

Jakob Bernoulli proposed, as a sequel to his solution of the brachistochrone problem, further problems of this type, namely the so-called *isoperimetric problems.* In the case of the brachistochrone, the class of curves considered consists of the curves passing through A and B. In isoperimetric problems one considers curves with prescribed length. For instance, it could be asked to find the curve through A and B with length l and comprising, together with the segment AB, the largest area (see figure 2.8.5). Jakob Bernoulli made much progress in finding methods to solve this type of problem. Euler unified and generalised these methods in his treatise *1744a*, thus shaping them into a separate branch of analysis. Lagrange contributed to the further development of the subject in his *1762a*, in which he introduced the concept of *variation* to which the subject owes its present name—the calculus of variat ons. On its history, see especially Woodhouse *1810a* and Todhunter *1861a*.

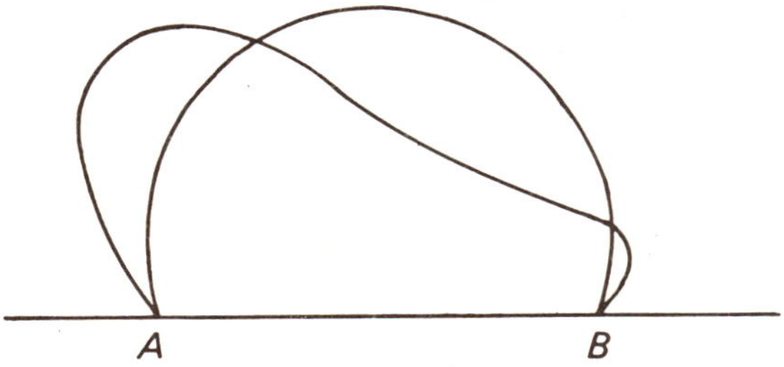

Figure 2.8.5.

2.9. *Rational mechanics*

The catenary and brachistochrone problems were two problems whose solution was made possible by the new methods. There were many more such problems, and their origins were diverse. The direct observation of simple mechanical processes suggested the problems of the form of an elastic beam under tension, the problem of the vibrating string (which Taylor, Daniel Bernoulli, d'Alembert, Euler and many others studied ; see section 3.3) and the problem of the form of a sail blown by the wind (discussed by the Bernoulli brothers in the early 1690s).

More technologically involved constructions suggested the study of pendulum motion (which Huygens initiated), the path of projectiles, and the flow of water through pipes. Astronomy and philosophy suggested the motion of heavenly bodies as a subject for mathematical treatment. Mathematics itself suggested problems too : special differential equations were generalised, types of integrals were classified (for example, elliptic integrals), and so on. Certain types of problems began rather quickly to form coherent fields with a unified mathematical approach: the calculus of variations, celestial mechanics, hydrodynamics, and mechanics in general. Somewhat later, probability (on which Jakob Bernoulli wrote a fundamental treatise *Ars conjectandi* (' The art of guessing '), which was published posthumously as *1713a*), joined this group of mathematicised sciences, or sub-fields of mathematics.

Something more should be said here about the new branches of mechanics (or ' rational mechanics ' as it was then called, to distinguish it from the study of machines), which acquired its now familiar mathematicised form in the 18th century. The basis for this mathematicisation was laid by Newton in his *Philosophiae naturalis principia mathematica* (*1687a*), in which he formulated the Newtonian laws of motion and showed that the supposition of a gravitational force inversely proportional to the square of the distance yields an appropriate description of the motion of planets as well as of the motion of falling and projected bodies here on earth. He gave here (among many other things) a full treatment of the motion of two bodies under influence of their mutual gravitational forces, several important results on the ' three-body problem ', and a theory of the motion of projectiles in a resisting medium. However, a great deal in the way of mathematicisation of these subjects still had to be done after the *Principia*. Though Newton made full use of his new infinitesimal methods in the *Principia*, he found and presented his results in a strongly geometrical style. Thus, although implicitly he set up and solved many differential equations, exactly or by approximation through series expansions, one rarely finds them written out in formulas in the *Principia*. Neither are his laws of motion expressed as fundamental differential equations to form the starting-point of studies in mechanics.

In the first half of the 18th century, through the efforts of men like Jakob, Johann and Daniel Bernoulli, d'Alembert, Clairaut and Euler, the style in this kind of study was further mathematicised—that is, the methods were transformed into the analytical methods—and they were unified through the formulation of basic laws expressed as mathematical formulas, differential equations in particular. Other fields were also tackled in this way, such as the mechanics of elastic bodies (on which Jakob Bernoulli published a fundamental article *1694a*) and hydrodynamics, on which father and son Johann and Daniel Bernoulli wrote early treatises (*1743a* and *1738a* respectively).

Great textbooks of analytic mechanics, such as Euler's *Mechanica* (*1736a*), d'Alembert's *1743a* and Lagrange's *1788a*, show a gradual process of mathematicisation of mechanics. Though Euler's *Mechanica* was strongly analytical, the formulation of Newton's laws in terms of differential equations (now termed ' Newton's equations ') occurred for the first time only in a study of Euler published in 1752 (see Truesdell *1960a*). These branches of rational mechanics were very abstract fields in which highly simplified models of reality were studied. Therefore, the results were less often applicable than one might have hoped. These studies served to develop many new mathematical methods and theoretical frameworks for natural science which were to prove fruitful in a wider context only much later. Still, the interest in the problems treated was not entirely internally derived. Thus the projectiles of artillery suggested the study of motion in a resisting medium, while the three-body problem was studied by Newton, Euler and many others, especially in connection with the motion of the moon under the influence of the earth and the sun, a celestial phenomenon which was of the utmost importance for navigation as good moon tables would solve the problem of determining a ship's position at sea (the so-called ' longitudinal problem '). Indeed, Euler's theoretical studies of this problem, combined with the practical astronomical expertise of Johann Tobias Mayer, gave navigation, in the 1760s, the first moon tables accurate enough to yield a sufficiently reliable means for determining position at sea.

Central problems in hydrodynamics were the efflux of fluid from an opening in a vessel, and the problem of the shape of the earth. The latter problem was of philosophical as well as practical importance, because Cartesian philosophy predicted a form of the earth elongated along the axis, while Newtonian philosophy, considering the earth as a fluid mass under the influence of its own gravity and centrifugal forces through its rotation, concluded that the earth should be flattened at the poles. In practice, the deviation of the surface of the earth from the exact sphere form has to be known in order to calculate actual distances from astronomically determined geographical latitude and longitude. Several expeditions were held to measure one degree along a meridian in different parts of the earth, and the findings of these expeditions finally corroborated the Newtonian view.

2.10. *What was left unsolved : the foundational questions*

The problem that was left unsolved throughout the 18th century was that of the foundations of the calculus. That there *was* a problem was well-known, and that is hardly surprising when one considers how obviously self-contradictory properties were claimed for the fundamental concept of the calculus, the differential. According to l'Hôpital's first postulate, a differential can increase a quantity without increasing it. Nevertheless, this postulate is necessary for deriving the rules of the calculus, where higher-order differentials (or powers or products of differentials) have to be discarded with respect to ordinary differentials, and similarly ordinary differentials have to be discarded with respect to finite quantities (see (2.5.1)). Also, when Bernoulli takes the differential of the area $\mathcal{Q}$ to be equal to $y\,dx$ he discards the small triangle at the top of the strip (like MmR in figure 2.5.2) because it is infinitely small with respect to $y\,dx$. Thus the differentials have necessary but apparently self-contradictory properties. This leads to the foundational question of the calculus as many mathematicians since Leibniz saw it :

FQ 1 : *Do infinitely small quantities exist ?*

Most practitioners of the Leibnizian calculus convinced themselves in some way or other that the answer to FQ 1 is 'yes', and thus they considered the rules of the calculus sufficiently proved. There is, however, a more sophisticated way of looking at the question, a way which for instance Leibniz himself adopted (see Bos *1974a*, 53–66). He had his doubts about the existence of infinitely small quantities, and he therefore tried to prove that by using the differentials as possibly meaningless symbols, and by applying the rules of the calculus, one would arrive at correct results. So his foundational question was :

FQ 2 : *Is the use of infinitely small quantities in the calculus reliable ?*
He did not obtain a satisfactory answer.

In Newton's fluxional calculus (see section 2.2) there also was a foundational problem. Newton claimed that his calculus was independent of infinitely small quantities. His fundamental concept was the *fluxion*, the velocity of change of a variable which may be considered to increase or decrease with time. In the actual use of the fluxional calculus, the fluxions themselves are not important (in fact they are undetermined), but their ratios are. Thus the tangent of a curve is found by the argument that the ratio of ordinate to sub-tangent is equal to the ratio of the fluxions of the ordinate and the abscissa respectively : $y/\sigma=\dot{y}/\dot{x}$ ($\dot{y}$ is the fluxion of y, $\dot{x}$ the fluxion of x ; see figure 2.10.1).

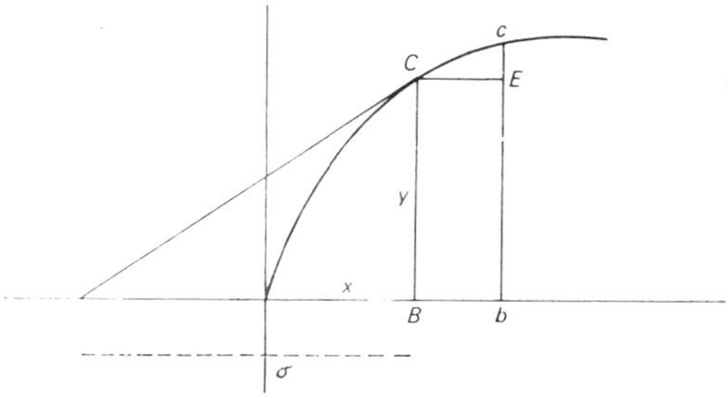

Figure 2.10.1.

He explains that the ratio of the fluxions $\dot{y}/\dot{x}$ is equal to the 'prime' or 'ultimate' ratio of the augments or decrements of y and x (see Newton *1693a*; *Works*$_2$, vol. 1, 141). That is, he conceives corresponding increments Bb of x and Ec of y, and he considers the ratio Ec/CE for Ec and CE both decreasing towards 0 or both increasing from 0. In the first case he speaks of their *ultimate ratio* which they have just when they vanish into zero or nothingness; in the latter case he speaks about their *prime ratio*, which they have when they come into being from zero or nothingness. The ratio $\dot{y}/\dot{x}$ is precisely equal to this ultimate ratio of evanescent augments, or equivalently to this prime ratio of 'nascent' augments.

Obviously there is a limit-concept implicit in this argument, but it is also clear that the formulation as it stands leaves room for doubt. For as long as the augments exist their ratio is not their ultimate ratio, and when they have ceased to exist they have no ratio. So here too is a foundational question, namely:

FQ 3: *Do prime or ultimate ratios exist?*

2.11. *Berkeley's fundamental critique of the calculus*

Most mathematicians who dealt with calculus techniques in the early 18th century did not worry overmuch about foundational questions. Indeed, it is significant that the first intensive discussion on the foundations of the calculus was not caused by difficulties encountered in working out or applying the new techniques, but by the critique of an outsider on the pretence of mathematicians that their science is based on secure foundations and therefore attains truth. The outsider was Bishop George Berkeley, the famous philosopher, and the target of his critique is made quite clear in the title of his tract *1734a*: 'The Analyst; or a Discourse Addressed to an Infidel Mathematician Wherein It Is Examined Whether the Object, Principles, and Inferences of the Modern Analysis are More Distinctly Conceived, or More Evidently Deduced, than Religious Mysteries and Points of Faith'.

As we have seen, Berkeley indeed had a point. In sharp but captivating words he exposed the vagueness of infinitely small quantities, evanescent increments and their ratios, higher-order differentials and higher-order fluxions (*1734a*, para. 4):

> Now as our Sense is strained and puzzled with the perception of Objects extremely minute, even so the Imagination, which Faculty derives from Sense, is very much strained and puzzled to frame clear Ideas of the least Particles of time, or the least Increments generated therein: and much more so to comprehend the Moments, or those

Increments of the flowing Quantities in *statu nascenti*, in their very first origin or beginning to exist, before they become finite Particles. And it seems still more difficult, to conceive the abstracted Velocities of such nascent imperfect Entities. But the Velocities of the Velocities, the second, third, fourth and fifth Velocities, &c. exceed, if I mistake not, all Humane Understanding. The further the Mind analyseth and pursueth these fugitive Ideas, the more it is lost and bewildered ; the Objects, at first fleeting and minute, soon vanishing out of sight. Certainly in any Sense a second or third Fluxion seems an obscure Mystery. The incipient Celerity of an incipient Celerity, the nascent Augment of a nascent Augment *i.e.* of a thing which hath no Magnitude : Take it in which light you please, the clear Conception of it will, if I mistake not, be found impossible, whether it be so or no I appeal to the trial of every thinking Reader. And if a second Fluxion be inconceivable, what are we to think of third, fourth, fifth Fluxions, and so onward without end ?

Further on comes the most famous quote from *The analyst* : ' And what are these Fluxions ? The Velocities of evanescent Increments ? And what are these same evanescent Increments ? They are neither finite Quantities, nor Quantities infinitely small nor yet nothing. May we not call them the Ghosts of departed Quantities ? ' (para. 35). Berkeley also criticised the logical inconsistency of working with small increments which first are supposed unequal to zero in order to be able to divide by them, and finally are considered to be equal to zero in order to get rid of them.

Of course Berkeley knew that the calculus, notwithstanding the unclarities of its fundamental concepts, led, with great success, to correct conclusions. He explained this success—which led mathematicians to believe in the certainty of their science—by a *compensation of errors*, implicit in the application of the rules of the calculus. For instance, if one determines a tangent, one first supposes the characteristic triangle similar to the triangle of ordinate, sub-tangent and tangent, which involves an error because these triangles are only approximately similar. Subsequently one applies the rules of the calculus to find the ratio dy/dx, which again involves an error as the rules are derived by discarding higher-order differentials. These two errors compensate each other, and thus the mathematicians arrive 'though not at Science, yet at Truth, For Science it cannot be called, when you proceed blindfold, and arrive at the Truth not knowing how or by what means' (*1734a*, para. 22).

2.12. *Limits and other attempts to solve the foundational questions*

Berkeley's critique started a long-lasting debate on the foundations of the calculus. Before mentioning some arguments in this debate, it may be useful to recall how in modern differential calculus the foundational question is solved. Modern calculus concerns *functions* and relates to a function f its derivative f', which is again a function, defined by means of the concept of limit :

$$f'(x) \underset{\mathrm{Df}}{=} \lim_{h \to 0} \left(\frac{f(x+h)-f(x)}{h} \right). \tag{2.12.1}$$

The preliminaries for this approach were worked out in the 18th and 19th centuries ; they played different roles in the various approaches to the foundational questions which were adopted in that period. It is instructive to list the preliminaries. They are :

(1) the idea that the calculus concerns *functions* (rather than variables) ;

(2) the choice of the *derivative* as fundamental concept of the differential calculus (rather than the differential) ;

(3) the conception of the derivative as a function ; and

(4) the concept of *limit*, in particular the limit of a function for explicitly indicated behaviour of the independent variable (thus explicitly $\lim_{h \to 0} (p(h))$, rather than merely the limit of the variable p).

Of the various approaches to the questions raised by Berkeley's critique, we have already seen the one adopted by Euler : he did conceive the calculus as concerning functions, but for him the prime concept was still the differential, which he considered as equal to zero but capable of having finite ratios to other differentials. Obviously this still leaves the foundational question QF 3 of section 2.10 unanswered. In fact, it does not seem that Euler was too much concerned about foundational questions.

Berkeley's idea of compensating errors was used by others to show that, rather than proceeding blindfold, the calculus precisely compensates equal errors and thus arrives at truth along a sure and well-balanced path. The idea was developed by Lazare Carnot among others. Another approach was due to Joseph Louis Lagrange, who supposed that for every function f and for every x one can expand $f(x+h)$ in a series

$$f(x+h) = f(x) + Ah + Bh^2 + Ch^3 + \ldots \,. \tag{2.12.2}$$

So Lagrange defined the ' derived function ' $f'(x)$ as equal to the coefficient of h in this expansion. The idea, published first in *1772a*, became somewhat influential later through Lagrange's *Théorie des fonctions analytiques* (*Functions*). As a solution of the foundational questions the idea is unsound (not every $f(x+h)$ can be so expanded, and even so there would be the question of convergence), but in other ways this approach was quite fruitful ; it conceived the calculus as a theory about functions and their derived functions, which are themselves again functions. For more details on Carnot and Lagrange, see sections 3.3 and 3.4.

Eventually the most important approach towards solving the foundational questions was the use of limits. This was advocated with respect to the fluxional calculus by Benjamin Robins (see his *1761a*, vol. 2, 49), and with respect to the differential calculus by d'Alembert. Robins and d'Alembert considered limits of variables as the limiting value which these variables can approach as near as one wishes. Thus d'Alembert explains the concept in an article *1765a* on ' Limite ' in the *Encyclopédie* which he edited with D. Diderot : ' One magnitude is said to be the *limit* of another magnitude when the second may approach the first within any given magnitude, however small, though the first magnitude may never exceed the magnitude it approaches '.

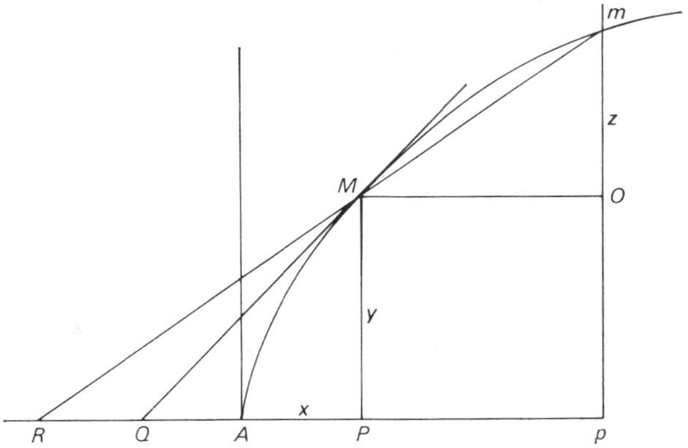

Figure 2.12.1.

In the *Encyclopédie* article ' Différentiel ' (*1764a*) d'Alembert gave a lengthy explanation, with the parabola $y^2 = ax$ as example. His argument can be summarised as follows. From figure 2.12.1 it follows that MP/PQ is the limit of mO/OM. In formulae, $mO/OM = a/(2y+z)$, and algebraically the limit of $a/(2y+z)$ is easily seen to be $a/2y$. One variable can have only one limit, hence $MP/PQ = a/2y$. Furthermore, the rules of the calculus also give $dy/dx = a/2y$, so that we must conceive dy/dx not as a ratio of differentials or as $0/0$, but as the limit of the ratio of finite differences mO/OM.

Robins and d'Alembert were not the first to formulate the concept of limit ; in fact it occurs already implicitly in ancient Greek mathematics, and later Simon Stevin for instance came very close to formulating it (see his *Works*, vol. 1, 229–231). For a very long time after Robins and d'Alembert propagated the use of this concept to solve the foundational questions, the limit approach was just one among many approaches to the problem. The reason why it took so long until the value of the limit approach was recognised lay in the fact that Robins and d'Alembert considered limits of *variables*. In that way the concept still involves much unclarity (for details, see Baron and Bos *1976a*, unit 4) which could only be removed once the limit concept was applied to *functions* under explicitly specified behaviour of the independent variable.

2.13. *In conclusion*

In the century which followed Newton's and Leibniz's independent discoveries of the calculus, analysis developed in a most impressive way, despite its rather insecure foundations, thus making possible a mathematical treatment of large parts of natural science. During these developments analysis also underwent deep changes ; for Newton and Leibniz did not invent the modern calculus, nor did they invent the same calculus. It will be useful to recall, in conclusion, the main features of both systems, their mutual differences, and their differences from the forms of calculus to which we are now used (compare Baron and Bos *1976a*, unit 3, 55–57).

Both Newton's and Leibniz's calculi were concerned with *variable quantities*. However, Newton conceived these quantities as *changing in time*, whereas Leibniz rather saw them as *ranging over a sequence of infinitely close values*. This yielded a difference in the fundamental concepts of the two calculi ; Newton's fundamental concept was the *fluxion*, the finite velocity or rate of change (with respect to time) of the variable, while Leibniz's fundamental concept was the *differential*, the infinitely small difference between successive values in the sequence.

There was also a difference between the two calculi in the conception of the *integral*, and in the role of the *fundamental theorem*. For Newton integration was *finding the fluent quantity of a given fluxion* ; in his calculus, therefore, the fundamental theorem was implied in the definition of integration. Leibniz saw integration as *summation* ; hence for him the fundamental theorem was not implied in the definition of integration, but was a consequence of the inverse relationship between summing and taking differences. However, the Bernoullis re-interpreted the Leibnizian integral as the converse of differentiation, so that throughout the 18th century the fundamental theorem was implied in the definition of integration.

Both Newton and Leibniz worked with *infinitely small quantities* and were aware of the logical difficulties inherent in their use. Newton claimed that his calculus could be given a rigorous foundation by means of the concept of *prime and ultimate ratio*, a concept akin to (but certainly not the same as) the concept of limit.

Leibniz valued *notation* very much, and his choice of symbols for the calculus proved to be a happier one than Newton's. His use of separate letters, 'd' and '$\int$', indicated the role of differentiation and integration as operators ; moreover, his symbols were incorporated into complicated formulas much more easily than were Newton's. In general, Leibniz's calculus was the more analytical ; Newton's was nearer to the geometrical figures, with accompanying arguments in prose.

These are the principal differences between the two systems. If we compare them with the modern calculus, we note three further differences. Firstly, whereas Newton's and Leibniz's calculi were concerned with *variables*, the modern calculus deals with *functions*. Secondly, the operation of differentiation is defined in the modern calculus differently from in the 18th century ; it relates to a function a derived function, or derivative, defined by means of the concept of limit. Thirdly, unlike 18th-century calculus, modern analysis has a generally accepted approach to the problem of the foundation of the calculus namely, through a definition of real numbers (instead of the vague concept of quantity which had to serve as a basis for analysis before the 1870s) and through the use of a well-defined concept of limit. The next chapter describes much of this future progress.

Chapter 3

The Emergence of Mathematical Analysis and its Foundational Progress, 1780–1880

I. Grattan-Guinness

3.1. *Mathematical analysis and its relationship to algebra and geometry*

In this book we assume that the reader has an understanding of the essentials of its subject-matter though not of its historical development. I shall make use of this fact in this opening section by describing mathematical analysis in the form often taught at first degree level (in contrast to more modern treatments, which are used in higher levels of teaching and research but are not mentioned here). By thus specifying the ' target ' straightaway I hope to make it easier for the reader to recognise as such the rather complicated steps which led to its achievement.

Textbooks of this type in mathematical analysis usually begin with a chapter on the structure of the real line (including a definition of irrational numbers), the definition and basic properties of limits, and the arithmetic of inequalities. Sometimes they also include topological ideas, such as neighbourhoods, accumulation points, open and closed sets, and covering theorems ; but as the history of these ideas belongs largely to the later chapters of this book I shall not be much concerned with them here. The textbooks then proceed to the definition of the continuity of a function, and to basic theorems on functions (of one and several variables). The differential calculus is treated extensively, starting with the definition of the derivative as the limit of the difference quotient and proceeding to mean value theorems, Taylor's series with remainder terms, and so on. The integral is interpreted as an area and defined as the limiting value of a sequence of partition-sums. An exegesis of mean value theorems, the inverse relationship with differentiation, and many other results, then follows. There may also be a section on the convergence of infinite series : its definition, necessary and sufficient conditions for convergence, and some convergence tests.

An important refinement concerns series of functions as opposed to series of constants : uniform convergence is defined, and its bearing on validating the processes of analysis is discussed. A discussion may be included of Fourier series, a particularly important example of series of functions ; their coefficients are calculated, their manner of representing a function described, and tests for their convergence established. The Fourier integral is often also introduced and its inversion theorem proved.

A prominent feature of this kind of mathematical analysis is its apparent *autonomy* of algebra and geometry. Although algebraic formulae are manipulated and geometrical diagrams drawn, they are not essential to the rigour or justification of the subject. This is to be found in the theory of limits, and in proof-methods largely based on the manipulation of (small) quantities according to the arithmetic of inequalities. Another feature of mathematical analysis is the *unification* which the underlying theory of limits brings to otherwise somewhat disparate branches of mathematics.

In this chapter I shall outline the emergence of this autonomous and unified theory in the 19th century, starting with the first gropings in the late 18th century and continuing the story up to about 1880. I have also included a few later developments, which the accounts of integration and set theory in chapters 4 and 5 considerably supplement ; but the progress of mathematical analysis from about 1860 on is so many-sided and complicated that for reasons of space I have indicated only the principal ideas which these developments took as their source. In preface, the next section discusses motivations to the subject and its development in different countries.

3.2. *Educational stimuli and national comparisons*

In chapter 2 we saw that differential equations and their solution became a very important use of the calculus. This feature was maintained in the 19th century, especially with an increasing use of partial differential equations with boundary conditions. Another stimulus to mathematical analysis, though of a different kind, was mathematical education. The calculus was taught in textbook style from the early days of l'Hôpital's textbook *1696a*, and the close association with education continued throughout the broadening of the subject into mathematical analysis. In section 0.1 I mentioned that the founding in 1795 of the *Ecole Polytechnique* in Paris was an important event in science education. Analysis was well represented in the mathematics teaching there (and also at the Paris *Ecole Normale*), and Laplace and Lagrange gave major courses. Among their assistants was Riche de Prony, a moderately good mathematician whose chief contributions to science lay in engineering. Here is a quotation from the introduction to his lecture course in ' pure and applied analysis ' at the *Ecole Polytechnique* in 1798, showing algebra, series and the calculus to be components of the subject matter (*1798a*, 215–216) :

> Mathematical analysis is divided into two parts : one which we may call *analyse déterminée,* considers the relationship between *unknown* but *invariable* quantities, and other quantities given in the question : the relationships between one and the other are made by means of *equations,* and one will explain to you all the aspects of the theory which were not demanded in the entrance examination.
>
> These researches lead us to the doctrine of series, which from many points of view is closely linked with the first part, but is so closely related to the second, which we call *analyse indéterminée,* that from a certain point of view it may be seen as the transition from one to the other.
>
> This second part considers the relationships between the quantities given by the question, and other quantities liable to an infinity of values subject to common law . . .
>
> The course in pure analysis will be concluded with differential and integral calculus, without which one cannot hope successfully to apply mathematics to the phenomena of nature.

However, it is not to Lagrange, Laplace or de Prony that we owe the synthesis of subject-matter of which I spoke earlier. This was due largely to Augustin-Louis Cauchy, also in lectures given at the *Ecole Polytechnique.* His *Cours d'analyse* (*1821a*) was based on his lecturing there ; its first (and only) published volume was sub-titled ' Algebraic analysis ', and characterised the subject in its introduction as follows :

> . . . I treat successively diverse aspects of real and imaginary functions, convergent or divergent series, the resolution of equations and of the decomposition of rational fractions. In speaking of the continuity of functions I could not dispense with making known the principal properties of infinitely small quantities, properties which serve as the basis of the infinitesimal calculus . . .
>
> As for methods, I have sought to give them all the rigour that we require in geometry, so as never to appeal to arguments drawn from the generality of algebra. Arguments of this kind, although commonly admitted, especially in the passage from convergent series to divergent series and from real quantities to imaginary expressions, are, it seems to me, only considered as inductions suitable to sound the truth sometimes but which accord little with the exactitude vaunted so much in the mathematical sciences.

This volume covered limits, the theory of continuous functions, and series. (There were some other topics there, which I shall not consider.) The calculus was reserved for his 1823 *Résumé*, where we read (*1823a*, introduction):

> This work, undertaken at the request of the council of instruction of the Ecole royale polytechnique, offers the resume of the lectures on the infinitesimal calculus which I have given at this school ... The methods that I have followed differ in several respects from those which are expounded in works of the same kind. My principal aim has been to reconcile the rigour, of which I had made myself a law in my *Cours d'analyse*, with the simplicity which results from the direct consideration of infinitely small quantities.

These and Cauchy's other textbooks set the style of mathematical analysis. They also continued the tradition that Lagrange and Laplace had started and which still holds today, whereby prominent French mathematicians present their Paris teaching in volumes of *Cours* or *Traités*. The quality of Cauchy's was so high that for some decades his successors only provided variations and expansions on his theme.

Meanwhile, pre-eminence in mathematics was passing from France to Germany. Many of the classic French texts were translated into German (especially by one C. H. Schnuse), and the next major advances in mathematical analysis came from Karl Weierstrass, both in his own work and especially in that of the students who attended his Berlin lectures from the early 1860s. It is due to this school, rather than to Cauchy's period, that mathematical analysis began to take the form that the modern textbooks now convey in their opening chapters.

How did the British fare in this subject? It was pointed out in section 2.2 that they stubbornly adhered to Newtonian fluxional methods until the early 19th century. The beginning of a change may be seen with Robert Woodhouse, but even his *Principles of analytical calculation* (*1803a*) showed a limited awareness of the foundational problems in mathematical analysis. The change to Leibniz's calculus came with a change of view among Irish mathematicians, and especially with the 'Analytical Society' founded at Cambridge in the 1810s by three undergraduates, later to become significant scientists: Charles Babbage William Herschel and George Peacock (see Dubbey *1963a*). One of their most useful contributions was to publish in 1816 an English translation of a textbook on the calculus by Sylvestre-François Lacroix. Like de Prony, Lacroix was a moderately good mathematician, and he specialised in writing widely used textbooks. To a restricted extent he formulated a theory of limits, but even this was too much for his translators, who stated in their preface (Lacroix *1816a*, iii–iv):

> The work of Lacroix, of which a translation is now presented to the Public, . . . may be considered as an abridgement of his great work on the Differential and Integral Calculus [*Treatise*], although in the demonstration of the first principles, he has substituted the method of limits of D'Alembert, in the place of the most correct and natural method of Lagrange, which was adopted in the former . . . The first twelve of the Notes were written by Mr. Peacock and were principally designed to enable the student to make use of the principle of Lagrange, adopting those statements and examples of our author, which do not involve the theory of limits.

In other words, the Analytical Society wanted to lift analysis into a state which the French were attempting to leave ! The Continental analysis did not make much impact in Britain until the early 1840s, when William Thomson, later to become Lord Kelvin but then still a teenager, began to study Fourier series and integrals. Even then British interest lay chiefly in applications to mathematical physics, where the achievements were very brilliant (see Burkhardt *1908a*, chs. 13 and 14 *passim*), rather than in foundations ; such studies concentrated more on the calculus of differential operators and the laws of their operation (see Koppelman *1972a*). Not until the work in the early years of this century by G. H. Hardy and W. H. Young were foundational studies brought fully into British education and research.

Finally in this section, a few remarks need to be made about Italy. As in Britain, Italian mathematicians tended to follow behind the results achieved in France and Germany. Prominence came to them only in the late 1870s, when Ulisse Dini and then Giuseppe Peano wrote significant textbooks, and Peano led a group of Italian mathematicians who participated in the exegesis of Weierstrassian analysis by bringing to it the new techniques of mathematical logic. We shall see evidence of the work of Peano and his school in this and the following chapters of this book.

3.3. *The vibrating string problem*

After these preliminary sections, it is time now to pick up the threads of Euler's calculus as they were left in section 2.7. We saw there that he brought considerable improvements and extensions to the Leibnizian calculus, although its foundations were still not completely clear. One of the foundational difficulties was the nature of the functions to be used in the calculus, and this problem was very prominent in the analysis of the vibrating string. To summarise briefly a rather extensive discussion and even polemic in the mid-18th century (for more details, see Truesdell *1960b*, pt. 3, or Grattan-Guinness *1970a*, ch. 1), the partial differential equation to represent the motion was agreed to be the ' wave equation '

$$\frac{\partial^2 y}{\partial x^2}=\frac{1}{c^2}\frac{\partial^2 y}{\partial t^2}, \tag{3.3.1}$$

where y is the transverse displacement at time t of the point x of a uniform string held at two points distant π apart on the x-axis. The solution was

$$y=f(x+ct)+g(x-ct), \tag{3.3.2}$$

where f and g were determined by the initial conditions. But the scope of the functions allowable in (3.3.2) was controversial ; Euler insisted that functions with corners be allowed in order to represent the plucking of the string (or rather, the mathematical idealisation of this event), but d'Alembert countered by saying that the second differential would not exist at such a point and so (3.3.1) would not apply there. However, Euler maintained that ' completely arbitrary ' functions could be admitted for f and g in (3.3.2) ; they did not have to be defined by an algebraic expression or mechanical law.

Then with *1753a* Daniel Bernoulli entered the discussion, arguing on the grounds of superposing the harmonic components of the vibration of the string that at any given time the string shows a combination of sinusoidal vibrations, so that its shape is given not by (3.3.2) but by

$$f(x)=\sum_{r=1}^{\infty} a_r \sin rx. \tag{3.3.3}$$

Bernoulli did not include terms in t, and so reduced the effectiveness of his claim. In only using sine terms he also limited the generality of his solution, and he offered no mathematical means of calculating the coefficients in (3.3.3).

In *1759a* Lagrange, at the beginning of his career, introduced a fresh analysis of the problem, defending (3.3.2) and Euler's interpretation of it but obtaining it by a different method. It is too long for description here (at one point it obtains formulae which are superficially like Bernoulli's (3.3.3)), but it makes extensive use of the manipulation of series.

Thereafter the discussion turned more to restatements of positions and to polemics, although some interesting additional factors were considered. The whole discussion is of considerable historical importance because it was a significant ' test case ' in the formation and solution of a partial differential equation to represent a physical problem. Let me mention a few aspects. Firstly, the wave equation (3.3.1) was not disputed, but it was derived in different ways. d'Alembert's and Euler's original derivations were obtained by analysing the motion of an infinitesimal slice of the string ; but Lagrange (and also Euler at one stage) used the ' n-body model ' method, where the string is interpreted as n equal and equally spaced masses joined by *weightless* cords, the motion of each mass is determined, and n is taken to infinity. Secondly, the functional solution (3.3.2) was obtained by different methods ; d'Alembert's original derivation involved a transformation of the independent variables x and t to $(x-ct)$ and $(x+ct)$, but later he also considered separation of variables. Thirdly, the definability of the functions in (3.3.2) in algebraic or geometrical terms was hotly disputed, with no clear explanation of the legitimate generality of the functions so defined being provided by anybody. Lastly, a variety of physical consequences were drawn from the solutions.

3.4. *Late-18th-century views on the foundations of the calculus*

In returning to the foundations of the calculus, I take up again the work of Lagrange. Before writing his first paper *1759a* on the vibrating string problem, he had been impressed by Berkeley's idea of the compensation of errors (see section 2.11), but from his *1772a* onwards he asserted that every function can be expanded in a ' Taylor series '

$$f(x+i)=a_0+a_1i+\frac{1}{2!}a_2i^2+\ldots, \tag{3.4.1}$$

and that its differential coefficients (which he called ' derived functions ') were *defined* as the coefficients a_0, a_1, a_2, . . . in the expansion (3.4.1). (The series is in Taylor *1715a*, 21–23 ; so named by MacLaurin in *1742a*, it was known to Newton, Leibniz and others.) He also said that they could be calculated by means which are ' redeemed from all considerations of the infinitely small, from vanishing [quantities], from limits and from fluxions, and reduced to the algebraic analysis of finite quantities '. This quotation comes from the continuation of the title of his principal textbook on analysis, his *Théorie des fonctions analytiques* (*Functions*), and shows that he thought that these coefficients were obtainable by orthodox algebraic means. The second edition of 1813 opens with these words of characteristic optimism :

> We call *function* of one or several variables any expression of calculation in which these quantities enter in any manner . . . The word *function* has been used by the first analysts to denote in general the powers of one quantity. Since then, the meaning of this word has been extended to any quantity formed in any manner from another quantity . . . When we assign any increment to the variable of a function by adding an indeterminate quantity, we can, if the function is algebraic, expand it in powers of this indeterminate

[quantity] by the ordinary rules of Algebra. The first term of the expansion will be the proposed function, which we shall call the *primitive function* ; the following terms will be formed of different functions of the same variable, multiplied by successive powers of the indeterminate [quantity]. These new functions depend uniquely on the primitive function from which they derive and may be called *derived functions* . . . We shall see in this Work that the Analysis which is popularly called *transcendental* or *infinitesimal* is at root only the Analysis of primitive and derived functions, and that the differential and integral Calculi are, speaking properly, only the calculation of these same functions.

Lagrange proved the basic theorems of the calculus, but could obtain the ' derived functions ' by algebraic means only for simple functions. Although his emphasis on power series was influential (see Yushkevich *1959a*) and his terminology ' derived function ' and notations ' $f'(x)$ ', ' $f''(x)$ ', . . . for them were widely adopted, few major mathematicians seem to have adopted his viewpoint (see Dickstein *1899a* and Yushkevich *1974a*, and note from section 3.2 above that the Analytical Society were supporters). Perhaps Lagrange himself was not completely satisfied, for during the period of his directorship of the mathematical section of the Berlin Academy a prize problem was proposed in 1784, calling for ' a clear and precise theory of what is called the infinite in mathematics ' :

> It is widely known that advanced geometry regularly employs the *infinitely great* and the *infinitely small* . . . The Academy, therefore, desires an explanation of how it is that so many correct theorems have been deduced from a contradictory supposition, together with . . . a truly mathematical principle that may properly be substituted for that of the *infinite* . . .[1]

The prize was awarded in 1786 to Simon Lhuilier, with a thoughtful if rather laborious study *1786a* of limits and their bearing on the calculus. He began by defining a limit in d'Alembert's style (see section 2.12 above) as the value from which a variable can differ by an arbitrarily small amount, and proved the product and division theorems :

$$\lim_{n\to\infty} (p_n/q_n) = (\lim_{n\to\infty} p_n)/(\lim_{n\to\infty} q_n), \quad \lim_{n\to\infty} (p_n q_n) = (\lim_{n\to\infty} p_n)(\lim_{n\to\infty} q_n) \qquad (3.4.2)$$

(assuming that the limits exist : pp. 18–24, 204–206). He also introduced the notation ' lim ', and in a revised version of his essay he mentioned the obvious though apparently hitherto unremarked fact that the passage towards the limit need not be monotonic (*1795a*, 17–21).

[1] *Nouveaux mémoires de l'Académie Royale des Sciences, Arts et Belles-Lettres*, (1784 : publ. 1786), 12–13.

Lhuilier's treatment of the differential calculus in *1786a* is very interesting, for he defined the derivative in the modern style: dy/dx is the limit of the difference quotient, and 'dy/dx' is to be read as a single sign, *not* as a ratio (pp. 24, 44). Ironically, he called it a 'differential ratio', which leads to a rather confusing clash of definition and terminology. (Similar approaches were to be adopted by the British mathematicians decades later; for example, Whewell (*1838a*, 110) and De Morgan (*1842a*, 58, 50) both defined the 'differential coefficient' this way, and took 'dy/dx' to be a whole symbol.) Lhuilier also defined the integral as the inverse of the differential ratio, and called it the 'integral ratio'.

Altogether the turgidity of much of Lhuilier's presentation greatly offsets the clarity of his doctrine of limits, and his work does not seem to have had as much influence as it deserved. More widely read at the time was a monograph by the French mathematician and engineer Lazare Carnot (father of the physicist Sadi Carnot). Curiously, the first version of this essay was also submitted for the 1786 Berlin prize (see Yushkevich *1971a*), but since Lhuilier won that competition it was published only very recently (see Carnot *1785a*). It was the revised and extended versions *Reflections*$_1$ (1797) and especially *Reflections*$_2$ (1813) which became very popular; the three are compared in Yushkevich *1971b*.

Carnot surveyed various known methods of founding the calculus, including Berkeley's doctrine of the compensation of errors and Euler's view on calculation with zeros (see sections 2.11 and 2.7 respectively). He supported the compensation of errors, but his most valuable discussions concerned the definition and use of infinitesimals, differentials and higher-order differentials. Here are some sentences from his *Reflections*$_2$: 'I call *infinitely small quantity*, any quantity which is considered as continually decreasing, such that it may be rendered as small as we want without us being obliged by that to vary those [other quantities] of which we seek the relation' (art. 14). 'We understand by the word *differential* the difference of two successive values of one variable, when we consider the system to which it belongs, in two or several successive states, of which one is regarded as fixed and the others as drawing simultaneously and continuously to the first until differing from it by as little as we wish' (art. 46). 'Instead of considering the system of variable quantities in two consecutive states we can consider it successively in two, three, four or a larger number of consecutive states, each infinitely little different from each other' (art. 68). Higher-order differentials were defined similarly, in terms of successive states of dx.

These (and other) statements seem to involve quantities being taken to limiting values; yet in *Reflections*$_2$ Carnot appears to give no formal definition of a limit, though in the previous version ' a *limit* is nothing other than a designated quantity by which an auxiliary quantity is supposed to draw indefinitely near so that it may differ from it by as little as we wish, and that its ultimate ratio is a ratio of equality ' (*Reflections*$_1$, art. 17). In such ways Carnot's work, like Lhuilier's, is fitful in achievement, and its advances are not sufficient to mark a radical improvement.

I turn now to another prize problem. In 1787, after Euler's death but doubtless under the influence of his mathematical work while Director there, the St. Petersburg Academy proposed this question on functions :

> If the arbitrary functions at which one arrives by integrating [differential] equations in three or several variables represent any curves or surfaces, either algebraic or transcendental, either mechanical, discontinuous, or produced by a voluntary movement of the hand ; or if these functions comprise only continuous curves represented by an algebraic or transcendental equation ?[1]

The phraseology of the question shows well the chaotic state of the theory of functions at that time ; a collection of terms drawn from a variety of sources—mechanics, geometry, algebra, and the still somewhat incoherent subject known as ' analysis '. The choice offered in the question is also interesting : do we have to wallow in this sea of confusion, or can we retreat to the safety of ' continuous curves represented by an algebraic or transcendental equation ' ? The prize was won by Louis Arbogast, who supported Lagrange's foundations for the calculus (see his *1789a*) and who in his essay *1791a* took the hint in a very interesting way. We have seen in chapters 1 and 2 that the calculus tended to operate with continuous objects, curves, values, functions, and if possible, differentials also. Arbogast extended the realm of ' safe ' continuity by distinguishing between two ways in which it can be broken (*1791a*, 9–10) :

> 1. The function can change form, that is to say, the law according to which the function depends on the variable can suddenly change. A curve formed of a collection of several portions of different curves is in this case. Let the curve be a portion of a parabola from A to B, from B to C a portion of an ellipse, from C to D a portion of a circle ; the continuity will be broken at the points B and C . . .
>
> We shall call *discontinuous curves* as much those which are formed by the joining of several portions of curves as those which, traced by the free movement of the hand, are not submitted to any law for any part of their course ; provided that all parts of the curves join

[1] *Nova acta Academiae Scientiarum Petropolitanae,* 5 (1787 : publ. 1789), pt. 1, 4–5.

together without interruption . . . By *discontinuous functions* we mean functions which represent this sort of curve . . .

2. The law of continuity is again broken, when the different parts of a curve do not join together . . .

We call curves of this type, *discontiguous curves*, because all their parts do not join up, or are not contiguous together, and we give the name of *discontiguous functions* to functions which are supposed to correspond to curves of this nature.

Arbogast concluded his essay with the opinion that ' in general, all arbitrary functions . . . are not subject either to the law of continuity or to that of contiguity ' (*1791a*, 96).

We must watch the terminology here. ' Discontinuous ' curves are so because they are defined by different laws (or functions) or by no law at all, but they are continuous to the modern view as they do join up. Our conception of discontinous functions, functions with jumps, is called ' discontiguous ', and for Arbogast continuity is broadened to include *contiguity*, connectedness—including, one presumes, vertical lines joining jumps.

The geometrical character of the definitions is clear, as also are the reasons for it. Algebraic function theory had become impossibly complicated (see Yushkevich *1976a*) : one expression could embody a corner or even a cusp in its path (consider, for example, $x^{2/3}$ around the origin), two expressions could have the same tangent at their join (for example, x^2 for positive x and x^3 for negative x), and so on. This move towards geometry was an important intermediate step in the progress towards mathematical analysis, and we shall see it also in the work of our next figure—Joseph Fourier.

3.5. *The impact of Fourier series on mathematical analysis*

Like de Prony (see section 3.2 above) Fourier assisted Lagrange and Laplace in the early years of the *Ecole Polytechnique*, but after that he had to devote his career to public duties of various kinds. He was a civilian member of the Egyptian campaign of 1798–1801, and from 1802 to 1815 he was prefect of the department of Isère on the Italian border.

It was during this latter period that he developed the mathematics that will interest us in this chapter. He first presented his results to the *Institut de France* (as the *Académie des Sciences* was called at the time) in December 1807 in the form of a large monograph on heat diffusion. There was controversy over the legitimacy of some of his results, and in compromise a prize problem in heat diffusion was set for 1811. He won it in 1812 with an expanded version of the earlier work, but no prospect of publication emerged and so he began a third version as a book, which appeared eventually as *Théorie analytique de la chaleur* (' Analytical theory of heat ' : Fourier *1822a*). The 1811 paper was published soon afterwards, but the 1807 monograph remained in manuscript. An edition of it is included in Grattan-Guinness and Ravetz *1972a*.

Much of the interest in his work was in his mathematicisation of a new area of physical phenomena, but I concentrate on the results pertaining to mathematical analysis. The equation to represent heat diffusion took different forms according to the shape of the body and the consideration or ignorance of external heat diffusion, but basically the ' diffusion equation ' satisfied by the temperature y of a point x of a one-dimensional body at time t had the form

$$\frac{\partial^2 y}{\partial x^2}=\frac{\partial y}{\partial t}. \tag{3.5.1}$$

By separating variables, solving the resultant equations, and inserting the initial condition that $y=f(x)$ when $t=0$, he obtained the ' Fourier series '

$$f(x)=a_0+\sum_{r=1}^{\infty}(a_r \cos rx+b_r \sin rx), \quad x\in[0, 2\pi]. \tag{3.5.2}$$

Fourier now proceeded to a general solution of the diffusion equation. For certain bodies and initial conditions only the sine terms appeared in (3.5.2) (the ' Fourier sine series ', of the kind which Bernoulli had urged in (3.3.3) in the vibrating string problem), and in other configurations only the cosine terms and the constant a_0 were present. In those cases the interval of x was of length π.

Fourier surpassed Bernoulli by calculating the coefficients in (3.5.2) :

$$a_0=\frac{1}{2\pi}\int_0^{2\pi} f(u)\, du, \quad a_r=\frac{1}{\pi}\int_0^{2\pi} f(u) \cos ru\, du,$$

$$b_r=\frac{1}{\pi}\int_0^{2\pi} f(u) \sin ru\, du. \tag{3.5.3}$$

One of his methods of calculation was by integration term-by-term ; but in this he was not novel, for Euler had obtained the cosine coefficients this way in his *1777a*. Fourier seems to have found (3.5.3) independently, and he certainly surpassed Euler in a most important aspect of the series—the manner of their representation of a function.

The main problem with competing solutions to a differential equation is to establish their generality, or lack of it. Now one of Euler's objections to Bernoulli's trigonometric series (3.3.3) was that the periodicity and oddness of its sine functions would forbid a general function from being represented : ' all the curves contained in that equation [(3.3.3)], although one increases the number of terms to infinity, have certain characteristics which distinguish them from all other curves ' (Euler *1755a*, art. 9). But solutions of the wave equation (3.3.1) have to represent the function only over a finite interval of x, so that this criticism is incorrect ; but it would be historically unwise to dismiss it as a minor slip, for it is a substantial conceptual point issuing deep from the algebra-oriented calculus which Euler did so much to develop.

In such a calculus, grounded basically on algebraic expressions and operations upon them, functions would normally be used over the whole of their (finite or infinite) range of definability. As we saw in section 3.4, discontinuous functions could be composed out of ' continuous ' portions ; but the interpretation of Bernoulli's (3.3.3) or Fourier's (3.5.2) demands in addition that only the part of the (' continuous ' or ' discontinuous ') f over the finite interval in question was needed, and therefore that either ' = ' in (3.4.2) be restricted to that interval or else f be redefined as the function repeating periodically over the rest of the real line. Euler recognised the geometrical manner of representation of f (*1755a*, art. 36), but, as we saw, he denied that a function could be represented by a trigonometric series. I am sure that his doubts were concerned with the question of periodicity and generality, and not entirely with the calculation of the coefficients ; for, as mentioned above, he did calculate the cosine coefficients in *1777a* in another context but did not relate the new discovery to the vibrating string problem. The manner of representation was fully clarified only by Fourier in his 1807 monograph. A Fourier series differs from its (normally non-periodic) defining function outside the interval of definition ; it also usually differs from the sine and cosine series of the function, and they differ from each other.

Now ideas such as these are far removed from an algebraic conception of the calculus : indeed, like Arbogast's distinctions between discontiguity and discontinuity in the last section, they require a change to geometry. The clearest explanation of them that I know occurs in Fourier's 1807 monograph. Figure 3.5.1 follows his original diagrams of the sine and cosine series of $y = \frac{1}{2}x$ over $[0, \pi]$, (and their extensions along the real line), with the defining function also shown as a dotted line in the first case, while figure 3.5.2 shows his representation of the sine series over $[0, \pi]$ and the full series over $[0, 2\pi]$ of functions which are constant over part of the defining interval and zero over the rest—a type of function whose representation by a trigonometric series Euler had denied (Grattan-Guinness and Ravetz *1972a*, 221, 226, 230, 271). Modern textbooks rarely explain the representability of Fourier series

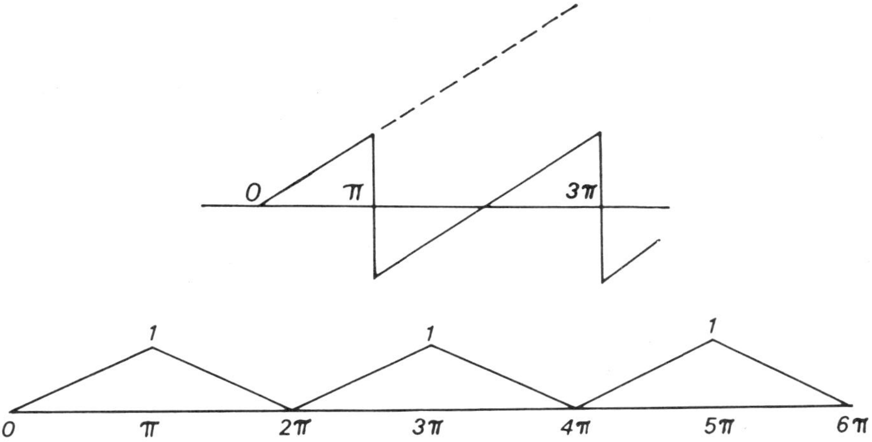

Figure 3.5.1.

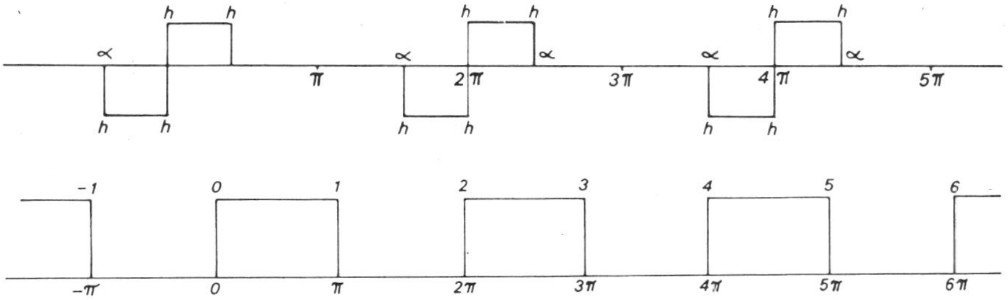

Figure 3.5.2.

with this care, and students often find the problem difficult to understand. Personally I suspect that the reason for their difficulties lies in the quasi-18th-century calculus taught and learnt at schools ; it leads students to approach Fourier series at university with a Eulerian cast of mind.

Figures 3.5.1 and 3.5.2 show other of Fourier's conceptions. Like Arbogast, he connected jumps with vertical lines (which he also mentioned in his verbal descriptions of his diagrams), presumably for the geometrical reasons that the partial sums of the series were continuous and so the limit curve should be continuous also—which means contiguity when necessary. He also urged the interpretation of the integral in his formulae (3.5.3) for the coefficients as an area (and also as a sum in the Leibniz–Euler sense when necessary) rather than the inverse of the differential coefficient (Grattan-Guinness and Ravetz *1972a*, 213–215). I think that the influence here was not Arbogast but Gaspard

Monge, to whom Fourier was deeply attached and who brought his geometrical style of mathematics to bear in his work on differential equations (*ibid.*, 133–134).

Another fundamental question concerning Fourier series was their convergence. Fourier produced rigorous proofs for individual series (*ibid.*, 161, 168, 219), but in 1807 he did not offer any kind of general proof. However, this defect does *not* seem to have underlain the ensuing controversy. The representability problem appears to have been more important, and it is noteworthy that Fourier removed his clear 1807 diagrams of representation in the later, published, versions of his work.

Who were the participants in the controversy ? As far as can be determined, the principal critic was Lagrange, who had supported Euler's criticism of Bernoulli's trigonometric solution in his own work on the vibrating string problem and seems to have repeated them in some form to Fourier (*ibid.*, 169–172). He was supported by Siméon-Denis Poisson and Jean-Baptiste Biot, professional Paris scientists of Fourier's generation who were also interested in heat diffusion and saw rivalry looming. But Fourier also had supporters, including Monge, Lacroix and especially Laplace, the only Parisian mathematician of comparable stature to Lagrange.

The contrast of personalities between Lagrange and Laplace is striking. Lagrange had the habit of deciding early on in his career how a branch of mathematics should be pursued and holding to his view for the rest of his career ; his belief in power-series in analysis is only one example. Thus he was a sort of mathematical *statesman*, and indeed by the 1800s he was the *doyen* of mathematicians. By contrast, Laplace was pragmatic and impressionable in his mathematical as well as his public life (though he was unremitting in his advocacy of Newtonian principles in physics and astronomy), and he acted much like a mathematical *politician*. Thus when Fourier series made their surprising re-appearance in mathematics, he was attracted by the new treatment and took up a problem left unsolved by Fourier in his 1807 monograph.

There *is* a valid criticism of the periodicity of a trigonometric series ; it can represent a function only over a finite interval. Fourier could not see an alternative for the infinite interval, but in *1809a* Laplace gave a clue by producing an *integral* solution to the diffusion equation (3.5.1) over an infinite range of values of x :

$$y=\frac{1}{\sqrt{\pi}}\int_{-\infty}^{\infty} f(x+2u\sqrt{t})\exp(-u^2)\,du. \tag{3.5.4}$$

Fourier then looked for integral solutions of his own, and after a highly unrigorous manipulation of differentials he found the ' Fourier integral formula ', which has forms such as

$$f(x)=\frac{1}{\pi}\int_{-\infty}^{\infty}\int_{0}^{\infty} f(u)\cos(q(u-x))\,du\,dq. \tag{3.5.5}$$

A discussion of this type of solution formed an extra section in his 1811 prize essay, and received substantial analysis in his book (*1822a*, arts. 342–385). It was not only of great significance for the furtherance of mathematical physics but also a sign of the times for mathematical analysis, for it involved a type of function which was becoming prominent : a function defined by an integral representation. Such functions were also an early stimulus for our next major figure—Augustin-Louis Cauchy.

3.6. *Cauchy's analysis : limits, infinitesimals and continuity*

Cauchy's interest in mathematical analysis began in the early 1810s, when he wrote a major paper *1814a* which inaugurated his ' residue calculus ' of functions of a complex variable. Adrien-Marie Legendre was one of the examiners of the paper when it was submitted to the *Académie des Sciences*, and he had an interesting dispute with Cauchy over the evaluation of the integral

$$F(a)=\int_{0}^{\infty}\frac{x\cos ax}{\sin bx}\frac{dx}{1+x^2} \tag{3.6.1}$$

(for details, see Grattan-Guinness *1970a*, ch. 2). The matter is of relevance here because, like Fourier's inversion formula (3.5.5), it involves a function defined by an integral representation. In the immediately following years Cauchy obtained integral solutions of one sort or another to differential equations, and in *1817a* he discovered Fourier's integral formula (3.5.5) for himself. But his considered thoughts on the foundations of real-variable analysis came only in 1820s, when he published his *Cours d'analyse* (*1821a*) and other textbooks based on his teaching at the *Ecole Polytechnique*. In this and the next two sections I shall discuss these contributions, apart from his treatment of the integral, which will be found in section 4.3 of the next chapter on integration.

In section 3.2 I cited Cauchy's *Cours d'analyse* as the book in which we could see mathematical analysis, in the sense now understood by the term, come into fruition. He expounded the theory of limits in much more detail than anyone before him, basing it on the following formulation of the definition of a limit : ' When the values successively attributed to a particular variable approach indefinitely a fixed value, so as to finish by differing from it by as little as one wishes, this latter is called the *limit* of all the others ' (*1821a*, 4 ; *Works*, 19).[1] One of the uses made of this definition was the following definition of an infinitesimal : ' We say that a variable quantity becomes *infinitely small*, when its numerical value decreases indefinitely so as to converge towards the limit zero ' (*1821a*, 26 ; *Works*, 37).

These two definitions raise some important questions about Cauchy's analysis. Nowadays the word ' variable ' is taken to name a symbol ; the variable x is a symbol which refers indifferently to any one of a collection of values, but not to a variable quantity, for there is no such object. But this practice has been introduced only in the study of the foundations of mathematics, described in chapter 6 below. Earlier mathematical literature is often unclear on the matter, and Cauchy's definition of an infinitesimal seems to be an example (as Bolzano pointed out in his *1851a*, art. 12) ; there does seem to be a variable *quantity* whose numerical *values* tend to zero. This is in contrast to the view adopted from the Weierstrassians onwards (and occasionally earlier), where an infinitesimal is a variable with limit zero and talk of the infinitely small does not presuppose the existence of infinitely small values. Cauchy's phrases ' as little as one wishes ' and ' decrease indefinitely so as to converge to the limit zero ' in his definitions leave open the question of whether or not there actually are infinitely small values through which the variable may pass (compare section 5.13 on this point). In other words, these phrases fail to make explicit the kind of independent variable with respect to which the sequence of values passes to its limit. (Possibly Cauchy gave these formulations to cover passage to the limit over both the real numbers ($\lim_{x\to a} f(x)$) and the integers ($\lim_{n\to\infty} s_n$).) The matter is of importance because these later views share with Cauchy's conception (and basically with Newton's in section 2.2) a ' dynamic ' characterisation of infinitesimals in terms of values passing to zero as limiting value. By contrast, in the ' static ' type of infinitesimal of Leibniz and Euler dx is a variable in the sense of a *fixed* if indeterminate value (possibly dependent on x and other variables, and their differentials).

The next basic definition for our attention is that of the continuity of a function $f(x)$: ' *The function $f(x)$ will remain continuous with respect to x between the given limits, if between these limits an infinitely small increase of the variable always produces an infinitely small increase of the function itself* ' (*1821a*, 34–35 ; *Works*, 43). Setting aside the possible meanings of ' infinitely small ', other features are of interest. In

[1] Cauchy's *Cours* has to be cited by page numbers, and I have given the numbers of both the original edition and the one in Cauchy's *Works*. For brevity I omit mentioning the relevant volume of the *Works*, which is ser. 2, vol. 3.

referring to (all) x ' between the given limits ' the definition seems to follow the tradition of defining continuity globally, and the examples which he gives of continuous functions are all algebraic expressions ; but the defining condition is local, describing the behaviour of $f(x)$ around x, a situation which Cauchy describes *separately* in his text. Again, the talk of ' increase of the function ' instead of change (either way) in its value exemplify the habits (which still endure !) of concentrating on monotonically increasing functions (or, equivalently, of failing to take the positive values of increments) and of identifying a function with its value. However, Cauchy mentions absolute (for him, ' numerical ') values just prior to the above quotation ; and he is clear, unlike many of his 18th-century predecessors, that a function had to be single-valued (*1821a*, 34 ; *Works*, 43).

A final feature of these definitions, and of all aspects of Cauchy's analysis, is that they do not rely on geometrical considerations. By using the theory of limits as the source of definitions of basic properties, and the arithmetic of inequalities as the chief device in proofs, Cauchy was able to bring to mathematical analysis an autonomy from both geometry and algebra. A striking feature of the *Cours d'analyse*, and the *Résumé* on the calculus, is that not a single diagram is used, not even for illustrative purposes.

3.7. *On Cauchy's differential calculus*

Cauchy discussed his differential calculus in detail first in his *Résumé* (*1823a*) of his calculus teaching at the *Ecole Polytechnique*. The introduction contains a tactfully worded criticism of Lagrange, then dead ten years, in which he consigned to the same fate Lagrange's belief in the Taylor series as a foundation for the calculus. He mentioned the danger of taking its convergence for granted, but the chief reason for his rejection of Lagrange's views was to be found in lecture 38 (and first in his paper *1822a*) ; that the function $\exp(-1/x^2)$ and all its derivatives are zero when $x=0$, so that it does not have a Taylor expansion there.

As might be expected, Cauchy constructed a system based on his limit-oriented ideas given in the *Cours* ; but the system he produced was not at all like the modern treatment. Let $f(x)$ be continuous and i an infinitesimal ; then by the definition of continuity $(f(x+i)-f(x))$ is also an infinitesimal, and if their ratio tends to a limit, then this limit is Lagrange's derived function $f'(x)$:

$$f'(x) \underset{\mathrm{Df}}{=} \lim_{i\to 0} \{(f(x+i)-f(x))/i\} \tag{3.7.1}$$

(*1823a*, lecture 3). He also defined the ' differential ' $df(x)$ of $f(x)$ as

$$df(x) \underset{\text{Df}}{=} \lim_{\alpha \to 0} \{(f(x+\alpha h)-f(x))/\alpha\}, \tag{3.7.2}$$

where α is infinitesimal but h is finite (*1823a*, lecture 4).[1] This is a type of differential that is still used and taught (see A. Taylor *1974a*), but it differs markedly from the normal Leibnizian sense, for it can take *finite* as well as ' infinitely small ' values. Cauchy gave this example : putting

$$i=\alpha h \tag{3.7.3}$$

in (3.7.1) yields from (3.7.2)

$$df(x)=hf'(x). \tag{3.7.4}$$

In the particular case of $f(x)=x$ (3.7.4) becomes

$$dx=h, \tag{3.7.5}$$

so that this differential takes the finite value h (*1823a*, lecture 4).

This curious mixture of the ' new ' style of limits and echoes from from the past make Cauchy's differential calculus unexpected in form (see Robinson *1966a*, 259–276). For example, putting (3.7.5) into (3.7.4) produces

$$df(x)/dx=f'(x), \tag{3.7.6}$$

which is a *theorem*, not a definition of either $df(x)/dx$ or $f'(x)$: it says that the derivative $f'(x)$ equals the ratio of the differentials $df(x)$ and dx. Hence we have the derivative defined by (3.7.1) as the limit of the difference quotient, but we do *not* have ' dy/dx ' (where $y=f(x)$) used in Lhuilier's or the modern sense as a whole symbol. Higher-order derivatives are handled by Cauchy in the same way ; in lecture 12, for example, the second derived function $f''(x)$ is defined from (3.7.2) as the derived function of $f'(x)$, the second differential ddy is given by $f''(x)(dx)^2$, and d^2y/dx^2 is the ratio of ddy and $(dx)^2$.

I shall now describe some of the theorems which relate to Cauchy's differential calculus. An immediate problem is the question of whether or not a continuous function is differentiable. To us the answer is undoubtedly in the negative, for a function with a corner is continuous but not differentiable there. But in the unrigorous systems of the 18th and early 19th centuries the properties of continuity and differentiability were not so clearly defined anyway, and infinitesimals might allow a corner to be interpreted as an infinitely tight (but differentiable) loop.

[1] This discussion differs slightly from that in Grattan-Guinness *1970a*, 58–60, which is marred by an unfortunate printing mistake : the limit in the defining expression for $df(x)$ was taken there over h instead of α. Cauchy's definition of the differential was anticipated by the obscure Portugese mathematician J. A. da Cunha ; see Yushkevich *1973a*.

For example, Lagrange's belief in the expandability of a function in a Taylor series (see section 3.4 above) has the differentiability of a continuous function as an obvious consequence, for the power series expansion (3.4.1) of $f(x+i)$ in i is necessarily continuous (unless an infinity in the function occurs at x, when a negative power of i will be involved) and the derived function $f'(x)$ can assuredly be calculated. Andre-Marie Ampère, who is best remembered now for his work in electricity and magnetism, was an acute commentator on analysis and tried to improve this type of reasoning by an argument which, though very hazy, set a new style of thinking. For convenience I shall use the notation '$DQ(p, q)$' to represent the difference quotient $(f(p)-f(q))/(p-q)$ over (p, q). Ampère assumed that $DQ(x, x+i)$ was not infinite or zero in value except perhaps 'at particular and isolated points' of the interval $[a, k]$. He then took a uniform partition $\{x_r\}$ of $[a, k]$, and asserted that $DQ(a, k)$ was less than $DQ(x_j, x_{j+1})$ for some interval (x_j, x_{j+1}) defined by the partition, and greater than $DQ(x_m, x_{m+1})$ for another interval. This led him to conclude that $DQ(x, x+i)$ passes from a value less than $DQ(a, k)$ to a value greater than it as x traverses $[a, k]$, and so equals it for some $x \in [a, k]$. He also decided that $DQ(x, x+i)$ cannot 'augment or diminish indefinitely' but instead that, if

$$f'(x)+I=(f(x+i)-f(x))/i, \tag{3.7.7}$$

then I and i tend to 0 together.

Cauchy made use of Ampère's approach in his own proof of the mean value theorem of the differential calculus. For example, he refined Ampère's use of inequalities into lemmas of this kind: let $a_1, \ldots a_n$ and $\alpha_1, \ldots \alpha_n$ be two sets of real numbers, the latter set sharing the same sign, and let the minimum and maximum of the a_r/α_r be m and M. Then there exists an h between m and M such that

$$\sum_{r=1}^{n} a_r = h \sum_{r=1}^{n} \alpha_r \tag{3.7.8}$$

(*1821a*, 16–17, 447–449; *Works*, 28–29, 368). He applied it to the mean value theorem as follows: let $f(x)$ be continuous over $[x_0, X]$, and partition that interval into sub-intervals by points $x_0, x_1, \ldots x_n(=X)$. Then from (3.7.8),

$$\frac{f(X)-f(x_0)}{X-x_0} = \frac{\sum_{r=1}^{n} (f(x_r)-f(x_{r-1}))}{\sum_{r=1}^{n} (x_r-x_{r-1})} = h, \tag{3.7.9}$$

where h lies within the minimum and maximum of $DQ(x_r, x_{r+1})$,

$1 \leqslant r \leqslant n$ (Cauchy *1823a*, lecture 7). Now when the sub-intervals are small ($<\delta$) $DQ(x_r, x_{r+1})$ is ' always ' within a small ϵ of $f'(x_r)$ for each r. Hence h also lies between the minimum and maximum of $f'(x)$ over the interval. Further, if $f'(x)$ is continuous over the interval, then—by appeal to the intermediate value theorem, ' proved ' in the *Cours* with some unavoidable lack of rigour (*1821a*, 43–44, 460–462 ; *Works*, 50–51, 378–380)—it must take the value h in (3.7.9) for at least one value of x within that interval. Hence, as required,

$$f(X)-f(x_0)=(X-x_0)f'(x_0+\theta(X-x_0)), \quad 0<\theta<1. \qquad (3.7.10)$$

In an ' Addition ' inserted at the end of the *Résumé* Cauchy proved in broadly similar style his generalised mean value theorem :

$$\frac{f(X)-f(x_0)}{F(X)-F(x_0)}=\frac{f'(x_0+\theta(X-x_0))}{F'(x_0+\theta(X-x_0))}. \qquad (3.7.11)$$

The limitations in the generality of Cauchy's theorem lay chiefly in the fact that his proof assumed (consciously) that $f'(x)$ has to be continuous, and (unconsciously) that if the difference quotients in $(3.7.9)_2$ were ' always ' within ϵ of $f'(x_r)$ then their convergence to the derived functions would be *uniform* over $[x_{r-1}, x_r]$, which itself effectively assumes the theorem to be proved (see Flett *1974a*, 68). (The Weierstrassians' revision of the theorem is described in section 3.14 below, and uniform convergence is analysed in section 3.12.) The limitation in rigour is due to a lack of structure in the real line in ' proving ' existence theorems such as the intermediate value theorem. All these criticisms apply in some form also to Ampère's reasoning, which is further vitiated by an inexplicit theory of limits.

Much more could be said about the proofs and theorems of Cauchy's differential calculus, and the results and their presentation that he gave in later papers and books. But the discussion above should convey his style. I shall conclude this section with some remarks on the confusing variety of terms and notations that have accumulated in the course of the development of the differential calculus.

In the modern approach (and also in Lhuilier's) the value of the derivative of the function $y=f(x)$ is defined as the limiting value of the difference quotient $(f(x+\delta x)-f(x))/\delta x$ as $\delta x \to 0$, and is denoted either by ' $f'(x)$ ' (if $f(x)$ is in mind) or by the *whole symbol* ' dy/dx ' (if y is referred to). dy and dx do not exist ; in the integral $\int f(x)\,dx$, ' dx ' appears as part of the collective notation ' $\int dx$ '. In Cauchy's system $f'(x)$ is the value of the derivative of $f(x)$ but is also sometimes called (after Lagrange) the ' derived function ' of $f(x)$. dy and dx do exist, after (3.7.2).[1] With Lagrange himself the derived function $f'(x)$, allegedly not obtained by taking limits or using infinitesimals, was a development of Euler's theory of differential coefficients

$$dx = p\ dt, \quad dp = q\ dt, \ldots \tag{3.7.12}$$

(see section 2.7). For both Euler and Leibniz differentials like dx did exist in some sense as infinitesimal but non-zero increments dx on x. dy/dx was not the single-symbol derivative but the ratio of the differentials dy and dx, $d^2y/(dx)^2$ was the ratio of ddy and $(dx)^2$, and so on. The same applies to Cauchy, with his type of differential. For all these writers ' differential equations ' literally means equations relating differentials.

This variety of definitions and interpretations of notations is very complicated to interpret or even follow historically. The situation becomes even more complicated when functions of several variables are considered, with their partial derivatives, partial differentials, total differentials, and the differentiation of functions of functions (for a survey of the notations that were introduced, see Cajori *1929a*, 196–241). Additional definitions and notations arose in connection with the calculus of variations, and in the calculus of differential operators, where ' dy/dx ' denoted the operation d/dx on y (see Burkhardt *1908a*, chs. 13 and 14 *passim*, and Koppelman *1972a*). And still more complications were caused by the frequent practice until the late 19th century of using ' dy/dx ' to denote partial as well as ordinary derivatives (or ratios of differentials).[2] Further, mathematicians interested in applications of the calculus tended to use calculus learnt in their youth rather than the most ' modern ' conception. For example, an Eulerian-style calculus can be found used throughout 19th-century mathematical physics.

[1] They exist even in ' $\int f(x)\, dx$ ', for while Cauchy explained that the integral is the limit of the sum $\Sigma f(x)\Delta x$ as Δx ' becomes infinitely small ' and that ' $\int$ ' ' does not indicate any longer a sum of products . . . but the limit of a sum of this type ', dx was still a differential and could be substituted via (3.7.5) to produce ' $\int f(x)h$ ' as an alternative expression of the integral (*1823a*, lecture 30). I have no idea what this means.

[2] An example is Fourier, and I have indicated the importance of reading his (Eulerian) calculus correctly in Grattan-Guinness *1975b*. The effect of notations and their related definitions on the development of the calculus is little studied. For example, Truesdell *1960b*, a fine account of the progress in rational mechanics from Galileo to Lagrange, modernises all these notations and speaks of ' derivatives ', although none of the historical figures used derivatives in this modern sense.

We saw in chapters 1 and 2 the importance of using symbols as free variables as well as unknown constants in the development of the (algebraic) calculus. Textbooks rarely explain this distinction well, and students are often confused about it. Briefly: in $y=f(x)$, x is a free variable (so is y) ; in $0=f(x)$, x is an unknown constant, taking the zeros (if any) of $f(x)$ as its values ; in the property $\phi x \underset{\mathrm{Df}}{=} f(x)=0$, x is a free variable for the propositional function ϕ (for more on this idea, see section 6.7), but still an unknown constant relative to the mathematical function f.

Thus a schism appeared between the theory and the practice of the calculus as the level of rigour in the calculus was raised : the foundationalists had one set of rules, the practitioners another. The situation has persisted to this day, with quite unfortunate and unnecessary confusions for students. It is a common experience for them to learn in the calculus lectures that infinitesimally small differentials do not exist, but to use them constantly in the mathematical physics lectures. While the Eulerian calculus is not rigorous, it should be taught for what it is : a powerful tool for the analysis of physical and geometrical phenomena, which has left its considerable mark on the conceptions, terminology and notations of later presentations of the subject. As things are, the treatment in textbooks is unsatisfactory. Some basically follow Cauchy's practice of notating the derivative by ' f' ' and defining the differential by some equivalent of his (3.7.2), while others notate the derivative by the *single symbol* ' dy/dx ' and omit differentials altogether ; and neither treatment warns the reader of the existence of the other. Further, both treatments give a prime place to limits without explaining why the standard of rigour and generality obtainable from this very difficult concept is desirable in the first place, or what kinds of less rigorous approach are being superseded.

3.8. *Cauchy's analysis : convergence of series*

I turn now to Cauchy's view on the convergence of series, a singularly important aspect of his mathematical analysis in that it brought series for the first time into full unity with the calculus under the common ground of limits. A few remarks on the prehistory of Cauchy's work are necessary.

It is easy to underrate the treatment of convergence in the 18th century. In those days (and much earlier), the convergence of some simple series was proved, and the basic interpretation of a series as the ordered summation of its terms was accepted. Beyond that, however, there were many important unclarities. The word ' convergent ' was applied to series if their n-th term, or n-th remainder term, tended to zero as n increased, and ' divergent ' series included those which took an infinite sum, or oscillated finitely or infinitely. Both words would sometimes be used for the same series. For example, the modern literature refers occasionally to ' d'Alembert's ratio test ', which is very lucky for d'Alembert ; for although he considered the ratio of successive terms of the series expansion of $(1+\mu)^m$ he said that, when $m=-2$ and $\mu=99/100$, the series was divergent up to the 99th term but convergent thereafter because the ratio was greater than 1 up to there and less than 1 afterwards (d'Alembert *1768a*, 171, 176)—which is hardly a convergence test as we now use it.

But a more important difficulty concerned the development of non-orthodox methods of summing series, especially power-series, for which the ' sum ' was definable as the limiting value (if it existed) of the corresponding sequence of partial sums. Unfortunately the ingenuity of summation was not matched by a corresponding reflection on the interpretation of the results obtained—that a sum of the series is defined relative only to the method of summation involved, so that the same series can have different sums (or no sum at all) relative to different methods. Euler was particularly talented in devising summations (see Hofmann *1959a*) and has left us a mass of results in summability and ' divergent ' series, and such studies continued in the 19th century (see Burkhardt *1910a*) ; but the theory within which they need to be interpreted did not begin to emerge until the end of the 19th century (see Tucciarone *1973a*).

In such circumstances it is very understandable that Cauchy demanded the study of infinite series to be preceded by an analysis of their convergence. The tendency towards such a view can be found already in Lacroix, Fourier and Gauss among others (see Grattan-Guinness *1970a*, 171), but Cauchy explored the definition of convergence and its consequences to new levels of detail. ' *A divergent series does not have a sum* ', he wrote on the introduction to the *Cours d'analyse*, perhaps with his professional contemporaries as much as his students in mind (although, like them, he seemed to include finite and infinite oscillation under the heading of ' divergent '). In the body of the book he explained that ' if, for always increasing values of n, the [n-th partial] sum s_n draws indefinitely near to a certain limit s, the series will be called *convergent*, and the limit in question will be called the *sum* of the series. On the other hand, if the sum s_n does not approach any fixed limit while n increases indefinitely, the series will be *divergent*, and will no longer have a sum ' (*1821a*, 123 ; *Works*, 114). And a page later he was claiming a remarkable theorem : for a series to be convergent ' it is necessary and it suffices that, for infinitely large values of the number n, the sums

$$s_n, s_{n+1}, s_{n+2}, \text{\&c.} \ldots$$

differ from the limit s, and consequently between themselves, by infinitely small quantities '.

The importance of this result is in the fact that the condition which is expressed by the words ' and consequently between themselves ' does not mention the sum of the series, which in practice may not even be guessed at. A rigorous proof of the sufficiency of the condition is hard, since it asserts the existence of a limit and so needs knowledge of the structure of the real line. Cauchy did not really attempt a proof.[1]

An important example of convergence which Cauchy studied was that of Taylor's series. Attempts to formulate an expression for the remainder term had been made already (see Pringsheim *1900a*), but Cauchy's rejection of Lagrange's faith in the series lent special im portance to the study of its convergence. He found the integral form of the remainder term by using integration by parts :

$$f(x+h)-f(x)=\int_0^h f'(x+h-z)\,dz$$

$$=hf'(x)+\frac{1}{2!}h^2f''(x)+\ldots+\frac{1}{(n-1)!}h^{n-1}f^{(n-1)}(x)$$

$$+\int_0^h \frac{1}{(n-1)!}z^{n-1}f^{(n)}(x+h-z)\,dz, \qquad (3.8.1)$$

as long as each derivative is continuous. By appeal to mean value theorems of the kind stated in (3.7.10) he converted the integral remainder term in (3.8.1) to the differential form

$$f^{(n)}(x+\theta h)\int_0^h \frac{1}{(n-1)!}z^n\,dz, \quad \text{or} \quad \frac{1}{n!}h^n(x+\theta h), \quad 0<\theta<1 \qquad (3.8.2)$$

(Cauchy *1823a*, lecture 36).

Recall of Cauchy's mean value theorem brings back also Ampère, whose *1806a* was mentioned in the last section. As part of his attempt to improve on Lagrange's foundation to the calculus he searched for a remainder term to Taylor's series there, and in his *1826a* he returned to the theme with a similar argument which is worth describing as an alternative approach of that time. His reasoning was given for a function of several variables, but for convenience I present a one-variable form. By appeal to his earlier paper he claimed that

$$P(\alpha)\underset{\mathrm{Df}}{=}\{f(x+h)-f(x+\alpha h)\}/(1-\alpha) \qquad (3.8.3)$$

was in general neither zero nor infinite. Rewriting (3.8.3) as

$$f(x+h)=f(x+\alpha h)+(1-\alpha)P(\alpha) \qquad (3.8.4)$$

[1] This condition and an attempted proof are to be found also in Bolzano *1817a*, and are two of several similarities with Cauchy's *Cours d'analyse*. Others include the primacy of limits, the definitions of continuity of a function and of convergence, the intermediate value theorem, and proofs by the partition of intervals (hence the so-called ' Bolzano–Weierstrass theorem ', mentioned in section 3.14 below), and some are so close that they need the full sophistications of Weierstrassian analysis (see sections 3.12 and 3.13 below) to distinguish them apart.

and differentiating successively with respect to α, he obtained

$$0=\frac{df\,(x+\alpha h)}{d\alpha}+(1-\alpha)P'(\alpha)-P(\alpha), \tag{3.8.5}$$

. . . .

$$0=\frac{1}{n}\frac{d^n f\,(x+\alpha h)}{d\alpha^n}+\frac{1-\alpha}{n}P^{(n)}(\alpha)-P(\alpha). \tag{3.8.6}$$

Replacing $P(\alpha)$ in (3.8.4) by itself from (3.8.5), and then $P'(\alpha)$ by itself from (3.8.6) when $n=2$, then $P''(\alpha)$ when $n=3$, and so on, he found that

$$f(x+h)=f(x+\alpha h)+(1-\alpha)\frac{df\,(x+\alpha h)}{d\alpha}$$
$$+\frac{(1-\alpha)^2}{2!}\frac{d^2 f\,(x+\alpha h)}{d\alpha^2}+\ldots+\frac{(1-\alpha)^n}{n!}P^{(n+1)}(\alpha), \tag{3.8.7}$$

whose last term gives the remainder in a differential form based on (3.8.3).

I return to Cauchy, and to his inauguration of convergence tests. Nowadays these tests are often presented as an unconnected list of useful if boring results ; but in the 1820s the logical relationship between tests was one of the stimuli to their development. A test is a sufficient condition for the convergence or divergence of an infinite series, and it is limited in use by being restricted to certain types of series and (usually) by requiring a function of the n-th (and perhaps other) terms to converge to a limit, which does not equal 1, as n increases. If known tests cannot determine the behaviour of a particular series, then a new test is needed, and certain tests may imply others for particular types of series. All this was well understood during their development in the quarter-century from the first ones in the *Cours d'analyse* (see the detailed description in Grattan-Guinness *1970a*, appendix). The proofs were broadly similar to the modern ones (including allowance for the distinction between absolute and conditional convergence), except that, following Cauchy, the upper limit of a sequence was rather vaguely defined by phrases such as ' the limit of the greatest values ' of the appropriate function of the terms.

Cauchy proved these tests in the *Cours d'analyse*, where appropriate for series of the same and of mixed terms : root (in terms of the behaviour of $\sqrt[n]{|u_n|}$ as $n\to\infty$) ratio ($|u_{n+1}/u_n|$), condensation ($\sum_{r=1}^{\infty} u_r$ and $\sum_{r=0}^{\infty} 2^r u_{2^r-1}$ converge or diverge together), logarithmic ($\log u_n/\log 1/n$), product (if $\sum_{r=0}^{\infty} u_r$ and $\sum_{=0}^{\infty} v_r$ converge absolutely, then

$\sum_{r=0}^{\infty} \sum_{i=1}^{r} u_i v_{r-i}$ converges to their product), and alternating (the u_r decrease in absolute value to zero and alternate in sign). Later tests included Abel's (if $\sum_{r=0}^{\infty} u_r$ converges and v_n is bounded, then $\sum_{r=0}^{\infty} u_r v_r$ converges also), Cauchy's integral (if $u(x)$ decreases monotonically to zero, then $\int_0^{\infty} u(x)\,dx$ and $\sum_{r=0}^{\infty} u(r)$ converge or diverge together), Raabe's $(n|(u_n/u_{n+1})-1|)$, and various forms of comparison test. Then there were a number of iterative tests, in which a sequence of expressions could be calculated and successively tested. As an example of motivation of tests, Cauchy introduced his integral test to deal with certain series for which his root test was indecisive.

Of these later results Abel's test is of special interest, for it appeared in an important paper of 1826 which explored the convergence and other properties of the binomial series. As part of his groundwork he proved results following Cauchy's principles of the type now called 'Abel's limit theorem'. One of the prominent features of the rather unrigorous manipulations of infinite series had been to assume that what was true up to the limit was true there also, and thus to insert the limiting value of a variable into a series. But now Abel, having read Cauchy's *Cours d'analyse*, knew better.

Abel stated his first theorem as follows (*1826a*, theorem 4):

> If the series
>
> $$f(\alpha)=v_0+v_1\alpha+v_2\alpha^2+\ldots+v_m\alpha^m+\ldots \qquad [(3.8.8)]$$
>
> is convergent for a certain value δ of α, it will also be convergent for every value less than δ, and, for always decreasing values of β, the function $f(\alpha-\beta)$ approaches indefinitely the limit $f(\alpha)$, presuming that α is equal to or smaller than δ.

His proof was very sloppy, for he did not follow Cauchy's usual care in taking positive values of expressions and so his manipulation of inequalities was unrigorous (see Grattan-Guinness *1970a*, 82–84), but his result drew attention to a technique which had hitherto been used uncritically. In theorem 5 of his paper he generalised this theorem to the case where the v_m are continuous functions of a variable x, and in a footnote he made an important criticism of a theorem in Cauchy's *Cours*.

Cauchy's theorem had stated (*1821a*, 131–132; *Works*, 120):

> When the different terms of the series $\left[\sum_{r=0}^{\infty} u_r\right]$ are functions

of the same variable x, continuous with respect to that variable in the vicinity of a particular value for which the series is convergent, the sum s of the series is also a continuous function of x in the vicinity of that particular value.

Abel pointed out in his footnote that the Fourier sine series of $\frac{1}{2}x$ over $[0, \pi]$ would be a counter-example to this theorem, for ' it is discontinuous for all values $(2m+1)\pi$ of x, m being a whole number '. Now we recall from section 3.4 that Fourier did *not* have jumps in his representation but joined them with vertical lines ; and if these lines could be interpreted as being infinitely steep, then the series and the theorem would be reconciled. However, virtually no such argument seems to have been offered at the time ;[1] and if Abel's interpretation of the Fourier series as being discontinuous when $x=(2m+1)\pi$ is upheld, then Cauchy's theorem, and especially the proof, needs examination.

In his proof Cauchy claimed that the n-th partial sum s_n was continuous, and so by his definition of continuity (discussed in section 3.6 above) suffered a small increment when x received a small increment α. The remainder term r_n was small for large n by convergence of the series, and the increment on r_n due to α ' becomes insensible at the same time ' as r_n. Thus s also has a small increase with α and so is continuous, as was to be proved (*1821a*, 131 ; *Works*, 120).

The quoted phrase ' insensible at the same time ' is crucial. What does it mean ? We have to bear in mind : (1) the smallness of $s_n(x+\alpha) - s_n(x)$ with small α due to the continuity of the function $s_n(x)$ around x, a local property although x can be any point in an interval (see section 3.6 above) ; (2) Cauchy's claimed smallness of $r_n(x+\alpha)-r_n(x)$ with $r_n(x)$; and (3) the smallness of $r_n(x)$ with large n. How do these small quantities relate to each other ? Is α independent of $r_n(x)$ (and thus of the largeness of n) or dependent on it ? And does the interval over which continuity of $s_n(x)$ is defined play a role ? There are various combinations of answers to these questions and, as we shall see in section 3.12, each combination defines a different mode of *uniform* convergence. Suffice it to say here that the technicalities involved are considerable, and that Cauchy's theorem, like his analysis in general, is *intrinsically vague* with respect to them. It is noteworthy that he used the symbols ' s_n ', ' r_n ' and ' s ' in his proof *without* symbolising their functional dependence on x.

Thus we have found a major difficulty in Cauchy's analysis, a setback to put against the many achievements and improvements described

[1] Such possibilities seem available in the ' new ' infinitesimals of non-standard analysis (see Cleave *1971a*). I shall not pursue this point here, for this highly sophisticated theory has virtually no bearing on the history of mathematical analysis (see Bos *1974a*, 81–86), though its teaching possibilities are very interesting (see Keisler *1976a*).

already. We shall see it now at work on a major problem of its time; the convergence of Fourier series.

3.9. *The general convergence problem of Fourier series*

This problem was an exciting challenge for mathematical analysis of this period, not only due to the importance that Fourier analysis was achieving in applied mathematics but also because of the mixture of functions, limits, convergence and integration that their convergence involved. I mentioned in section 3.5 that Fourier had proved the convergence of particular series in his 1807 monograph but had not obtained a general proof. In this section I shall briefly summarise the early general proofs (for more details, see Grattan-Guinness *1970a*, ch. 5).

The first proof was published in 1820 by Poisson, who brought to the problem his facility for manipulating power-series. He presented his method in various forms over a number of years, but basically it did not change, and I shall describe the version in his *1823a*. He took the formula

$$\frac{1-p^2}{1-2p\cos(x-\alpha)+p^2}=1+2\sum_{r=1}^{\infty}p^r\cos(r(x-\alpha)),\quad |p|<1, \qquad (3.9.1)$$

and multiplied through it by $f(\alpha)$ and integrated over $[-\pi, +\pi]$ of α to obtain:

$$\int_{-\pi}^{+\pi} f(\alpha)\frac{1-p^2}{1-2p\cos(x-\alpha)+p^2}\,d\alpha=$$
$$\int_{-\pi}^{+\pi} f(\alpha)\left\{1+2\sum_{r=1}^{\infty}p^r\cos(r(x-\alpha))\right\}d\alpha. \qquad (3.9.2)$$

Then he let p go to 1. The integrand on the left hand side of (3.9.2) took the indeterminate form $0/0$ when $\alpha=x$, but after granting himself some rather unrigorous manipulations he obtained

$$2\pi f(x)=\int_{-\pi}^{+\pi} f(\alpha)\left\{1+2\sum_{r=1}^{\infty}\cos(r(x-\alpha))\right\}d\alpha, \qquad (3.9.3)$$

which becomes the required full Fourier series of $f(x)$ over $[-\pi, +\pi]$ when $\cos(r(x-\alpha))$ is expanded and terms are rearranged:

$$f(x)=\frac{1}{2\pi}\int_{-\pi}^{+\pi} f(\alpha)\,d\alpha+\frac{1}{\pi}\sum_{r=1}^{\infty}\left\{\cos rx\int_{-\pi}^{+\pi} f(\alpha)\cos r\alpha\,d\alpha\right.$$
$$\left.+\sin rx\int_{-\pi}^{+\pi} f(\alpha)\sin r\alpha\,d\alpha\right\}. \qquad (3.9.4)$$

Some of the manipulations are merely disregarding inconvenient powers of the infinitesimal $(1-p)$ and retaining the useful ones, and these processes can mostly be reworked more rigorously. But there are also two more serious difficulties. One is allowing p to take its limiting value 1, a process which Poisson retained even after Abel's limit theorems in *1826a*. The other is the fact that no restriction seems to be placed on $f(x)$; (3.9.4) appears to be valid for 'any' function $f(x)$. These two points severely undermine Poisson's argument, although he did allow the summation of the series by other means and the 'Poisson integral' on the left hand side of (3.9.2) was to be of considerable interest for both real- and complex-variable analysis.

Cauchy published an attempted proof in *1826a*, using the following theorem from his complex variable calculus, which still often took the form of putting $i\ (=\sqrt{-1})$ into a real-variable integral: Let $f(t)$ be a finite-valued function of a real variable t, and b be a positive real number. Then

$$\int_0^a \exp(\pm ibt)f(t)\,dt = \mp i \int_0^\infty \{\exp(\pm iab)f(a\pm iu) - f(\pm iu)\} \times \exp(-bu)\,du. \qquad (3.9.5)$$

Cauchy used this theorem to transform each term of the Fourier series (3.9.4) from the form of the left hand side of (3.9.5) to its right hand side and then manipulate the series into a multiple of the (convergent) Fourier series of $\frac{1}{2}(\pi - x)$ over $[-\pi, +\pi]$. But his proof involved integrating a series term-by-term and rearranging the order of the terms (and thus assuming the convergence to be proved), it did not show that the series actually summed to $f(x)$, and, like Poisson's proof, it seemed to be valid for 'any' function $f(x)$. Cauchy may have become dissatisfied with his proof, for he offered another in *1827a* which also was based on his complex variable calculus; it appears to make some assumptions of convergence and is very restricted in the types of function to which it can apply.

There is another error in Cauchy's first proof which I mention only now because it helped to motive the first proof of convergence which was to do justice to the problem. Cauchy assumed there that if $(u_n - v_n) \to 0$ as $n \to \infty$ and if $\sum_{r=0}^{\infty} u_r$ is convergent, then so is $\sum_{r=0}^{\infty} v_r$. However, he was in error, for $(-1)^n n^{-1/2}$ and $(n^{-1} + (-1)^n n^{-1/2})$ satisfy the property required of u_n and v_n, but the first generates a convergent series while the other creates a divergent one. This was pointed out by a young German mathematician, Peter Lejeune-Dirichlet, in a paper *1829a* in which he presented his own treatment of the problem.

Like many young scientists of his age, Dirichlet spent part of his early career in Paris, then the scientific capital of the world. He became closely attached to Fourier, and developed his own proof as a (considerable) refinement of a sketched proof in Fourier's book. By means of the trigonometric identity

$$\sum_{r=-n}^{n} \cos ru = \cos nu + \sin nu \frac{\sin u}{1-\cos u} \tag{3.9.6}$$

Fourier had converted the n-th partial sum (that is, the first $(2n+1)$ terms) of his full series (3.9.4) into

$$\frac{1}{2\pi}\int_{+\pi}^{+\pi} f(\alpha)\cos(n(x-\alpha))\,d\alpha + \frac{1}{2\pi}\int_{-\pi}^{-\pi} f(\alpha)\sin(n(x-\alpha))\frac{\sin(x-\alpha)}{1-\cos(x-\alpha)}\,d\alpha. \tag{3.9.7}$$

He then interpreted each integral in (3.9.7) as an area, when the oscillation of $\cos(n(x-\alpha))$ and $\sin(n(x-\alpha))$ suggests that each integral decomposes into areas of equal magnitude and opposite sign and so has zero as value. However, in the second integral, the fraction $\sin(x-\alpha)/[1-\cos(x-\alpha)]$ becomes $0/0$ when $\alpha = x$, and so needs special treatment. As in Poisson's case, an unrigorous use of infinitesimals 'showed' that the value there was $f(x)$, so that convergence was established (*1822a*, art. 423).

Fourier's standard of rigour here is about the same as Poisson's, and he shared with Poisson (and Cauchy) a belief in the validity of his proof for 'any' function $f(x)$. But he gave Dirichlet the clue both to understand the convergence problem and to offer a solution: seek *sufficient conditions on* $f(x)$ from which convergence can be demonstrated, and start from a trigonometric identity. In his *1829a* Dirichlet selected, instead of (3.9.6), the identity

$$\sum_{=-n}^{n} \cos ru = \frac{\sin((n+\frac{1}{2})u)}{\sin(\frac{1}{2}u)}. \tag{3.9.8}$$

Thus, instead of Fourier's (3.9.7), he obtained

$$s_n(x) = \frac{1}{\pi}\int_{-\pi}^{+\pi} f(\alpha)\frac{\sin((n+\frac{1}{2})(x-\alpha))}{2\sin(\frac{1}{2}(x-\alpha))}\,d\alpha, \tag{3.9.9}$$

which is the form of the integral now known as the 'Dirichlet integral'

$$\int_0^h f(u)\frac{\sin mu}{\sin u}\,du, \quad 0 < h \leqslant \frac{\pi}{2}. \tag{3.9.10}$$

Dirichlet's treatment of (3.9.10) was geometrical, like Fourier's, but done in the style of Cauchy's analysis. It was a rigorous version of the following intuitive argument. The oscillation of sin mu suggests a breakdown of $(0, h)$ into sub-intervals $(0, \pi/m)$, $(\pi/m, 2\pi/m)$, . . ., over each of which sin mu is monotonic. If $f(x)$ decreases monotonically, then the integral in (3.9.10) should break down into a series of areas of alternating sign and decreasing magnitude and so (by the alternating series test) converge as $n \to \infty$. When $u=0$ the integrand becomes 0/0, but putting $mu=v$ in (3.9.10) shows that it tends to

$$f(0) \int_0^\infty \frac{\sin v}{v}\, dv, \quad \text{or} \quad \frac{\pi}{2} f(0). \tag{3.9.11}$$

Hence from (3.9.10) and (3.9.11),

$$\lim_{m\to\infty} \int_0^h f(u) \frac{\sin mu}{\sin u}\, du = \frac{\pi}{2} f(0). \tag{3.9.12}$$

In order to establish conditions for convergence of the Fourier series Dirichlet proved some easy consequences of this result. (3.9.12) is true if $f(x)$ increases monotonically to zero or is constant, and therefore if $f(x)$ changes sign within the interval (for then we apply (3.9.12) to the functions $(c+f(x))$ and c for a c large enough to make $(c+f(x))$ positive over $(0, h)$ and subtract the results). (3.9.12) is also true for the interval $(0, g)$ if $0<g<h$, and so by subtraction the Dirichlet integral is zero over (g, h). Finally, if $f(x)$ is discontinuous at g and/or h, then we use $f(g+0)$ and $f(h-0)$ in the proof instead of $f(g)$ and $f(h)$. (In a reworking *1837a* of *1829a*, he actually introduced these notations for the left- and right-hand limiting values of $f(x)$ at x.)

In order to apply these results to (3.9.9) Dirichlet assumed that within the interval $[-\pi, +\pi]$ of α, $f(\alpha)$ took discontinuities or turning values only at the finite number of points $d_1, d_2, \ldots d_s$, and he split up $(-\pi, +\pi)$ into the sub-intervals $(-\pi, d_1)$, (d_1, d_2), . . . (d_N, x), (x, d_{N+1}), . . . (d_s, π). (He actually specified closed sub-intervals.) Now the interval $(0, h)$ of u in (3.9.12) corresponds to the sub-interval (x, d_{N+1}) of $(-\pi, \pi)$ of α in (3.9.9). Thus when $n \to \infty$ there, (x, d_{N+1}) contributes

$$\frac{1}{\pi}\left(\frac{\pi}{2} f(x+0)\right)$$

—that is, $\frac{1}{2}f(x+0)$—to the value of $s_n(x)$. Similarly, (d_N, x) contributes $\frac{1}{2}f(x-0)$. The other sub-intervals of $(-\pi, \pi)$ contribute 0 to $s_n(x)$ by the extension of (3.9.12) to (g, h) of u. Hence under the sufficient conditions for $f(x)$ stated above, its Fourier series converges to $\frac{1}{2}(f(x-0)+f(x+0))$. In a later paper Dirichlet extended the conditions on $f(x)$ to include a finite number of infinities for which its indefinite integral remains continuous (*1837b*, addition).

I have discussed this proof in some detail not only because of its importance but also to convey the contemporary difficulties involved in ' doing ' mathematical analysis. It is one of the most important proofs in the history of the subject, for it both showed how mathematical analysis should be done and formulated an important task : to try to obtain more general sufficient conditions for the convergence of Fourier series. This would clearly mean considering functions with an *infinitude* of turning values, discontinuities and/or infinities within a finite interval, and we shall see such studies occur quite prominently in the later development of mathematical analysis. Dirichlet concluded his *1829a* with a particularly prophetic remark, defining a function which did *not* satisfy his conditions :

> $f(x)$ equals a definite constant c when the variable takes a rational value, and equals another constant d when this variable is irrational. The function thus defined has finite and definite values for every value of x, and meanwhile one cannot substitute it into the [Fourier] series, seeing that the different integrals which enter in that series lose all significance in this case.

In his *1837a* he urged a very general conception of a function as not tied to any analytic expression (on this idea, compare Lacroix *Treatise*$_1$, vol. 1 (1797), 1), and clearly thought that this function (effectively the characteristic function of the rationals) was an example of such generality. Such ideas as these form an important part of the next chapter of this book ; the history of 19th-century integration.

In his *1829a* Dirichlet also promised to prove ' some other quite remarkable properties ' of Fourier series and discuss the ' fundamental principles of infinitesimal analysis ', but unfortunately he seems never to have made them fully public, not even in his lectures. He is best remembered today for his work on number theory ; but his contributions to mathematical analysis were considerable, the best since Cauchy and certainly more important than those of Abel, whose name is usually bracketed with Cauchy's. I shall conclude this section with his proof of Abel's limit theorem, which I stated in section 3.8. Dirichlet appreciated the need of the theorem, but he may have felt dissatisfied with Abel's vague proof, for he reformulated the theorem and proved it afresh as follows in his posthumous *1862a*. Let

$$f(h) = \sum_{r=0}^{\infty} v_r h^r \tag{3.9.13}$$

when $0 \leqslant h < 1$, and

$$A = \sum_{r=0}^{\infty} v_r. \tag{3.9.14}$$

Then the theorem asserts that $A = f(1-0)(= \lim_{h \to 1^-} f(h))$.

Dirichlet's proof begins by showing that, if s_n is the n-th partial sum of A, then by a valid rearrangement of terms in (3.9.13),

$$f(h) = (1-h) \sum_{r=0}^{\infty} s_r h^r. \tag{3.9.15}$$

For a fixed n, $0 < |s_r| < K$ for some finite K for all $r \leqslant n$, so that

$$\left| (1-h) \sum_{r=0}^{n-1} s_r h^r \right| < (1-h) \sum_{r=0}^{n-1} K h^r < (1-h) nK. \tag{3.9.16}$$

Further, by an intermediate value argument after the style of Cauchy's (3.7.8),

$$(1-h) \sum_{r=n}^{\infty} s_r h^r = (1-h) P_n \sum_{r=n}^{\infty} h^r = P_n h^n, \tag{3.9.17}$$

where P_n lies between the maximum and minimum of $s_n, s_{n+1}, \ldots$. Now he split up the series in (3.9.15) into its partial sum and remainder, and let $h \to 1$ throughout the equation. $f(h) \to f(1-0)$; by (3.9.16), the partial sum $\to 0$; and by (3.9.15) the remainder term $\to P_n$. Now let $n \to \infty$ in each of these three expressions. The only one affected is P_n, which by definition and (3.9.14) takes A as its limiting value, as was to be proved.

Dirichlet was careful here in his handling of double limits, as he was in his study of the convergence of Fourier series (where $x \to a$ (for some value of a) and then $n \to \infty$). It was in this area, the handling of multiple limits, that Cauchy's analysis was not sufficiently rigorous. The next section describes some of the stumbling advances that were made in the direction which Dirichlet presaged.

3.10. *Some advances in the study of series of functions*

Dirichlet's conditions for the convergence of Fourier series sharpened the apparent clash with Cauchy's theorem in the *Cours d'analyse* on the continuity of the sum-function of a convergent series of continuous functions (see section 3.8 above), for they included the possibility that the represented function could be discontinuous (at a finite number of points). Cauchy did not seem impressed, for he repeated his theorem word for word in 1833 in his *Résumés analytiques* (*1833a*, 46); but Phillip Seidel, one of Dirichlet's students, took the matter up in his *1848a*.

Unlike Cauchy, Seidel did symbolise the argument variable in this problem (though I shall not use his notations), which he approached by proposing the following distinction: the series $\sum_{r=0}^{\infty} u_r(x)$ converges 'arbitrarily slowly' or 'not arbitrarily slowly' if n_0, the smallest value of n for which

$$|r_n(x+h)| < \epsilon, \qquad (3.10.1)$$

where $n \geqslant n_0$ and $(x+h)$ is in the neighbourhood of x, respectively does not or does pass to infinity as ϵ tends to zero. Now consider the equation

$$s(x+h)-s(x) = \{s_n(x+h)-s_n(x)\}+\{r_n(x+h)-r_n(x)\}. \qquad (3.10.2)$$

If h is small, then, by the continuity of $s_n(x)$, $\{s_n(x+h)-s_n(x)\}$ will be small also; and both $r_n(x+h)$ and $r_n(x)$ are small for large n. All this applied in Cauchy's proof also (see section 3.8 above), but now Seidel utilised his new definition (3.10.1). Let the increment h' on x decrease from h to 0 inclusive, and watch the effect of n_0 in (3.10.1). If n_0 achieves only a *finite* maximal value, then the convergence is not arbitrarily slow, $\{r_n(x+h)-r_n(x)\}$ remains small, and hence, from (3.10.2), the sum function is continuous as in Cauchy's theorem. However, if n_0 passes to infinity, then the convergence is arbitrarily slow and Cauchy's theorem does not hold.

I have simplified Seidel's very laborious reasoning; for clarity of language he is little better than Cauchy's 'insensible at the same time'. Indeed, in an important respect he marks a *retreat* in standards, for in requiring h to pass to zero he spoils the limit-avoiding character of Cauchy's mathematical analysis, in which we take small *but non-zero* increments of variables and explore the manner of their functional dependence via the arithmetic of inequalities. The same applies to a proof of a similar theorem by G. G. Stokes, produced almost simultaneously but apparently independently, who defined the properties of 'infinitely slow convergence' and 'not infinitely slow convergence' for an infinite series. The only difference between his definition and Seidel's (3.10.1) is that the inequality is assumed to hold only when $n = n_0$ and not necessarily for all larger values of n (Stokes *1849a*, art. 39).

Mention of Stokes shows a British mathematician having something significant to say at last about the foundations of mathematical analysis. Within a year another Briton made a useful contribution which, even more than with Stokes, seems not to have had the influence that it deserved. In section 3.5 above I noted that Fourier represented the discontinuities in a function by joining them up contiguously with vertical lines. In 1848 Henry Wilbraham considered Fourier's proof of the convergence of the series

$$\cos x - \tfrac{1}{3}\cos 3x + \tfrac{1}{5}\cos 5x - \ldots, \tag{3.10.3}$$

which converges to $\pi/4$ over $(-\pi/2, +\pi/2)$, and to $-\pi/4$ over $(\pi/2, 3\pi/2)$ (Fourier *1822a*, art. 179 ; although it has only cosine terms, it is really a sine series in its mode of representation, as can be seen by changing the origin to $x = -\pi/2$ by the substitution $2y = 2x + \pi$). Fourier obtained the expression

$$\int_0^x \frac{\sin 2nv}{2\cos v}\, dv \tag{3.10.4}$$

for the remainder after n terms, but Wilbraham queried the behaviour of this term as n tends to infinity. He took (3.10.4) for $2n$ rather than n terms, and then substituted $4nv = 2n\pi - u$ to obtain for the remainder term

$$r_{2n}(x) = -\int_{2n\pi-4nx}^{2n\pi} \frac{\sin u}{8n\sin(u/4n)}\, du. \tag{3.10.5}$$

He noted that the integrand took large values when $u/4n$ was small, and so changed the denominator to $8n(u/4n)$, or $2u$. He also changed the upper limit of the integral to ∞ (but left the lower limit alone !) and so obtained

$$r_{2n}(x) = -\int_{2n\pi-4nx}^{\infty} \frac{\sin u}{2u}\, du. \tag{3.10.6}$$

When x is distant from $\pi/2$, (3.10.6) is zero ; when $x = \pi/2$, it is $-\pi/2$; but when it is close to $\pi/2$ to an order within $1/n$, it oscillates in a manner governed by $\sin nx$.

This result was sadly overlooked for half a century, when it was rediscovered and mis-named ' the Gibbs phenomenon ' (see Carslaw *1925a*). There were differences between Wilbraham's and the later results (of J. W. Gibbs and others) ; Wilbraham took his work to clarify the behaviour of the series, whereas his successors recognised the distinction between the series (which has a jump at this point, by Dirichlet's conditions) and its geometrical interpretation as a limit curve (for which Wilbraham had refined Fourier's geometrical reasoning). Further, Gibbs did not discuss the oscillation close to the vertical line but the limiting case (see figure 3.10.1). In terms of double limits, the expression (3.10.5) is a function of both n and x ; while Dirichlet's conditions relate to the behaviour of the *repeated* limit $\lim_{n\to\infty}\lim_{x\to a}\{r_n(x)\}$, Wilbraham is concerned with the converse repeated limit $\lim_{x\to a}\lim_{n\to\infty}\{r_n(x)\}$, and Gibbs with the double limit $\lim_{x\to a,\, n\to\infty}\{r_n(x)\}$.

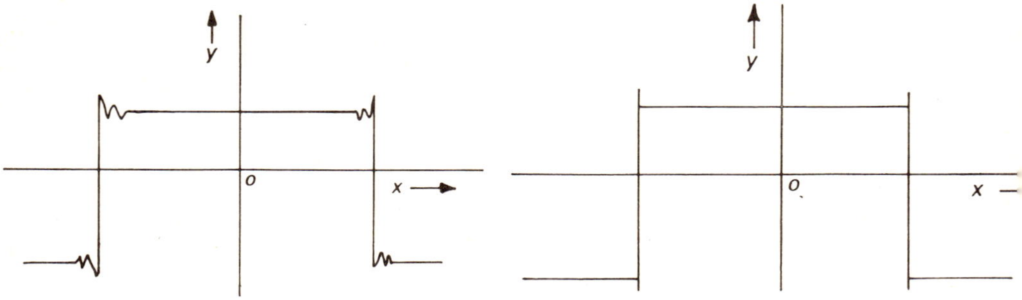

Wilbraham's interpretation. Gibbs's interpretation.

Figure 3.10.1.

Perhaps the interest which was being showed in Fourier series in the late 1840s caused Cauchy to look again at this theorem of the *Cours* on the sum function of a convergent series of functions. During the intervening period he had written some new papers on analysis, possibly motivated by his researches on the existence theorems for differential equations (see Dobrovolski *1971a*), and his follower the Abbé Moigno was publishing textbooks on the calculus (and other branches of mathematics) following Cauchy's principles (see Moigno *1840–1844a*) ; but no fundamental revisions to his system had occurred. However, in a paper *1853a* Cauchy admitted at last that the Fourier sine series of $\frac{1}{2}(\pi - x)$ over $[0, \pi]$ was a counter-example to his old theorem. He reformulated both his theorem and his necessary and sufficient condition for convergence so as to include what we now recognise as uniform convergence over an interval : $|r_n(x)|$ is ' always ' small—that is, it is small for all x in the interval. Thus in (3.10.2) the continuity of $s(x)$ is established by now familiar reasoning, which is much clearer than Seidel's and also than Cauchy's own ' insensible at the same time ' of the *Cours d'analyse* thirty years earlier (compare section 3.8 above) ; $\{r_n(x+h) - r_n(x)\}$ is now definitely small with h, and thus the left hand side of (3.10.2) is also small, as required.

So we find Cauchy making yet another contribution to mathematical analysis. But this one was not a fundamental breakthrough, for in his paper he did not explore the further consequences of his new definition, not even the other places in his system where uniformity of convergence had been taken for granted. For such advances we have to turn from France to Germany, whither the centre of mathematical attention had been moving.

3.11. *The impact of Riemann and Weierstrass*

In the investigations described in the last section we begin to see emerging the kind of rigour to which the modern textbooks accustom us. However, its eventual achievement is an extremely complicated story which would need a book of its own for the telling, and so I need to indicate how much of it will be conveyed in this chapter and the way in which it is treated.

In this section I shall describe the historical circumstances of the work of the principal figures, and in the rest of the chapter discuss in some detail the contributions made by them and their followers to the convergence of series of functions, to the extension of the theory of functions, and to the refinement of the differential calculus. For reasons of space I have considered only the most important definitions and techniques, and have not described the reworking of *every* theorem discussed earlier in this chapter. However, by way of compensation I have occasionally gone beyond the upper time limit of 1880 when some later developments could be concisely conveyed in a convenient context.

The most important contribution to Fourier series after Dirichlet's establishment of his conditions was Bernhard Riemann's *Habilitationsschrift* (second doctorate) of 1854 on trigonometric series. He never finished it, but after his death in 1866 his friend Dedekind recognised its importance and published it in 1868. To emphasise the date of its composition I shall cite it as Riemann *1854a*.

Riemann's essay falls into three parts: a history of trigonometric series, an analysis of the integral (which is discussed in section 4.3), and a miscellany of comments on functions and their representability by series. He had discussed series personally with Dirichlet, and took up some problems which Dirichlet had left. For example, he reported that Dirichlet had been partly motivated to Fourier series by the discovery that the sum of a conditionally convergent series could be changed by re-ordering its terms (compare Dirichlet *1837c*), and he generalised this result by showing that the series could be reordered so as to converge to *any* value (*1854a*, art. 5).

Riemann also constructed functions with infinitely many discontinuities or oscillations, and studied their integrals and possibility of representation by a Fourier series. But he seems to have felt the extension of Dirichlet's sufficient conditions for convergence to be too difficult, and so he examined the problem from the other end by seeking *necessary* conditions for convergence and analysing the circumstances under which the n-th term of the series would tend to 0 as $n \to \infty$. To do this he considered the function $F(x)$ obtained by twice formally integrating the trigonometric series term-by-term :

$$F(x) = C + C'x + \tfrac{1}{2}a_0x^2 - \sum_{r=1}^{\infty} (a_r \cos rx + b_r \sin rx)/r^2 \qquad (3.11.1)$$

and proved theorems on the behaviour of

$$\lim_{\alpha\to 0,\ \beta\to 0} \{(F(x+\alpha+\beta) - F(x+\alpha-\beta) - F(x-\alpha+\beta) + F(x-\alpha-\beta))/4\alpha\beta\} \qquad (3.11.2)$$

for convergent trigonometric series (*1854a*, arts. 8–12). In the process he not only proved his so-called ‘ localisation theorem ’, that the convergence of the series at a point depends only upon the behaviour of the function in the neighbourhood of that point (art. 9, where it takes a much more general form) but also posed many new problems. He suggested that two functions which differed at a finite number of points took the same series, which raised the questions of whether an infinity of such points can be allowed, how one is to understand the apparent insensitivity of the series to such exceptional points of non-convergence to the function, whether the convergence of the n-th term to 0 as $n \to \infty$ implied the convergence of a_n and b_n to 0, and whether every trigonometric series was a Fourier series (that is, whether there was always an integrable function to serve as the sum of a convergent trigonometric series). All these questions and some others (including extending Dirichlet's conditions) were a continual source of problems ; some are not even completely resolved today. They make Riemann's paper one of the most seminal in the history of mathematical analysis.

But what about the techniques to solve these problems, and the foundational questions posed by these techniques ? Here the principal figure was Karl Weierstrass. Weierstrass's career falls into two sharply distinct but almost equal halves. For his first forty years he formed and shaped his mathematical skills while holding obscure teaching positions (see Mittag-Leffler *1923a*) ; but when he began to publish substantial papers in the mid-1850s he rapidly won the approbation of the scientific world. The second half of his life was spent at Berlin University, where he exerted enormous influence through his teaching from the late 1850s to the late 1880s (see Biermann *1966a*). It is to him and his students that we owe most of the rigorous foundations for mathematical analysis which are now taught.

Thus the history of mathematical analysis during the last third of the 19th century is in notable measure the story of mathematicians applying Weierstrassian techniques to Riemannian problems. However, the writing of this history presents formidable difficulties

(Pringsheim *1899a*, Hawkins *1970a* and Medvedev *1975a* and *1976a* are important sources), partly for the complexity of the developments involved and especially due to the fact that Weierstrass did not publish his lectures and seems to have discouraged even the taking of notes. The ' new word ' was spread as much by correspondence and teaching as by research papers and published accounts (of which Pincherle *1880a* was a significant early example, and Dugac *1973a* is a recent historical survey). Hence, somewhat unusually for the mathematics of the late 19th century, we can take some account of unpublished materials, confident that their content was being disseminated to the community of mathematicians by the means described above.

Although Weierstrass probably saw his contributions to the foundations of mathematical analysis as building on Cauchy's achievements, he seems to have begun his reconstructions before reading any of Cauchy's works and was at times critical of Cauchy's methods in both real and complex analysis. In all his work on analysis, especially elliptic functions and differential equations, he made such extensive use of series, especially power-series, of both real and complex variables that he seems to have regarded them as providing some kind of foundation for mathematical analysis (see Jourdain *1906–1909a*). In this general sense he continued the views of Lagrange, but of course he brought to the approach a much more sophisticated treatment of the problems of convergence and of the existence of limits. Let me illustrate some of these points by returning to the years of his obscurity and tracing the genesis of his contributions to the theory of convergence.

3.12. *The importance of the property of uniformity*

Weierstrass probably learnt of elliptic functions from their study by his teacher Christoph Gudermann (see Manning *1975a*, 363–364). The theory of elliptic functions involved infinite series and therefore was a branch of mathematical analysis where convergence was an important matter (see Brill and Noether *1894a*). Gudermann was interested in comparing the rates of convergence of a series for different values of its argument variable, and at one point he mentioned that, like infinite series, infinite products ' have a convergence [which] on the whole [is at the] same rate [einen im Ganzen gleichen Grad der Convergenz] ' (*1838a*, 251–252). The context shows that he was using a form of what we now recognise as uniform convergence ; the n-th partial sum (or product) is distant from its limiting value by the same amount for all (or nearly all) values of the argument variable. However, it seems doubtful that he saw the potential of his idea for mathematical analysis, or even that his readers understood it here, for he gave it no formal definition and did not mention it again in his writings. But probably it came into his discussions at Münster with his student Weierstrass, for in a manuscript of 1841 on power-series Weierstrass spoke informally of their ' uniform convergence '—his phrase (' gleichmässig Convergenz '), and now ours (*1841a*, 68–69).

Weierstrass discussed uniform convergence in detail at his Berlin lectures, and in a paper *1880a* on the definability of functions in terms of power-series he showed some of the relationships between different modes of convergence. Firstly, he defined uniform convergence in a region as follows (*1880a*, art. 1) :

> An infinite series $\sum_{\nu=0}^{\infty} f_\nu$, whose members are functions of arbitrarily many variables, converges uniformly in a given part (B) of their region of convergence if after the assumption of an arbitrarily small positive quantity δ a whole number m can be specified, so that the absolute value of the sum $\sum_{\nu=n}^{\infty} f_\nu$ is smaller than δ for each value of n which is $\geqslant m$, and for each collection of values of the variables belonging to the region B.

He then defined the different property of uniform convergence in the neighbourhood of a point :

> Further, let a positive magnitude ρ be assumed for a specific point a of this region, so that the series converges uniformly for the values of x corresponding to the condition
>
> $$|x-a| \leqq \rho,$$
>
> then I shall say, that the series converges uniformly in the neighbourhood of the point a.

It is obvious that if the series converges over a region then it converges in a neighbourhood of a ; but the converse result is not so evident, and Weierstrass offered the following interesting proof :

> The quantity ρ then has an upper bound ; let it be R, then the collection of values of x for which $|x-a| < R$ may be called—in relation to the considered series—the vicinity of a, and R its half-measure. One assumes any point one likes in this vicinity, then it is clear that the series also converges uniformly in the neighbourhood of the latter [point]. It follows from this, that the collection of points in whose neighbourhood the series converges uniformly is represented in the plane [Ebene] of the variable x by a

simple surface [Flache] but which can exist as several pieces separated from each other.

The last sentence suggests that Weierstrass might also have had another mode of uniform convergence in mind ; namely, over intervals, where

$$\left| \sum_{\nu=n}^{\infty} f_\nu \right| < \delta \tag{3.12.1}$$

when x belongs to any one of a number of parts of the region of convergence. In these pages Weierstrass also offered a test for uniform convergence which is now called ' Weierstrass's M-test '.

Weierstrass's proof contains a tacit assumption of a theorem known today as the ' Heine–Borel theorem ' ; that if a bounded closed set can be covered by an infinite collection of open sets, then it can be covered by a finite sub-collection of those sets. This theorem was proved in 1895 by Emile Borel (see section 4.6 below) and was given the name (by Arthur Schönflies in *1900a*, 119), because Eduard Heine had also tacitly assumed it in proving in his *1872a* that a function $f(x)$ which is continuous and finite-valued over a finite interval is also uniformly continuous over that interval. (Since the similarity between Heine's and Borel's theorems is superficial, the name is misleading.) Heine's theorem is worth studying as a typical early example of the emergence of Weierstrassian methods in print. He defines uniform continuity as follows (*1872a*, 184) :

> A function $f(x)$ is called . . . *uniformly continuous* from $x=a$ to $x=b$, if for any given quantity ϵ, however small, there exists a positive quantity η_0 such that for all positive values η which are less than η_0, $f(x\pm\eta)-f(x)$ remains less than ϵ. Whichever value one may give to x, assuming only that x and $x\pm\eta$ belong to the region from a to b, *the same* η_0 must effect the required [property].

Notice Heine's use of symbols to help clarify the functional dependence of increments on the variables, in contrast to Cauchy's purely verbal definition of continuity in section 3.6 above. However, Heine failed to specify the positive value of $(f(x\pm\eta)-f(x))$—another characteristic of this time.

Heine's proof proceeds by partitioning the interval $[a, b]$ with points $x_1, \ldots x_{n-1}$ such that, for a given ϵ,

$$\left.\begin{array}{r} |f(x_1)-f(a)|=3\epsilon \;.\; \& \;.\; x\in[a, x_1] \;.\; \rightarrow \;.\; |f(x)-f(a)| \leqslant 3\epsilon, \\ |f(x_2)-f(x_1)|=3\epsilon \;.\; \& \;.\; x\in[x_1, x_2] \;.\; \rightarrow \;.\; |f(x)-f(x_1)| \leqslant 3\epsilon, \\ \cdots\cdots\cdots\cdots\cdots\cdots \\ x\in[x_{n-1}, b] \;.\; \rightarrow \;.\; |f(x)-f(x_{n-1})| \leqslant 3\epsilon. \end{array}\right\} \tag{3.12.2}$$

There may be an infinity of the x_r; but if so, then they have a limiting value X within $[a, b]$.[1] Let η_0 have the property that

$$0<\eta<\eta_0 \,.\ \rightarrow .\ |f(X)-f(X-\eta)|<\epsilon. \tag{3.12.3}$$

Now $x_n, x_{n+1}, x_{n+2}, \ldots$ all lie within $[X-\eta, X]$ for a suitably large n, so that, from (3.12.1) and (3.12.3), for them

$$|f(x_{n+1})-f(x_n)|<2\epsilon, \ldots, \tag{3.12.4}$$

infringing their definition in $(3.12.2)_1$. Hence there is only a finite number of x_r (this is the 'Heine–Borel' step), so that the theorem is proved from the definition of uniform continuity and (3.12.2) (Heine *1872a*, 188).

I shall now make some comments on the formulation of definitions of properties which involve uniformity. A series $\sum_{r=1}^{\infty} u_r(x)$ is *uniformly convergent over* $[a, b]$ if $|r_n(x)|$ is small when n is large for all x in the interval. In symbols, this definition may be written:

$$\epsilon>0\ ;\ \exists N\{\text{if } x\in[a, b] \text{ and } n\geqslant N, \text{ then } |r_n(x)|<\epsilon\}. \tag{3.12.5}$$

As we have seen, Cauchy and Weierstrass independently defined this mode of convergence, and Seidel struggled towards it.

The series is *uniformly convergent in the neighbourhood of a point* x_0 if there is a neighbourhood $[x_0-\delta, x_0+\delta]$ of x_0 around which it is uniformly convergent in the sense of (3.12.5). As we saw, Weierstrass had the conception of this mode. It is to be distinguished from *uniform convergence at the point* x_0, which is defined by the condition:

$$\epsilon>0\ ;\ \exists\delta, N\{\text{if } x\in[x_0-\delta, x_0+\delta] \text{ and } n\geqslant N, \text{ then } |r_n(x)|<\epsilon\}. \tag{3.12.6}$$

The difference between this mode of convergence and its predecessor is that the δ now depends on the ϵ and so decreases with it, hence giving uniform convergence *at* x_0, whereas in the previous definition the δ is independent of the size of ϵ.

In addition, there is a parallel sequence of modes of convergence

[1] Heine justifies this step by a definition of irrational numbers and a proof of the intermediate value theorem for a continuous function; but I omit them, as they correspond closely to Cantor's definition of irrationals and Weierstrass's proof of that theorem (sections 5.2 and 3.14 below respectively).

I use some of the notations listed in section 0.5. However, I have not given a *fully* symbolic form to these definitions. Such a version of (3.12.5), for example, would read:

$$(\forall\epsilon)(\exists N)(\forall x)(\forall n)\{\epsilon>0 \,.\, \& \,.\, n\geqslant N \,.\, \& \,.\, x\in[a, b] \,.\, \rightarrow \,.\, |r_n(x)|<\epsilon\}.$$

The distinction between the 'closed' form of this definition, where all the variables are quantified, and the 'schematic' character of (3.12.5), where some variables are free, is important with respect to the formal language being used; but it is often not observed for either definitions or axioms in any branch of mathematics.

called ' quasi-uniform convergence ', in which the condition that $|r_n(x)|$ is small for all values of $n \geqslant N$ is replaced by smallness for an infinity of values of n. As an example I shall formulate *quasi-uniform convergence in the neighbourhood of* x_0, for Stokes's ' not infinitely slow ' convergence (see section 3.10 above) is closely related to it :

$$\exists\delta[\epsilon, N > 0\ ;\ \exists n_0\{\text{if } n = n_0 > N \text{ and } x \in [x_0 - \delta, x_0 + \delta], \text{ then } |r_n(x)| < \epsilon\}]. \qquad (3.12.7)$$

With each of these modes there is also its negative (non-uniform and quasi-non-uniform convergence), and also more sophisticated modes such as uniform convergence by intervals, which I suggested earlier in this section may have been within Weierstrass's purview.

I shall not describe the history of the introduction of these modes, many of which came in only after 1880 (see Hardy *1918a*). My main purpose is to point out the *fundamental reworking* of mathematical analysis that the new distinctions involve. Their importance is often *not* well conveyed in modern textbooks, for they are treated as apparently minor refinements to a scattering of theorems ; but it *is* well treated in some of the older ones (see, for example, Carslaw *1930a* and Knopp *1951a*), and was emphasised by Weierstrass in his own teaching (according to Stolz *1881a*, 256). One aspect of their importance is the need for a large measure of symbolism.

While on the subject of the convergence of series of functions, I shall return briefly to the convergence of Fourier series and to a paper which presaged the Riemann–Weierstrass period and exerted some influence on it. In his doctoral thesis *1864a*—written after the drafting of Riemann's projected thesis but before its publication—Rudolph Lipschitz took up the extension of Dirichlet's conditions on to an infinity of oscillations and/or discontinuities. In considering how they could be distributed along the real line he developed some primitive ideas of measure and point set theory (compare sections 4.5 and 5.5). He required discontinuities to lie within a finite union of intervals of arbitrarily small total length over which the contribution to the integral of $f(x)$ is arbitrarily small. This latter stipulation begs the question of extending the conditions, however, and he decided to ' set aside research in these cases '. But with oscillations he was more successful. As the most general case he saw the interval divided up into sub-intervals over each of which either an infinity of oscillations fell within a sub-sub-interval and a finite number outside, or else an infinity fell within every sub-sub-interval (when they would be ' dense ' over the sub-interval, to use the modern terminology). By an intricate argument closely following Dirichlet's proof-method (described in section 3.9 above) he showed that convergence of the Fourier series occurred if $f(x)$ satisfied the ' Lipschitz condition ' of order α ;

$$|f(x+h)-f(x)|<Bh^{\alpha},\ h,\ B,\ \alpha>0, \tag{3.12.8}$$

over the interval, and so could include an infinity of damped oscillations within $(x,\ x+h)$.

Still more general conditions for the convergence of Fourier series, including the possibility of their uniform convergence, were developed during the rest of the century (see Paplauskas *1966a*). A Weierstrassian who wrote prolifically on these questions and indeed on the whole range of Weierstrassian analysis (if not always to the required standard of rigour) was Paul du Bois Reymond. For example, in his *1876a* he applied Riemann's second difference formula (3.11.2) to both Fourier series and integrals, and in an appendix refuted a conjecture of Dirichlet and Riemann that every continuous function has a convergent Fourier series. In his *1874a* he discussed the difference between repeated and double limits in considering the discontinuity or contiguity (he did not use this term) of functions across jumps ; indeed, had he not handled Dirichlet's partial sum formula (3.9.8) rather carelessly he would have obtained the ' Gibbs phenomenon ' which we saw Wilbraham examining in section 3.10 above. Some of his and much of others' work on Fourier analysis involved the more general definitions of the integral which will be discussed in chapter 4, and forms of summation other than term-by-term addition. But I shall set these developments aside, and turn to another important aspect of the Riemann–Weierstrass style of analysis : the extension of the class of definable functions.

3.13. *The post-Dirichletian theory of functions*

In his essay on trigonometric series Riemann accepted Dirichlet's very general conception of a function (see section 3.9 above), and he sought examples of functions which possessed an infinity of discontinuities or turning values but which also had a definable integral and/or a convergent trigonometric series. One of his examples is described in section 4.3.

The Weierstrassians took up the construction of analytic expressions which defined functions with infinities of oscillations and discontinuous within a finite interval, and they examined the consequences for mathematical analysis. A classification of these types of function was introduced by Hermann Hankel in his dissertation *1870a.* He considered functions $\phi(y)$ which have a convergent Taylor series over $[-1,\ +1]$ of y except when $y=0$, where the left- and right-hand limiting values are -1 and $+1$ respectively. He called such a point a ' singularity ', and then stated that such singularities could occur for infinitely many values of x in $f(x)$, where

$$f(x) = \sum_{r=1}^{\infty} \phi(\sin(r\pi x))r^{-s}, \quad s > 3. \tag{3.13.1}$$

He based his classification of functions on the ' condensation of singularities ' which they display (Hankel *1870a*, art. 4).

For continuous functions Hankel distinguished between those of decreasing and of constant amplitude, giving $x \sin 1/x$ and $\sin 1/x$ around $x=0$ as an example of each (art. 3; he mentioned Lipschitz's work here). For ' linear discontinuous functions ' he considered the ' collection [Schaar] ' of points in an interval at which ' a certain property ' (such as discontinuity or oscillation) occurred, and divided functions into three kinds: those for whose points ' no sub-interval [Intervall] however small can be specified in the interval [Strecke], in which at least one point of that collection does not lie ' ; those for which ' between any two arbitrarily near points of the interval a sub-interval can always be found in which no point of that collection lies ' ; and ' total linear discontinuous functions ', whose ' points with jumps which exceed a certain finite magnitude *fill up* the whole interval ' (art. 7).

As with Lipschitz, there is point set theory nascent here, which will be seen in chapter 5 to grow in Cantor's hands. Hankel devoted most of his paper to theorems related to his types of function and examples of them. Of particular interest is the totally discontinuous characteristic function of the irrationals, for which he provided the analytical definition

$$\sum_{r=1}^{\infty} r^{-s} \Big/ \sum_{r=1}^{\infty} r^{-s}(\phi(\sin r\pi x))^{-2}, \tag{3.13.2}$$

where $\phi(y)$ is the Fourier sine series of 1 over $[0, \pi]$ of y (art. 9), contrary to Dirichlet's apparent belief that this function could not be so represented (see section 3.9 above). Simpler forms of expression for this function, of which the most familiar today is

$$\lim_{m\to\infty} \lim_{n\to\infty} \{\cos(m!\pi x)^{2n}\}, \tag{3.13.3}$$

were found in the next twenty years (for some references, see Pringsheim *1899a*, 41).

Further such examples were studied in a long paper *1875a* on various aspects of Weierstrassian analysis and Riemannian problems by Gaston Darboux. His paper was of considerable importance in bringing the new Germanic ideas to the attention of French mathematicians, and we shall meet it again in section 4.4. He extended (and criticised) Hankel's handling of (3.13.2). Among other things, he stated that if

$$\phi(y)=y^2 \sin 1/y \tag{3.13.4}$$

and $\sum_{r=1}^{\infty} a_r$ is absolutely convergent, then

$$f(x)=\sum_{r=1}^{\infty} \pi a_r \phi'(\sin(r\pi x)) \cos(r\pi x) \tag{3.13.5}$$

is uniformly convergent even though its terms are discontinuous functions, and takes

$$F(x)=\sum_{=1}^{\infty} r^{-1} a_r \phi(\sin(r\pi x)) \tag{3.13.6}$$

as its indefinite integral. ‘ We thus obtain *a continuous* function of which the derivative is discontinuous for all commensurable [rational] values of x, but exists for all values of x ’, he concluded (p. 109). After considering this and other examples of functions constructed in his paper, he remarked understandably that ‘ it seems difficult to indicate a general character which allows the recognition of whether a function $f(x)$ has a primitive function, that is to say, if it is the derived [function] of another function ’ (p. 111). Note the durability of Lagrange's terminology, defined in section 3.4 above ; it persists in French mathematics to this day.

With functions such as these we see the question of continuity vis-à-vis differentiability arise in a stark form. Could there be a continuous function which is differentiable nowhere ? Weierstrass answered this question in the affirmative in a manuscript *1872a* with the example

$$\sum_{r=0}^{\infty} b^r \cos(a^r \pi x), \quad 2ab > 2+3\pi, \tag{3.13.7}$$

where a is an odd integer and $0<b<1$. He published it in his *1880a*, art. 6, but it appeared first in du Bois Reymond *1875a*, 29–31. An extensive discussion of such functions was carried on in print as well as in correspondence (see Hawkins *1970a*, 44–50), and a large number of continuous non-differentiable functions were invented over the following decades (see Boas *1969a*). Weierstrass also proved an important result for his belief in power series when in his *1885a* he proved his ‘ approximation theorem ’ (that any continuous function can be uniformly approximated over a finite interval by a sequence of polynomials, and that it could be so approximated by a sequence of finite sums of sines and cosines).

The French mathematician René Baire introduced some important refinements, especially in his *1899a* (see Dugas *1940a*, 64–71). He defined the properties of upper and lower semi-continuity, which related to the question of when a continuous function achieved the upper and lower bounds of its values over a closed interval of its argument. He also introduced a classification of functions, in which the lowest class was that of continuous functions, and each function in a higher class was defined as the limit function of a convergent sequence of functions of the immediately lower class. This classification led Baire's colleague Henri Lebesgue to study at length in *1905a* the scope of the analytical representability of functions.

3.14. *Refinements to proof-methods and to the differential calculus*

In this final section on Weierstrassian analysis I shall survey some of the refinements introduced to the differential calculus. In preface I shall comment briefly upon the standard of rigour which Weierstrass sought to see applied in all areas of mathematical analysis.

I mentioned in section 3.6 above that Cauchy's verbal formulation of a limit did not explicitly state the functional relationship between the variables involved. Weierstrass popularised the practice of expressing this relationship in the ' ϵ, δ ' form of the definition (for an early example, see Dugac *1973a*, 119). Augustus de Morgan anticipated this form of definition in his *1835a*, 154–155, but he did not develop it in his writings on analysis.

As part of his programme Weierstrass introduced a definition of irrational numbers.[1] One motivation for the definition was to make sure that infinitesimals could be dispensed with entirely, although some of the Weierstrassians affirmed forms of actual infinitesimals (see Vivanti *1891a*). More important was the need to avoid the question-begging proofs by his predecessors of theorems, such as the intermediate value theorem, on the existence of limits. An important lemma in the proof of such theorems was the ' Bolzano-Weierstrass theorem ', which asserted that an infinite bounded set (of real numbers) contains at least one limit-point. According to those who heard the theorem in his lectures, Weierstrass used in his proof the method of successively dividing up the set and homing in on the required point (see Pincherle *1880a*, 237), and they gave the theorem this name because of the (unrigorous) use of this method in Bolzano's proof of the intermediate value theorem in his *1817a*. The same kind of proof-method was also used to prove theorems such as that a continuous function achieves its upper and lower bounds over a finite interval (see Dugac *1973a*, 77, 120).

[1] As the definition is complicated I shall not describe it but mention the accounts given by Weierstrass's students in Pincherle *1880a*, Dugac *1973a* (an edition of manuscripts) and elsewhere. Sections 5.2 and 6.2 below include accounts of Cantor's and Dedekind's rather simpler theories, which were introduced for the same purposes. For a survey of these definitions, see Pringsheim *1898a*.

Weierstrass also made effective use of these theorems in the differential calculus. The following lemma is characteristic : If $f(x)$ and $f'(x)$ are both continuous and finite-valued over $[a, b]$ and $f'(x)$ is always positive, then $f(b) > f(a)$, and $f(b)$ and $f(a)$ are the largest and smallest values that $f(x)$ can take over that interval (so that $f(x)$ achieves its bounds within it ; see Dugac *1973a*, 118–122). The key equation is

$$f(x_0+h)-f(x_0)=h\{f'(x_0)+hh_1(h, x)\}, \tag{3.14.1}$$

where $h_1 \to 0$ as $h \to 0$ (thus correctly expressing the functional dependence of the ' error term ' on h). In a letter of 1870 to his fellow Weierstrassian Cantor, K. H. A. Schwarz used such results to prove that : if $f(x)$ is continuous and finite-valued over $[a, b]$, and for every x

$$\lim_{h\to 0}\{(f(x+h)-f(x))/h\}=0, \tag{3.14.2}$$

then $f'(x)$ exists and is always zero, so that $f(x)$ is constant. In his proof Schwarz considered the functions

$$f(x)-f(a)-k(x-a) \quad \text{and} \quad f(x)-f(a)+k(x-a),$$

where k is positive, and used Weierstrass's lemma to show that the former function is negative and has $-k$ as derivative and the latter is positive with derivative $+k$, and hence that

$$|f(b)-f(a)|<2k(b-a). \tag{3.14.3}$$

But k is arbitrary and so can be made as small as we like ; and b is arbitrary also. Hence from (3.14.3) the desired conclusions follow. ' The proof above seems to be completely rigorous ', Schwarz wrote proudly to Cantor, ' It is the foundation of the differential and integral calculus ' (Meschkowski *1964a*, 89).

From proofs such as these more rigorous treatments of the mean value theorem of the differential calculus were possible. For by considering the behaviour of the derivative f' of the continuous f over $[a, b]$ one could use Weierstrass's proof of the intermediate value theorem to prove the mean value theorem without using Cauchy's assumption that the derivative was always continuous (compare section 3.7 above). This line of reasoning was first produced by Ossian Bonnet, a French mathematician who made notable contributions to mean value theorems, and was published in Serret *1868a*, 17–23 ; Darboux and Schwarz were among those who developed extensions. Similarly, the convergence of Taylor's series was studied further (both for functions of one and of several variables), and various new expressions obtained for the remainder term obtained ; it was also applied to a correspondingly more advanced study of the maxima and minima of a function (see Voss *1899a*, 65–88).

Very many more results in the differential calculus could be described ; for example, the definitions of differentials and the theory of partial derivatives. But I shall conclude this section with the Weierstrassians' treatment of a pair of quartets—four derivatives and four limits. We bring in Dini, the first of the Italians to learn the new Weierstrassian techniques. He published a textbook *1878a* on the foundations of analysis which was so good that Cantor tried to organise a German translation (see Grattan-Guinness *1974a*, 115–116) ; eventually a German edition (prepared by others) was published as *1892a*.

Dini accepted the new types of function which had been introduced by Weierstrass and Riemann, and realised that, since their non-differentiability was due to their continually oscillatory character, the basic idea of defining the derivative of a continuous function as the limiting value of the difference quotient would need refining. For this purpose he took the intervals (a, x) and (x, b), and considered separately the behaviour of the difference quotient over each interval. Let l_x and L_x be the greatest lower bound and least upper bound of the values of the quotient over (x, b) :

$$l_x \underset{\text{Df}}{=} \inf \{(f(x+h)-f(x))/h \mid x+h \in [x, b)\}, \tag{3.14.4}$$

$$L_x \underset{\text{Df}}{=} \sup \{(f(x+h)-f(x))/h \mid x+h \in [x, b)\}. \tag{3.14.5}$$

x is fixed, and l_x and L_x are functions of b. Now reduce the size of the interval by bringing b towards x. h is correspondingly reduced in its range of values, and thus the quotient must take a smaller set of values. Hence l_x will increase in value (or remain constant) as b approaches x, and L_x will similarly decrease. Hence each will take a limiting value when b reaches x, which we now usually call the ' lower and upper right-hand derivatives ' of $f(x)$ at x and write as ' $D^+f(x)$ ' and ' $D_+f(x)$ '. From (3.14.4) we can see that the formal definition of $D^+f(x)$ is

$$D^+f(x) \underset{\text{Df}}{=} \lim_{b \to x^+} \sup \{(f(x+h)-f(x))/h \mid x+h \in [x, b)\} \tag{3.14.6}$$

(Dini *1878a*, esp. art. 145).

We can simplify this formulation by realising that the limit of the sequence of least upper bounds of the difference quotient as b tends down to x is the upper limit of the quotient as h decreases to zero. Hence we may replace (3.14.6) by

$$D^+f(x) \underset{\text{Df}}{=} \overline{\lim_{h \to 0^+}} \{(f(x+h)-f(x))/h\}. \tag{3.14.7}$$

We similarly define $D_+f(x)$ in terms of the lower limit of this difference quotient, and $D^-f(x)$ and $D_-f(x)$ as the upper and lower limits of the quotient $\{(f(x)-f(x-h))/-h\}$ (Dini *ibid.*). From the existence or non-existence of these derivatives and their mutual equality or inequality the differentiability of continuous functions could at last be studied adequately (see Hawkins *1970a*, 45–54).

A notable feature of this approach is the use that it makes of both greatest lower and least upper bounds (' inf ' and ' sup ' in (3.14.4) and (3.14.5) above) and also of lower and upper limits. I mentioned in section 3.8 above that Cauchy did not define the upper limit very clearly in his proofs of convergence tests, and until Weierstrass's lectures the distinctions between these types of limit were not always clearly understood. With their introduction the study of limits was very greatly enhanced (see Pringsheim *1898a*) ; but the literature is not easy to read, partly for the occasional remaining unclarities and also because of terminological difficulties. The greatest lower bound and least upper bound were usually called upper and lower ' Schrank ' or ' Grenze ', each of which can be translated as either ' limit ' or ' bound ' ! Lower and upper limits were called upper and lower ' Limes ' or ' Limite ', which are specialised technical terms which must be assigned to ' limit ' in translation ; and du Bois Reymond introduced in his *1870a*, art. 3 the ' limits of indefiniteness ' of a series, concepts which we now recognise as its lower and upper limits. Further, the notations were not standard among authors ; for example, only towards the end of the century were ' lim inf ' and ' $\underline{\lim}$ ', and ' lim sup ' and ' $\overline{\lim}$ ', introduced as symbols for lower and upper limits.

Another mathematician who studied the difference between these types of limit was our second Italian, Peano. Early on in his career he published a textbook *1884a* on the foundations of the calculus, which was of comparable importance to Dini's and also received in due course a German translation. It was an edition of the lectures of his teacher Angelo Genocchi prefaced by his own notes, which were the most important part of the book. Especially noteworthy is its attention to multi-variate analysis, including Taylorian expansions, the uniform continuity of functions of several variables, and an example of a function whose mixed partial derivatives are not equal.

Peano continued his interest in the foundations of analysis, and devoted several papers, especially his *1894a*, to the different kinds of limit and to the theory of multiple limits (see Cassina *1936a*). He also wrote on mean value theorems and the definition of the derivative. But perhaps he is most important in drawing attention to the symbolic language needed to express the subtleties of mathematical analysis. Throughout the century we see a gradually increasing use of symbolism in the subject, and Peano extended this practice to new lengths by using the notations that he had introduced in the course of his work in mathematical logic. If we say 'for all ϵ, then there is a δ . . .' then we ought to symbolise this part of the sentence as well as the succeeding inequalities, and it is Peano who paved the way for the kind of symbolisation of analysis which I have used in this chapter (especially in section 3.12).[1]

Peano was also significant in propagating the work of himself and the mathematicians around him in a journal which he started in the 1890s (initially called *Rivista di matematica* but varied in later volumes) and a textbook (called *Formulaire de mathématiques* at first but also changed in later editions). Thus it is appropriate for me to conclude my outline of the progress of mathematical analysis with Peano, for his and his followers' contributions to integration, set theory and the foundations of mathematics will find mention in the remaining three chapters of this book. But before moving forward, let us look back at the three chapters which are now completed.

3.15. *Unification and demarcation as twin aids to progress*

How do we measure progress in mathematical knowledge? The scope and generality of a mathematical theory is one criterion, which can be

[1] Sometimes the symbolisation is abused nowadays, so that the purpose of an epsilontic proof is not made clear. A proof often hinges on showing the smallness of some function F of the independent and dependent variables and constants of the problem when the suffix variables are large and the increments on the real variables are small. Hence *an order of smallness for F* is sought—$O(\epsilon)$ (sometimes $o(\epsilon)$), in Landau's notation. But textbook proofs are sometimes rigged so that they yield a naked 'ϵ' on the last line. I once saw a 'proof' which early on read something like:

$$\text{`If } n > N, \text{ then } |r_n(x)| < (7\pi^2\sqrt{\sigma})\epsilon/(2A+k^3)\text{'}, \tag{1}$$

so that the virginal 'ϵ' could be displayed at the end. This is mathematical education at its rabbit-out-of-hat worst; two of the constants hovering around the 'ϵ' in (1) were not even *defined* until several lines later!

Another criticism of textbook proofs in mathematical analysis concerns the presentation of inequalities. The proofs often read:

$$\left.\begin{aligned} \mathscr{A} &\leqslant \mathscr{B}_1 \\ &\leqslant \mathscr{B}_2 \\ &\leqslant \mathscr{B}_3 \ldots, \end{aligned}\right\} \tag{2}$$

where each expression $\mathscr{B}_i$ in the proof is related back to the original expression $\mathscr{A}$. But (2) does not show how the successive $\mathscr{B}_i$ relate to each other. It would be much clearer to the student if the signs were understood to relate to *successive* expressions, so that (2) would be replaced by the more informative

$$\left.\begin{aligned} \mathscr{A} &\leqslant \mathscr{B}_1 \\ &= \mathscr{B}_2 \\ &< \mathscr{B}_3 \ldots \end{aligned}\right\} \tag{3}$$

(say).

applied to its definitions and theorems, to the accumulation of solutions to special cases, to the results obtained for successively more complicated versions of the same type of problem, and so on. For example, we saw in chapter 2 that Leibniz's calculus, based on partitioning the range of the variable x into increments of (constant or variable) magnitudes dx, provided solutions to wider classes of problems than the various methods used earlier in the century, and that the inverse relationship between differentiation and integration vastly increased the scope of these methods. In this chapter we have seen the umbrella widen still further as the calculus became part of a limit-based mathematical analysis, and in the remaining chapters we shall see the same process of extension occur in definitions of the integral, in the consequences of the introduction of set theory, and in foundational studies in mathematics.

But much of our concern here, as also in the education and appreciation of mathematics, is with *foundations* and *rigour* of a mathematical theory ; and some special features occur here in the assessment of progress. I do not intend to dwell long on a difficult and complicated matter, but I shall exemplify a few features which have emerged during this chapter.

From the point of view of rigour, a theorem is only as sound as its proof allows it to be. Thus, strictly speaking, different proofs may prove different theorems, even if the same words and symbols are used in their statements. Now an important component in assessing the rigour of a proof is examining its *definitions*, for they are the places at which the chain of assumptions stops ; eventually we come to a definition whose defining expression uses necessarily undefined concepts. Thus rival mathematical theories can be assessed in terms of whether certain fundamental results are proved as theorems based on deeper definitions of other ideas, or whether they have to be assumed. For example, Cauchy's differential calculus has deeper roots than Lagrange's because Taylor's series is a theorem there whereas with Lagrange it has to be taken for granted. I use the word ' root ' quite deliberately, for its connotation of digging downwards for new foundations is exactly the sense in which foundations should be understood, as Gottlob Frege pointed out in the passage which I quoted at the head of chapter 0.

Definitions also play another important role ; they introduce *new distinctions*. Take, for example, the modes of uniform and non-uniform convergence described in section 3.12 above. Each definition there brings in a distinction which previously had been overlooked and with respect to which earlier work is incoherent. Thus Cauchy's ' false ' theorem from the *Cours d'analyse* is obscure there for want of these distinctions, and—relating this point to my earlier one that theorems change with their proofs—his theorem becomes different theorems in

Seidel's, Stokes's and his own later hands, because different new distinctions are introduced each time and thus affect the old proof in different ways.

Hence in foundational studies we see greater generality emerging from finer distinctions—the 'unification and demarcation' of the title of this section. I shall note a few examples from the first three chapters of this book, and leave as a hint for later reading that they also occur in the last three chapters.

Calculus techniques began as a rag-bag of partly geometrical techniques to which Leibniz's (and Newton's) calculi brought unity. However, Leibniz's calculus itself suffered ambiguity in the definition of higher-order differentials and the dependence of expressions using them on the particular infinitesimal partitions of the ranges of the variables. Thus it was superseded by Euler's version of the calculus, with one or several variables assigned as independent by the criterion of their uniform partition, and the other variables treated as functionally dependent upon them. Cauchy urged a new approach, based on the theory of limits; it united the calculus, the convergence of series and the theory of functions to form mathematical analysis. But it was at that stage only a *single*-limit analysis, whereas some of the major problems required a theory of *multiple* limits (or, more generally, *multi*-variate methods), before which it faltered. However, Dirichlet's work on Fourier series—the most prominent though by no means the only multiple-limit problem of the time—hinted at the way ahead to the multiple-limit analysis which the Weierstrassians were eventually to achieve, including the treatment of modes of convergence of series, attention to the structure of the real line, some new proof-methods (especially for existence theorems), and an extensive use of mathematical symbolism.

In emphasising the distinction between single and multiple limits I *do* intend to question the customary appelations attached to mathematical analysis. The term 'arithmeticisation of analysis' became widely used at the end of the last century because the definitions and proof-methods were reduced to real numbers and their arithmetic of inequalities. But the term hardly highlights the new features, for it could be applied with comparable strength to the manipulation of differentials in the previous century. More recently, the revolting word 'epsilontics' has been applied, in allusion to the '(ϵ, δ)' style of framing definitions. Now this is quite satisfactory for Cauchy's single-limit analysis, for he sometimes used exactly these symbols to denote small quantities related to increments on the independent and dependent variables—his proof of the mean value theorem described in section 3.7 above is an example. But otherwise 'epsilontics' is an unhelpful term, for it embraces indifferently both the (ϵ, δ) style of Cauchy's single-limit style and also the (ϵ, δ, N, . . . x, . . .) style of the multiple-limit analysis which the Weierstrassians were to introduce as such a fundamental reappraisal of the Cauchy period.

But, as we know from our textbooks, still finer distinctions are now used in mathematical analysis. With their introduction the rest of this book is much concerned.

Chapter 4

The Origins of Modern Theories of Integration

Thomas Hawkins

4.1. *Introduction*

At the beginning of the 20th century the French mathematician Henri Lebesgue created the theory of integration that became the paradigm for modern theories of the integral. The purpose of this chapter is to trace those developments in 19th-century analysis which provided the conceptual framework and the motivation for his work.

The starting-point is the changing conception of a function. In formulating his theory of integration, Lebesgue accepted the definition of a function of a real variable as any correspondence $x \to f(x)$ between real numbers, realised that the functions definable in this manner are so general that the meaning of the integral requires investigation, and sought to base that investigation upon the idea of the integral as an area. These considerations, which form the starting-point of his work, did not originate with him but in the first half of the 19th century. Before discussing the manner in which they arose, it will be helpful to recall the way in which functions and integration tended to be conceived in the 18th century.

During the second half of the 18th century, when the concept of a function became fundamental to mathematical analysis (see sections 2.7 and 3.3–3.4), functions and integration were regarded primarily in algebraic terms. A function $f(x)$ was some sort of an equation, possibly with an infinite number of terms ; and integration was the problem of determining a primitive function $F(x)$ which has $f(x)$ as its derivative, or, more generally, as the problem of determining an equation representing the solution of a differential equation. The algebraic orientation is not at all surprising, since the mathematical analysis of the 18th century originated with the transformation by Viète and Descartes of the geometrical analysis of the Greeks into a method of analysis by equations, which at the hands of Newton, Leibniz, Euler and others had proved to be an incredibly powerful device for solving diverse geometrical and mechanical problems.

Integration was a part of this essentially algebraic approach to problems. That integration was related to the determination of areas was naturally understood. Indeed, the Leibnizian notation $\int y\,dx$ was a constant reminder that the integral could be conceived of geometrically as the sum of infinitesimal rectangles $y\,dx$; and occasionally mathematicians would estimate the value of an integral by computing the area of an approximating polygonal figure. But the determination of areas represented no more than a single application of a much more generally applicable, and essentially algebraic, operation applied to algebraically conceived functions.

The conception of a function as a correspondence $x \to f(x)$ and of the integral as, in essence, an area, involved a return to a more geometrically oriented viewpoint. Actually that viewpoint already existed to a certain extent in the 18th century and was especially evident in the controversy over the vibrating string problem (see section 3.3). Recall that the form which d'Alembert gave to the solution to the wave equation prompted Euler to conclude that the functions which generate the solutions need not be ' continuous ', that is, given by a single equation. They could be ' irregular ' or ' absolutely arbitrary ' functions, by which he seems to have meant functions defined for different intervals of their domain by different equations and even functions not given by equations at all but by curves ' traced at will ' in the plane. These functions he called ' discontinuous '. Despite Euler's willingness to bring these more geometrically conceived functions within the purview of analysis, the result was not a reformulation of the foundations of analysis. The first steps towards a more geometrical approach to functions had been taken, but such an approach was not to bear fruit until it was re-introduced by Fourier at a time when a more critical attitude towards the foundations of analysis was emerging.

4.2. *Fourier analysis and arbitrary functions*

A side issue in the vibrating string controversy had been whether the arbitrary functions that, according to Euler, generated solutions are actually expressible as a series of sines. Daniel Bernoulli had argued on the basis of physical considerations that they must be, but no one else appears to have agreed with him. To Euler and Lagrange it seemed a contradiction in terms to assert that a completely arbitrary function is always expressible as a sum of sines over a finite interval, for such a representation appeared to delimit its arbitrariness. Fourier's investigation of the conduction of heat in solids, however, provided him with a strong mathematical reason for wanting Bernoulli's claim to be justified. To solve the boundary value problems to which his analysis led, Fourier used the now-familiar technique of separation of variables. The validity of the method depended upon the assumption that an 'arbitrary' function $f(x)$, which arises in the boundary conditions, can be represented by a trigonometric series. Specifically, if $f(x)$ is defined for the interval $[-l, l]$, then

$$f(x)=\tfrac{1}{2}a_0+\sum_{n=1}^{\infty}\left\{a_n\cos\left(\frac{n\pi x}{l}\right)+b_n\sin\left(\frac{n\pi x}{l}\right)\right\} \tag{4.2.1}$$

for all x in this interval.

Fourier was familiar with the 18th-century literature on the vibrating string problem and was made well aware of the lack of popularity of Bernoulli's position when Lagrange opposed acceptance of his first memoir (1807) by the Paris *Institut de France*, due, to some extent, to the fact that it depended upon the acceptance of Bernoulli's position. Space does not permit a detailed discussion of the case Fourier presented for the validity of (4.2.1). Judged by today's standards, some of his arguments are no more convincing than those of his 18th-century predecessors. However, he highlighted *for the first time* what would nowadays be assumed to be the heart of the matter: do the partial sums of the series in (4.2.1) converge to $f(x)$? He provided a strong case for an affirmative answer, although not an incontrovertible demonstration. From the standpoint of the history of the integral, however, one of his more 18th-century arguments is especially significant.

Fourier regarded (4.2.1) as an equation in an infinity of unknowns $a_0, a_1, a_2, \ldots, b_1, b_2, b_3, \ldots$. He showed that it could be 'solved' for these unknowns in the following manner. Consider, for example, a_0. From (4.2.1) it 'follows' that

$$\int_{-l}^{l} f(x)\,dx=\int_{-l}^{l}\tfrac{1}{2}a_0\,dx+\sum_{n=1}^{\infty}\int_{-l}^{l}\left\{a_n\cos\left(\frac{n\pi x}{l}\right)+b_n\sin\left(\frac{n\pi x}{l}\right)\right\}dx. \tag{4.2.2}$$

Since the integrals involving the sines and cosines are all zero, (4.2.2) yields

$$a_0=\frac{1}{l}\int_{-l}^{l} f(x)\,dx. \tag{4.2.3}$$

Here Fourier has taken for granted that the integral of an infinite sum is the sum of the integrals of its terms—an assumption that no-one as yet had seen reason to question. In other words, he had assumed that

$$\int_{-l}^{l} \{\lim_{n\to\infty} s_n(x)\}\,dx = \lim_{n\to\infty} \int_{-l}^{l} s_n(x)\,dx, \tag{4.2.4}$$

where

$$s_n(x) = \tfrac{1}{2}a_0 + \sum_{k=1}^{n} \left\{ a_k \cos\left(\frac{k\pi x}{l}\right) + b_k \sin\left(\frac{k\pi x}{l}\right) \right\}. \tag{4.2.5}$$

Using this assumption he obtained in similar fashion the formulas

$$a_n = \frac{1}{l}\int_{-l}^{l} f(x)\cos\left(\frac{n\pi x}{l}\right)dx, \quad b_n = \frac{1}{l}\int_{-l}^{l} f(x)\sin\left(\frac{n\pi x}{l}\right)dx. \tag{4.2.6}$$

It was the trigonometric series (4.2.1) with these values for the coefficients—the *Fourier series,* as we now say—that Fourier claimed converges to $f(x)$.

The definitive version of Fourier's case for the correctness of his Bernoulli-type claims appeared in his book *Théorie analytique de la chaleur* (*1822a*). By the time it appeared, Lagrange was no longer alive, and it is unclear whether Fourier ever convinced him. Fourier himself, however, was fully convinced that the ' discontinuous ' or ' arbitrary ' functions of Euler were, in fact, representable by trigonometric series. Fourier's arbitrary functions were actually more general than those that occur in the vibrating string problem. Since $y = f(x)$ represents the form of the string when $t = 0$ it must all be in one piece, whereas such a restriction was not necessary in the physical context of heat conduction ; so Fourier also considered ' arbitrary functions ' such as that sketched in figure 4.2.1. (Perhaps he included the solid vertical line to convey a semblance of continuity ; compare section 3.5.)

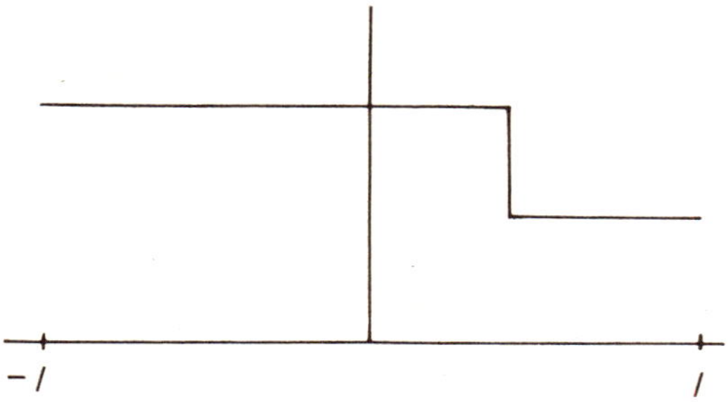

Figure 4.2.1.

The manner in which Fourier described his ' arbitrary functions ' is extremely important (*1822a*, art. 417) :

> In general the function $f(x)$ represents a succession of values or ordinates each of which is arbitrary. An infinity of values being given to the abscissa x, there is an equal number of ordinates $f(x)$. All have actual numerical values, either positive or negative or nul.
>
> We do not suppose these ordinates to be subject to a common law; they succeed each other in any manner whatever, and each of them is given as if it were a single quantity.

The conception of a function implied by these words is certainly far removed from the 18th-century idea of a 'continuous' function. Fourier's description is similar to Euler's description of 'discontinuous' functions, although it tends to emphasise more their arbitrariness. If it is taken literally, it implies the modern idea of a function as any well-defined correspondence $x \rightarrow f(x)$ between real numbers.

It is not clear how literally Fourier intended his description to be taken. The examples of arbitrary functions given in his book are all comprised of a finite number of 'continuous' pieces for $-l \leqslant x \leqslant l$. His remarks suggest he recognised the irrelevance of the law or laws (that is, equations) governing $f(x)$ for the validity of (4.2.1), although perhaps not all the implications of that irrelevance. In any case, he left it for others to develop those implications.

By venturing beyond the realm of 'continuous' functions, Fourier also had to give up the 18th-century conception of the integral. The fact that the coefficients in the representation (4.2.1) are given by (4.2.6), and therefore are definite integrals, required a reconsideration of the *meaning* of those integrals when $f(x)$ is 'arbitrary'. Obviously it was inappropriate, if not impossible, to speak of anti-derivatives of equations, and Fourier consequently resorted to a more geometrical interpretation: the integrals are to be conceived as areas. Since for each x an ordinate $f(x)$ exists, these ordinates determine a region in the plane, and he never doubted that this region had a definite area. He had unwittingly raised a significant mathematical question: precisely how does one define $\int_{-l}^{l} f(x)\,dx$ as an area when f is arbitrary? We shall now consider how three mathematicians, Cauchy, Dirichlet and Riemann, responded to it.

4.3. *Responses to Fourier, 1821–1854*

The first response to the question raised by Fourier's work came from Augustin-Louis Cauchy, a precocious mathematician who quickly achieved the prominent position of professor at the prestigious *Ecole Polytechnique* in Paris. In his *Cours d'analyse* (*1821a*) and *Résumé des leçons données à l'Ecole royale polytechnique sur le calcul infinitésimal* (*1823a*) Cauchy sought to develop the fundamental propositions of the calculus by methods as logically rigorous as those found in Greek geometry and to never have recourse 'to reasoning drawn from the generality of algebra' (see the quotation in section 3.2). The extent to which his treatises were influenced by Fourier's work is uncertain; he had other reasons for rejecting a formal, algebraic approach to the foundations of the calculus. Nevertheless, his treatises did represent a response to the questions tacitly raised by Fourier; for in striving to avoid the generalities of algebra Cauchy introduced a more geometrically inspired approach to analysis which provided, intentionally or not, an answer to Fourier's question about the meaning of the definite integral.

The starting point of Cauchy's reconstruction of analysis was the notion of a continuous function. As we saw in section 3.6, he dropped the 18th-century characterisation of continuity in favour of a more analytical one: a function $f(x)$ which is single-valued and finite-valued for x between a and b is continuous between these limits if the absolute value of $f(x+\alpha)-f(\alpha)$ 'decreases indefinitely with that of α' (*1821a*, 34–35; *Works*, 43). Because it is independent of the manner of representation of $f(x)$ as an equation, his definition is compatible with Fourier's conception of a function as a succession of ordinates.

In his *Résumé* (*1823a*) Cauchy proceeded to use his concept of a continuous function to present an approach to integration that was likewise compatible with Fourier's viewpoint. Restricting himself to continuous functions, Cauchy defined $\int_a^b f(x)\,dx$ as follows. Consider a partition P of $[a, b]$:

$$a=x_0<x_1<x_2<\ldots<x_n=b, \tag{4.3.1}$$

and the 'Cauchy sum'

$$S=\sum_{i=1}^{n} f(x_{i-1})(x_i-x_{i-1}). \tag{4.3.2}$$

By using the assumed continuity of f Cauchy was able to prove that for any two partitions P and P', the corresponding sums S and S' can be made to differ by an arbitrarily small amount provided the lengths of all sub-intervals $[x_{i-1}, x_i]$ in both partitions are sufficiently small.[1] Thus 'the value of S will obviously end up being constant . . . This limit is called a *definite integral*' (Cauchy *1823a*, lecture 21).

Cauchy had, in effect, provided an answer to the question raised by

[1] The proof uses continuity in the sense of uniform continuity. It is unclear what type of continuity Cauchy had in mind in his definition and likely that he did not grasp the distinction between pointwise and uniform continuity; compare section 3.12.

Fourier's remarks. His approach to the definite integral showed that its existence was not dependent upon the existence of an equation defining f and implied that $\int_a^b f(x)\,dx$ has a definite value for any ' arbitrary function ' *provided* that it is continuous in his sense. In fact, his definition and existence proof can readily be extended to cover the case of a bounded function with a finite number of points of discontinuity in $[a, b]$. In a later note (*1849a*) he proclaimed that such functions are precisely the kind of arbitrary functions that occur in the boundary value problems of mathematical physics. Perhaps Fourier would have agreed, had he been alive ; but Fourier's declarations in his *Théorie analytique de la chaleur*, if taken literally, posed an even more general problem than that resolved by Cauchy's definitions : can $\int_a^b f(x)\,dx$ be defined for *any* succession of ordinates $x \to f(x)$?

A negative answer was suggested by Dirichlet. In the 1820s, when Dirichlet was a student, the German universities had not yet become centres of mathematical research. They boasted one great mathematician, Gauss, who had a reputation for being relatively unapproachable. Dirichlet consequently decided to continue his mathematical education in Paris, which abounded with prominent mathematicians such as Laplace, Legendre, Poisson, Lacroix and Hachette, as well as Cauchy and Fourier, whose book on the theory of heat conduction appeared in the year (1822) that Dirichlet arrived in Paris. He remained there until 1825. He had little, if any, direct contact with Cauchy ; but he studied Cauchy's books on analysis, which also appeared around the time of his Paris sojourn, and accepted Cauchy's approach to analysis. Fourier was more accessible, and the mathematical questions raised by Fourier's work on heat became the subject of a ground-breaking memoir by Dirichlet (*1829a*), in which he provided answers to Fourier's questions by employing Cauchy's type of analysis (see section 3.9).

Dirichlet accepted a literal interpretation of Fourier's characterisation of an arbitrary function f and claimed that $\int_a^b f(x)\,dx$ need not have a definite value. He offered the following example. Let $f(x)=c$ if x is a rational number and let $f(x)=d$ $(d \neq c)$ if x is irrational. Thus for each x, an ordinate $f(x)$ has been specified ; f is an ' arbitrary function '. But, according to Dirichlet, $\int_a^b f(x)\,dx$ is meaningless. Probably Dirichlet was thinking in terms of Cauchy's definition of the integral. It is easily seen that a partition P of $[a, b]$ can be chosen with all $x_i - x_{i-1}$

arbitrarily small so that all the $x_i \neq a$ or b are irrational. The corresponding Cauchy sum S is then approximately equal to $d(b-a)$. On the other hand, partitions P' exist with all $x_i' - x_{i-1}'$ arbitrarily small and all rational $x_i' \neq a, b$. For such partitions, S' is arbitrarily close to $c(b-a)$. Thus S and S' are not approaching a *unique* limiting value : $\int_a^b f(x)\,dx$ does not exist in the sense of Cauchy.

Some restrictions on the nature of the arbitrary function thus seemed necessary to ensure that it could be integrated. Dirichlet claimed, however, that f need not be continuous or possess at most a finite number of points of discontinuity in order that $\int_{-\pi}^{\pi} f(x)\,dx$ exist. There could be an *infinity* of discontinuity points in $[-\pi, \pi]$; it was only necessary that ‘ if a and b denote two arbitrary quantities included between $-\pi$ and π, it is always possible to place other quantities r and s between a and b which are sufficiently close so that the function remains continuous in the interval from r to s ’ (*1829a*, 169). Expressed in modern terms, Dirichlet's condition is that the set of points of discontinuity be nowhere dense.

Dirichlet did not attempt to justify his assertion, because a clear proof ‘ requires some details connected with the fundamental principles of infinitesimal analysis, which will be presented in another note . . . ’. The promised note never appeared, and it seems likely that he discovered he could not carry through the demonstration. If he was thinking in terms of Cauchy's definition of the integral, his silence is understandable since the Cauchy sums need not approach a unique limiting value when the function satisfies Dirichlet's condition. Although inconclusive, Dirichlet's comments involved some bold ideas : a function is literally a succession of ordinates, and as such can be totally discontinuous in Cauchy's sense ; nevertheless, it should be possible to extend the definition of the definite integral to *some* functions with an infinite number of points of discontinuity in a finite interval.

As we saw in section 3.9, the main result of Dirichlet's *1829a* was that if $f(x)$ is defined and bounded on $[-\pi, \pi]$ and if it possesses a finite number of maxima and minima and is continuous except possibly at a finite number of points, then the Fourier series of f converges for all x to

$$\tfrac{1}{2}\{f(x-0)+f(x+0)\}, \tag{4.3.3}$$

where $f(x-0) = \lim_{h\to 0^+} f(x-h)$ and $f(x+0) = \lim_{h\to 0^+} f(x+h)$. (For $x = \pm\pi$, (4.3.3) must be specially interpreted.) The proof of this result, which develops ideas already found in Fourier's *Théorie*, does not directly depend upon the assumption of the continuity of $f(x)$; it was added simply to assure the meaning of the definite integrals (4.2.6) defining the Fourier coefficients. Thus Dirichlet's interest in extending the integral to more discontinuous functions was linked with his desire to confirm Fourier's claims about representing functions as trigonometric series for as large a class of arbitrary functions as possible.

The question of the integrability of highly discontinuous arbitrary functions which Dirichlet had left unresolved was taken up by Riemann, who had received his doctorate from the University of Göttingen in 1851. In order to lecture as an instructor at Göttingen, it was necessary to exhibit a further capacity for scholarship, to write a *Habilitationsschrift*. As his topic Riemann chose the problem of the representability of functions by trigonometric series ; and in the course of preparing his paper (*1854a*) he was encouraged and aided by Dirichlet, who was then a professor at the University of Berlin. The paper was presented to the Göttingen faculty in 1854.

Riemann had obviously picked a challenging problem, for it involved going beyond the results of Dirichlet : that is, it involved the consideration of more general 'arbitrary functions'. He was consequently forced to consider the question of when the definite integral of such functions can be defined. Riemann also accepted the definition of the integral given by Cauchy : a function f defined and bounded on $[a, b]$ is integrable if the 'Cauchy–Riemann' sums

$$S = \sum_{i=1}^{n} f(t_i)(x_i - x_{i-1}) \tag{4.3.4}$$

approach a unique limiting value as the $\{x_i - x_{i-1}\}$ approach zero, where $a = x_0 < x_1 < \ldots < x_n = b$ and $t_i \in [x_{i-1}, x_i]$. The limiting value is, by definition, $\int_a^b f(x)\,dx$.

Riemann formulated an important necessary and sufficient condition for the integral to exist, which will be considered in section 4.5 below. Using it he showed that Dirichlet's conjectured condition for integrability was actually not necessary : a function could be much more discontinuous than Dirichlet had imagined and still be integrable ; it can have an infinite number of points of discontinuity in any interval, no matter how small ! As he observed, 'since these functions have never yet been considered, it will be good to begin with a specific example'. The remarkable example he gave turned out to be consequential in more ways than he himself realised.

To construct his example Riemann began with the function $\phi(x)$ which is defined as follows : $\phi(x) = x - n$, where n is the integer closest to x ; for $x = \pm\frac{1}{2}, \pm\frac{3}{2}, \pm\frac{5}{2}, \ldots, \phi(x) = 0$. The graph of ϕ is indicated in figure 4.3.1. It is interesting and perhaps not entirely coincidental

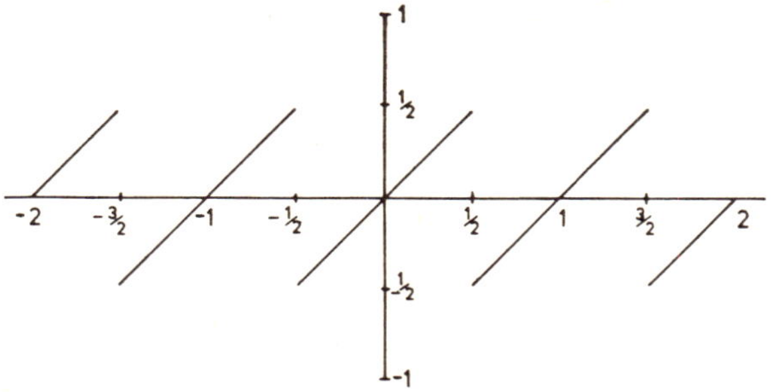

Figure 4.3.1.

that $\phi(x)$ is the Fourier series of $f(x)=x$ for the interval $[-\frac{1}{2}, +\frac{1}{2}]$. By Dirichlet's theorem the series converges for all x and the value, $\phi(x)$, of its sum when $x=\pm\frac{1}{2}, \pm\frac{3}{2}, \pm\frac{5}{2}, \ldots$ accords with (4.3.3).

From ϕ Riemann defines new functions ϕ_n by setting $\phi_n(x)=\phi(nx)$, $n=1, 2, 3, \ldots$. Now ϕ_n is discontinuous at $\pm 1/2n, \pm 3/2n, \pm 5/2n, \ldots$, and his idea is to add these functions together so as to obtain a highly discontinuous function. Thus f is defined by

$$f(x)=\phi_1(x)+\phi_2(x)/2^2+\phi_3(x)/3^2+\ldots+\phi_n(x)/n^2+\ldots. \quad (4.3.5)$$

The series converges uniformly, although this notion was not widely known in 1854 (see section 3.12) and Riemann never introduced it. He does use a consequence of uniform convergence, namely that

$$f(x\pm 0)=\sum_{n=1}^{\infty}\phi_n(x\pm 0)/n^2, \quad (4.3.6)$$

to deduce that f is discontinuous at all the points of discontinuity of the ϕ_n. That is, if x is any rational number of the form $p/2q$, where p is odd and p and q have no common divisor, then

$$f(x\pm 0)-f(x)=\mp\left(\frac{1}{2q^2}\right)(1+\tfrac{1}{9}+\tfrac{1}{25}+\ldots)=\mp\frac{\pi^2}{16q^2}. \quad (4.3.7)$$

Hence f is discontinuous at all x of the form $p/2q$, and there are infinitely many such x in any interval. But f is bounded and, as Riemann showed, integrable (*1854a*, art. 6).

Riemann's paper was first published in 1868, soon after his untimely death. As can be imagined, the unprecedented generality of his viewpoint and his striking example of an integrable function made a great impression upon mathematicians. It seemed to du Bois Reymond (*1883a*, 274) that Riemann had extended the concept of the integral to its outermost limits ; his integrability condition appeared to be the most general conceivable. (See also Weierstrass's remarks in Mittag-

Leffler *1923b*, 196.) The extent of the generality was forcibly exhibited by the above unprecedented example. It appeared impossible to conceive of the integrability and integral of a bounded function in any more general manner, for if the Cauchy–Riemann sums (4.3.4) failed to approach a unique limiting value, it did not seem to make much sense to speak of the area determined by its ordinates. The approach of Cauchy and Riemann was based upon a tradition that went all the way back to Archimedes.

Although the above attitude towards Riemann's theory of integration prevailed during the 19th century, various discoveries were made which, with the wisdom of hindsight—that is, from the vantage point of Lebesgue's work—can be regarded as revealing serious flaws in Riemann's theory, flaws which indicate that, appearances notwithstanding, Riemann's integrability condition was *not* sufficiently general. The discovery of some of these flaws will be sketched in the next section.

4.4. *Defects of the Riemann integral*

The discussion of the foundations of the integral calculus by Cauchy, Dirichlet and Riemann was but one aspect of the increasing concern with, and changing approach to, the foundations of mathematical analysis that occurred during the 19th century. Another aspect was a more critical attitude towards assumptions that were taken for granted or not proved by a standard of reasoning compatible with the new approach to analysis. Weierstrass's example of a continuous nowhere differentiable function (see section 3.13) typified the critical attitude. His example showed that the widely-held belief that continuous functions were differentiable in general, that is, except at a few exceptional points, could not be rigorously established. As we shall see, the critical approach to analysis also produced examples of functions that revealed that Riemann's definition of integrability was not as general as one would have wished.

The first treatise to synthesise the results of the new approach to analysis was Dini's *Fondamenti per la teorica delle funzioni di variabili reali* (' Foundations for the theory of functions of real variables ' : *1878a*). Dini naturally accorded a prominent place in his treatise to Riemann's theory of integration. One of the results presented by Dini was the following theorem, due to Gaston Darboux (*1875a*, 111–112) ; If a function f possesses a bounded, Riemann-integrable derivative f' on $[a, b]$ then for all $x \in [a, b]$,

$$\int_a^x f'(t)\, dt = f(x) - f(a). \tag{4.4.1}$$

Cauchy had established (4.4.1) for continuous derivatives, in accordance with his theory of integration. Darboux's generalisation was a significant triumph for Riemann's viewpoint, for it showed that the latter's much weaker integrability condition also sufficed to establish the fundamental relationship (4.4.1) of the calculus.

However, Dini perceived a less triumphal implication of Darboux's theorem. He observed that if f has the property that in every interval, no matter how small, points t exist for which $f'(t)=0$, then either f is a constant or f' cannot be integrable in Riemann's sense. Darboux's theorem implies that these are the only alternatives. For if f' possesses the above property and is bounded, then for each $x \in [a, b]$, $\int_a^x f'(t)\,dt = 0$: in the Cauchy–Riemann sums

$$S = \sum_{i=1}^{n} f'(t_i)(x_i - x_{i-1}) \tag{4.4.2}$$

defining this integral the t_i can always be chosen so that $f'(t_i)=0$; thus the only way these sums can approach a unique limit is if that limit is zero. It therefore follows from Darboux's theorem that *if* f' is integrable, then

$$f(x) - f(a) = \int_a^x f'(t)\,dt = 0, \tag{4.4.3}$$

which implies that $f(x)=f(a)$ for all $x \in [a, b]$.

Dini felt it was ' very likely ' that non-constant functions exist which have the above-mentioned property so that f' is not integrable (*1878a*, art. 100). An example confirming Dini's conjectures was provided in *1881b* by Vito Volterra. That is, he constructed an example of a non-constant function f with a bounded derivative f' which vanishes on a dense set of points. The example shows that within the context of Riemann's theory of integration the fundamental operations of differentiation and integration are not entirely reversible ; the process of differentiation could produce bounded functions f' which fail to be integrable in Riemann's sense and for which (4.4.1) is therefore meaningless.

Volterra pointed out that his example f showed that, in some cases at least, the approach to integration via anti-differentiation is more general, for f' possesses f as an anti-derivative so that its integral over $[a, b]$ could be taken as $f(b)-f(a)$. He did not intend his observation as a serious criticism of Riemann's approach to integration, and he hastened to add that it was superior to the anti-derivative approach because it at least provided necessary and sufficient conditions for a function to be integrable. At the time, Riemann's theory was the most satisfactory alternative, and, as noted at the end of the previous section, a more general alternative appeared impossible.

Other examples of the functions predicted by Dini were discovered later in the 19th century. The one given by the Swedish mathematician T. Brodén in 1896 is especially interesting, because it uses the same technique of 'condensation of singularities' (Hankel's term; see section 3.13) that Riemann had devised to illustrate how *general* his integrability condition is. Brodén's example indicates it was not general enough, although he himself did not draw this conclusion.

Brodén begins with the function $\phi(x) = x^{1/3}$ for $x \in [-1, 1]$. At $x = 0$ the graph of ϕ has a vertical tangent, so that $\phi'(0) = \infty$. For all other values of x, $\phi'(x)$ exists as a finite number. Let $\{a_n\} \subset [-1, 1]$ be a dense set, and define

$$\phi_n(x) = \phi(x - a_n) = (x - a_n)^{1/3}. \tag{4.4.4}$$

Thus $\phi_n'(x)$ exists for all $x \in [-1, 1]$, but $\phi_n'(x) = \infty$ for $x = a_n$. Define f on $[-1, 1]$ by

$$f(x) = \sum_{n=1}^{\infty} \phi_n(x)/2^n. \tag{4.4.5}$$

The analogy with the construction of Riemann's function should be clear. Moreover it is also the case that, for all $x \in [-1, 1]$,

$$f'(x) = \sum_{n=1}^{\infty} \phi_n'(x)/2^n, \tag{4.4.6}$$

which implies that $f'(x) = \infty$ for all x such that the right hand side of (4.4.6) is infinite. In particular, $f'(a_n) = \infty$ when $n = 1, 2, 3, \ldots$ Thus the graph of f possesses vertical tangents at a dense set of points on its graph.

The function f is easily seen to be strictly increasing, since this is true of ϕ and hence of all the ϕ_n. Thus f possesses a continuous inverse $g = f^{-1}$ defined on $[a, b] = [f(-1), f(1)]$. It is easily sensed from the geometrical relation of the graphs of f and g that the latter possesses horizontal tangents at a dense set of points. Expressed analytically, the result is that $g'(y) = 1/f'(x)$, where $y = f(x)$, and therefore $g'(b_n) = 0$ $n = 1, 2, 3, \ldots,$ where $b_n = f(a_n)$. Furthermore, a simple check shows that $f'(x)$ is bounded away from zero for $x \in [-1, 1]$, so that g' is bounded. Since g' vanishes on a dense set $\{b_n\}$ and is also strictly increasing (and hence not a constant), it must be non-integrable by virtue of Dini's observations.

The above example is a special case of an entire class of similar functions indicated by Brodén. Their significance to him was not that they displayed the limitations of Riemann's theory of integration: he never mentions Dini's observations and may have been unaware of them. For him and others who devised further examples of curves with the property of densely distributed horizontal tangents, these examples continued the trend initiated by Weierstrass's example of a continuous nowhere differentiable function. Weierstrass's function was not intuitable ; but the examples of Brodén and the others showed that even everywhere differentiable functions could be non-intuitable. They provided further confirmation of the Weierstrassian thesis that intuition could not be relied upon when establishing the fundamental principles of mathematical analysis.

The derivative of a function is obtained from it by a limiting process, and the above examples show (to us of the post-Lebesgue era) that the above defect in Riemann's theory stems from the fact that integrability in Riemann's sense is not preserved in limiting processes. This fact also revealed itself in another manner in the 19th century. We saw in section 4.2 above that Fourier had made the commonly accepted assumption that if $f(x)=\lim_{n\to\infty} f_n(x)$ for all $x\in[a, b]$, then

$$\int_a^b f(x)\,dx = \lim_{n\to\infty} \int_a^b f_n(x)\,dx$$
$$\text{or} \quad \int_a^b \{\lim_{n\to\infty} f_n(x)\}\,dx = \lim_{n\to\infty} \int_a^b f_n(x)\,dx. \tag{4.4.7}$$

Cauchy had in fact presented a proof of (4.4.7) under the assumption that the f_n are continuous in Lesson 40 of his *Résumé* (*1823a*), but it was inadequate because of his failure to distinguish between uniform and non-uniform convergence and continuity. In his lectures at Berlin, Weierstrass showed that (4.4.7) is valid provided the convergence is assumed to be uniform (see Heine *1870a*, 353).

The same result was proved in *1875a* by Darboux, who also gave a simple example showing that (4.4.7) need not hold when the convergence is not uniform (*1875a*, 77–84). He considered the sequence

$$f_n(x)=2n^2x \exp(-n^2x^2), \quad x\in[0, 1]. \tag{4.4.8}$$

For each $x\in[0, 1]$, $f(x)=\lim_{n\to\infty} f_n(x)=0$. Hence $\int_0^1 f(x)\,dx=0$, but

$$\int_0^1 f_n(x)\,dx=\{-\exp(-n^2x^2)\}_0^1=1-\exp(-n^2), \tag{4.4.9}$$

so that $\lim_{n\to\infty} \int_0^1 f_n(x)\,dx=1\neq 0$. He showed that the sequence f_n did not converge uniformly on [0, 1] by noting that $f_n(1/n)=2n/e$. Actually this shows that the sequence is not uniformly bounded. A sequence is *uniformly bounded* on $[a, b]$ if a positive number B exists such that $|f_n(x)| \leqslant B$ for all $x \in [a, b]$ and for all n. The concept of uniform boundedness had not been introduced when Darboux wrote his paper, and he apparently failed to realise its potential significance. He and other mathematicians of the period were so impressed by the discovery of the importance of uniform convergence that they overlooked the possibility of being able to assert (4.4.7) under more general conditions.

The oversight is consequential because it hides the fact that Riemann-integrability is not preserved in limiting processes. That is, if $f_n(x)$ converges to $f(x)$ uniformly on $[a, b]$ and if the f_n are Riemann-integrable, then so is f, as Darboux proved. But it was discovered by Cesare Arzelà (*1885a*) and W. F. Osgood (*1897a*) that uniform boundedness was the essential condition for (4.4.7) to hold. For example, Arzelà showed that (4.4.7) holds if it is assumed that f_n and f are Riemann-integrable and that the f_n are uniformly bounded. Thus, given a convergent, uniformly bounded sequence of integrable functions f_n, the only further condition for the validity of (4.4.7) was that the limit function be Riemann-integrable.

Arzelà did not produce an example of a non-integrable limit function, but a very simple one was inadvertently provided by René Baire in *1899a*. He considered the sequence $f_n(x)$ defined for $x \in [0, 1]$ as follows. Let $f_n(x) = 1$ if x is a rational number p/q, p and q having no common factors, and $q \leqslant n$. For all other $x \in [0, 1]$, $f_n(x) = 0$. Each function f_n is consequently zero except at a finite number of points in $[0, 1]$. Hence they are integrable and $|f_n(x)| \leqslant 1$, so they are uniformly bounded as well. It is easily seen that $f(x) = \lim_{n \to \infty} f_n(x)$ exists for all $x \in [0, 1]$, and that $f(x) = 1$ if x is rational and $f(x) = 0$ if x is irrational. The limit function f is therefore one of Dirichlet's examples of a function impossible to integrate (see section 4.3 above), and indeed it is not integrable in Riemann's sense.

When Lebesgue created his generalisation of the Riemann integral, he proved that (4.4.7) holds for any uniformly bounded convergent sequence ; the integrability, in Lebesgue's sense, of the limit function followed from the fact that it was such a limit function. In terms of Lebesgue's discoveries, Riemann's theory is thus revealed as insufficiently general ; but, prior to those discoveries, no one regarded the fact that such a limit function need not be Riemann-integrable as a serious setback for Riemann's approach to integration. For example, Baire made no references to integration theory when giving his example. He gave it to show that Dirichlet's functions are of the second ' Baire class '—that is, a limit of a sequence of functions which are themselves limits of sequences of continuous functions (see section 3.13). How could Riemann's theory of integration be criticised for not being sufficiently general in the respects discussed above when it seemed the most general conceivable ? By the close of the 19th century, however, Riemann's theory had been reformulated in terms of sets and their measure, a reformulation that provided new perspectives and, ultimately, the idea that a very natural generalisation is possible.

4.5. *Towards a measure-theoretic formulation of the integral*

I shall begin by briefly indicating the connection between the measure of sets and the integration of functions. Let E denote a subset of $[a, b]$. Two numbers $c_i(E)$ and $c_e(E)$, called, respectively, the inner and outer content of E, are defined in the following manner. Let $[a, b] = \bigcup_{k=1}^{n} I_k$ denote a partition P of $[a, b]$ into intervals I_k. Then

$$c_i(E) = \sup_P \sum_{I_k \subset E} l(I_k), \quad c_e(E) = \inf_P \sum_{I_k \cap E \neq \phi} l(I_k), \tag{4.5.1}$$

where $l(I_k)$ denotes the length of I_k. In other words, the inner content of E is obtained by considering, for a given partition P, those intervals entirely contained in E ; and the outer content is obtained by considering all intervals containing points of E. Clearly

$$\bigcup_{I_k \subset E} I_k \subset E \subset \bigcup_{I_k \cap E \neq \phi} I_k, \tag{4.5.2}$$

so that

$$\sum_{I_k \subset E} l(I_k) \leqslant c_i(E) \leqslant c_e(E) \leqslant \sum_{I_k \cap E \neq \phi} l(I_k). \tag{4.5.3}$$

The set E is defined to be measurable in the sense of Jordan if $c_i(E) = c_e(E)$, and in this case its content is denoted by $c(E)$.

These notions were introduced by Camille Jordan in *1892a* for reasons to be discussed further on. Using them, he showed that Riemann's condition for integrability can be reformulated in the following manner. Let $[a, b] = \bigcup_{k=1}^{n} E_k$ denote a partition P of $[a, b]$ into disjoint measurable sets E_k (that is, $E_i \cap E_j = \phi$ for $i \neq j$). The (generalised) lower and upper Riemann sums of a bounded function f corresponding to P are defined by

$$L(P) = \sum_{k=1}^{n} m_k c(E_k), \quad U(P) = \sum_{k=1}^{n} M_k c(E_k), \tag{4.5.4}$$

where m_k and M_k denote, respectively, the greatest lower bound (inf) and least upper bound (sup) of the numbers $f(x)$ for $x \in E_k$. The lower and upper Riemann integrals are then defined by

$$\underline{\int_a^b} f = \sup_P L(P) \qquad \overline{\int_a^b} f = \inf_P U(P), \tag{4.5.5}$$

and f is Riemann-integrable if and only if

$$\underline{\int_a^b} f = \overline{\int_a^b} f. \tag{4.5.6}$$

The characterisation of Riemann-integrability in terms of lower and upper integrals was introduced shortly after Riemann's *Habilitationsschrift* was published (posthumously) in 1868, and it was suggested by one of Riemann's own necessary and sufficient conditions for integrability. Prior to Jordan, however, the above notions had been developed without measure-theoretic concepts: the lower and upper Riemann sums, $L(P)$ and $U(P)$, were defined only for partitions P of $[a, b]$ into sub-intervals. By showing that more general types of partitions P are admissable in the definitions, Jordan unwittingly provided the idea for a fruitful way to generalize Riemann's theory of the integral. His measure-theoretic characterisation of Riemann-integrability implied that a generalisation of his theory of the measure of sets would bring with it a generalisation of the integral.

Suppose a *larger* class M of measurable sets can be defined, and a number, $m(E)$, associated with each E in M such that $m(E) = c(E)$ if E is measurable in Jordan's sense. Then Jordan's characterisation of Riemann-integrability immediately suggests the following generalisation. In the above definitions of P, $L(P)$ and $U(P)$ permit the sets E_k to be measurable in the generalised sense—that is, to belong to M. The resulting lower and upper integrals, $(*)\underline{\int_a^b} f$ and $(*)\overline{\int_a^b} f$, will then satisfy

$$\underline{\int_a^b} f \leqslant (*)\underline{\int_a^b} f \leqslant (*)\overline{\int_a^b} f \leqslant \overline{\int_a^b} f. \tag{4.5.7}$$

Thus if f is defined to be integrable when

$$(*)\underline{\int_a^b} f = (*)\overline{\int_a^b} f, \tag{4.5.8}$$

with this common value taken as the integral of f. (4.5.7) shows that every Riemann-integrable function is still integrable; we therefore obtain a generalisation of the Riemann integral. When M is the class of Lebesgue-measurable subsets of $[a, b]$, the generalised class of integrable functions is the class of bounded Lebesgue-integrable functions. It was by considerations such as these that Lebesgue was led to his theory of integration.

Now that the historical significance of the introduction of measure-theoretic ideas is clear, we must consider how and why it occurred. When Riemann's *Habilitationsschrift* on trigonometric series appeared in 1868, no one spoke explicitly of sets and their measure : such notions had not yet become a part of mathematics. But he had provided the following necessary and sufficient condition for the integrability of a bounded function : For every $\epsilon > 0$ and $\sigma > 0$ there exists a $\delta > 0$ such that, if P is any partition $a = x_0 < x_1 < \ldots < x_n = b$ with $\Delta x_i = x_i - x_{i-1} < \delta$ for all i, then $S(P, \sigma) < \epsilon$, where $S(P, \sigma)$ denotes the sum of all the Δx_i for which $M_i - m_i > \sigma$ (Riemann *1854a*, art. 5 ; as above, M_i and m_i denote the least upper bound and greatest lower bound of $f(x)$ for x in $[x_{i-1}, x_i]$). He introduced this condition to show that his example of a highly discontinuous function (4.3.5) is integrable. The discontinuities of this function are jump discontinuities, since the left-hand and right-hand limits $f(x \pm 0)$ exist at all points of discontinuity. Furthermore, there exist at most a *finite number* of points $x \in [a, b]$ at which the magnitude of the jump, $f(x-0) - f(x+0) = \pi^2/8q^2$ is greater than a given σ. Using this property of f, Riemann showed that $s(P, \sigma) < \epsilon$ for all P with Δx_i sufficiently small.

An important advance was taken in *1870a* by Hankel, who had studied under Riemann at Göttingen (compare section 3.13). He sought to devise a necessary and sufficient condition for integrability that would be directly applicable to Riemann's function by virtue of the above mentioned property of its jump discontinuities. He therefore focused his attention on *sets*, the sets of points x for which the jump is greater than positive σ's. Hankel realised that in general a function need not possess one-sided limits $f(x \pm 0)$ at a point x of discontinuity, and so he introduced a generalised definition of the jump of f at x (*1870a*, art. 7). His definition is similar to the more familiar notion of the oscillation of f at x, $\omega_f(x)$, which is defined as follows. For $\delta > 0$ let $I_\delta = (x - \delta, x + \delta)$ and $\omega_f(x, \delta) = M(\delta) - m(\delta)$, where $M(\delta)$ and $m(\delta)$ are, respectively, the least upper bound and greatest lower bound of the numbers $f(t)$ for $t \in I_\delta$. Then $\omega_f(x) = \lim_{\delta \to 0^+} \omega_f(x, \delta)$. A function f is continuous at x exactly when $\omega_f(x) = 0$.

Hankel focused attention on the sets S_σ of points $x \in [a, b]$ for which $\omega_f(x) > \sigma$ (using, of course, his slightly different definition of $\omega_f(x)$). Moreover, he clearly realised that a bounded f is integrable if and only if, for every $\sigma > 0$, the set S_σ can be enclosed in a finite number of intervals of arbitrarily small total length. The proof of this fact involves a straightforward generalisation of Riemann's proof that the function (4.3.5) satisfies his integrability criterion. In view of Hankel's realisation of this characterisation of integrability, it is natural to suppose that Hankel's work initiated not only the set-theoretic approach to the theory of integration but the *measure-theoretic* approach as well. In hindsight, it appears a small additional step to the criterion that a bounded function f is a Riemann-integrable on $[a, b]$ if and only if $c_e(S_\sigma)=\sigma$ for all $\sigma>0$. But measure-theoretic notions and this criterion were not introduced until the 1880s.

The question therefore arises as to why it took so long for this to happen. Historical questions do not usually have simple, irrefutable answers ; history is too complex—and interesting—for that to be the case. Nevertheless, the historian can isolate certain conditions or factors which would account for the historical phenomenon in question. In the case of the slow introduction of a measure-theoretic viewpoint, I would suggest that an important, perhaps decisive, role was played by the confusion that existed concerning measure-theoretic and topological characterisations of a negligible set.

Such confusion is evident in the following very nice, but false, theorem that Hankel thought he had proved : A bounded function f is integrable if and only if for every $\sigma>0$ the set S_σ is nowhere dense (*1870a*, art. 7). That is, the set S_σ must have the property introduced earlier by Dirichlet : between any two points an entire interval exists that is devoid of points of S_σ. Hankel termed such a set ' scattered '. He was familiar with Dirichlet's *1829a*, and indeed it seemed that he had confirmed Dirichlet's claim that the integral can be defined for any bounded function for which the set of points of discontinuity, D, form a nowhere dense set. That is, S_σ is a subset of D for all $\sigma>0$, so that if D is nowhere dense the sets S_σ are as well. By Hankel's ' theorem ' f is therefore integrable.

Hankel's ' theorem ' is only half true. If f is integrable then $c_e(S_\sigma)=0$ for all $\sigma>0$, which implies that S_σ is nowhere dense for all $\sigma>0$. (If S_σ is dense in some interval I, then $c_e(S_\sigma)\geqslant l(I)$.) The converse is incorrect. To establish it, he believed he had proved that a nowhere dense set can be enclosed in a finite number of intervals of arbitrarily small total length. Until it was discovered that his proof was invalid, it appeared that he had obtained a very nice topological characterisation (as we can now describe it) of the sets S_σ corresponding to an integrable function—a characterisation which directly confirmed Dirichlet's speculations on integrability conditions. There was consequently no need to develop a measure-theoretic characterisation of these sets, as long as it was not realised that topological negligibility (nowhere dense) and measure-theoretic negligibility (zero outer content) are *not* equivalent. The reason that the lack of equivalence went unnoticed for some time is not difficult to find : in the 1870s the theory of point sets had not yet been worked out (see chapter 5). Reasoning that involved infinite sets of points, when it did occur, remained on a very naive, careless level. No-one, including Hankel, fully realised the logical implications of formal definitions such as that of a nowhere dense set.

Cantor's first publication on point sets (*1872a*) actually had the effect of adding to the confusion. His investigation of the uniqueness of trigonometric series representations had led him to consider sets E with the property that the n-th derived set, $E^{(n)}$, is finite for some integer n. The set $E^{(n)}$ is defined as follows. The first derived set, E', is the set of limit-points of E, and, in general, $E^{(n)}=(E^{(n-1)})'$ (see section 5.2). A simple example of a set E such that $E^{(n)}$ is finite is the set of all numbers of the form

$$\frac{1}{m_1}+\frac{1}{m_2}+\ldots+\frac{1}{m_n}, \tag{4.5.9}$$

where the m_i denote positive integers. In this case, $E^{(n)}=\{0\}$. Following Cantor's later terminology, we shall refer to such sets as ' sets of the first species '. Cantor discovered that certain properties valid for finite sets carry over, by induction, to first species sets. In this manner it is easily seen that sets of the first species are nowhere dense and also that they can be enclosed in a finite number of intervals of arbitrarily small total length. Hence for these sets both the topological and the measure-theoretic characterisations of negligibility obtain. Several of the mathematicians interested in the theory of integration tended to think that first species sets were the only type of nowhere dense set possible. Thus nowhere dense sets (*qua* first species sets) were measure-theoretically negligible.

The special importance, for integration theory, of sets negligible in a measure-theoretic sense was not appreciated until, in the early 1880s, it became clear that nowhere dense sets exist which *cannot* be enclosed in intervals of arbitrarily small total length. The effect of this discovery was immediate and decisive. Special names were introduced for the special type of nowhere dense set which could be so enclosed ; and Cantor posed the problem of finding conditions that a nowhere dense set must possess in order to have zero content. Shortly thereafter, Cantor (*1884a*, art. 18) and Stolz (*1884a*) independently took the next obvious step and introduced the notion of the (outer) content of a set. Thus in 1884 the first theory of measure was finally created, thanks to the discovery of nowhere dense sets of positive outer content. Mathematicians had at last begun to think in measure-theoretic terms.

In view of the historical importance of the discovery of nowhere dense sets of positive outer content, I shall briefly indicate how they were discovered. Several mathematicians apparently discovered them independently,[1] but the basic idea behind their various constructions was the same. In seeking to conceive of the most general kind of nowhere dense set, mathematicians had previously proceeded inductively. A finite set E_1 is nowhere dense. If sequences of points converging to each element of E_1 are added, the resulting set E_2 is still nowhere dense. If now sequences of points converging to each element of E_2 are added, the resulting set E_3 is still nowhere dense, and so on. In this manner increasingly complex nowhere dense sets can be conceived. They are, however, of the first species, since $E_n^{(n-1)} = E_1$ is a finite set. The idea behind the construction of nowhere dense sets of positive outer content is to distribute *intervals* rather than points. That is, if a sequence of disjoint intervals I_n can be densely distributed in $[a, b]$ in the sense that any sub-interval of $[a, b]$ contains an I_n, then $E = [a, b] - \bigcup_{n=1}^{\infty} I_n$ is nowhere dense. If, in addition, $\sum_{n=1}^{\infty} l(I_n) < b - a$, then $c_e(E) > 0$.

The discovery of this type of nowhere dense set actually had a double significance for the history of the theory of integration, for these sets were also used by Volterra to construct examples of bounded derivatives which are not Riemann-integrable (see section 4.4 above). It is likely that Volterra got the idea of how to construct these sets from Dini's speculations concerning the existence of such derivatives. Although Dini could not construct an example, he conjectured that non-constant functions f exist with the property that between any pair of numbers $r < s$ there exists an interval on which f remains constant, so that $f'(x) = 0$ there. These intervals of constancy are densely distributed in $[a, b]$. If they are taken as open and denoted by $\{I_n\}$, then $E = [a, b] - \bigcup_{n=1}^{\infty} I_n$ is nowhere dense. Furthermore, $c_e(E) > 0$. The reason is that, assuming $f'(x)$ exists for all x and is bounded, it cannot be integrable on $[a, b]$ by virtue of Dini's corollary to the fundamental theorem of the calculus (4.4.1). Hence the set of points of discontinuity of f', D, must be such that $c_e(D) > 0$. (Otherwise f' would be integrable by the correct part of Hankel's 'theorem'.) But $D \subset E$ and hence $c_e(E) \geqslant c_e(D) > 0$. Although Dini himself failed to realise it, his hypothetical functions are intimately related to the existence of nowhere dense sets that are not negligible in terms of measure.

The measure-theoretic viewpoint, which had finally become a part

[1] See Smith *1875a* ; Volterra *1881a* ; du Bois Reymond *1880a*, and *1882a*, 188–189 ; and Cantor *1884a*, art. 18.

of Riemann's theory of the integral by the mid-1880s, received its definitive formulation in *1892a* at the hands of Camille Jordan. As noted at the beginning of this section, Jordan introduced the distinction between inner and outer content and the important concept of a measurable set. His work was motivated by his dissatisfaction with the way in which Riemann's theory of integration had been developed for functions of two or more variables. The integral of a function $f(x, y)$ was customarily defined over a region E of the plane which was regarded as bounded by a curve (see figure 4.5.1). To define the lower and upper

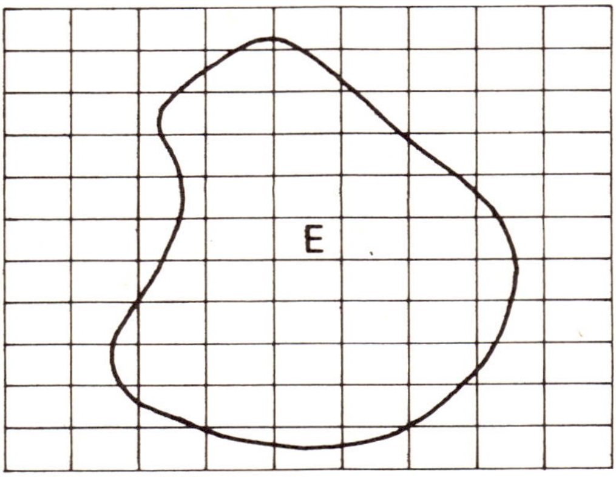

Figure 4.5.1.

Riemann sums or the Cauchy–Riemann sums, a partition of the plane into rectangles formed by lines parallel to the coordinate axes was considered. Such a partition divides E into pieces, most of which are rectangles. Exceptions occur, however, for rectangles meeting the boundary curve.

A certain amount of ambiguity or arbitrariness thus occurs in the definition of, for example, the Cauchy–Riemann sum corresponding to the partition. It could be defined as

$$\sum f(x_i, y_j)a(R_{ij}), \quad a(R_{ij}) = \text{area } R_{ij}, \tag{4.5.10}$$

where either (1) the summation is over all rectangles R_{ij} which are within E or (2) the summation is over all R_{ij} which simply contain points of E. (Non-rectangular boundary pieces cannot be used, since their area is not defined.) To justify the independence of the definition of $\int_E f(x, y)\, dE$ on the choice of option (1) or (2), it was usually noted that the sum of the areas meeting the boundary curve can be made arbitrarily small. Jordan himself had proceeded in this manner in the first edition of his *Cours d'analyse* (*1883a*). In *1890a*, however, Peano showed that the assumption about the boundary rectangles could not be taken for granted by constructing a continuous curve which passes through every point of the square $0 \leqslant x \leqslant 1$, $0 \leqslant y \leqslant 1$.

Perhaps Peano's space-filling curve prompted Jordan to dispense with the custom of regarding E as bounded by a curve. Instead he proposed in his *1892a* to treat the domain of integration E with the same degree of generality as Riemann had accorded to the functions to be integrated. This meant regarding E as a set of points, that is, treating E from the general standpoint of Cantor's theory of sets. The requirement that E be bounded by a curve is replaced by the requirement that E be measurable. (The definitions (4.5.1) of $c_i(E)$ and $c_e(E)$, and thence of measurability, extend immediately to points in the plane; partitions into intervals are simply replaced by partitions into rectangles.) Jordan defined the boundary of E to be the set of all points p such that every circular disc about p contains points in E and points not in E. He proved that E is measurable if and only if the boundary of E has zero outer content. Thus the measurability of E is precisely what is needed for the validity of the assumption made about the boundary curve of the domain of integration.

Having introduced the concept of a measurable set, Jordan proceeded to define $\int_E f$ for E measurable along the lines sketched at the beginning of the section. The consideration of partitions into arbitrary measurable sets E_k arises naturally from the fact that E is an arbitrary measurable set. Jordan developed his formulation of Riemann's theory of integration for functions of n variables. The form it took was motivated by considerations which arise in the case $n=2$, but the approach was developed for any n, including $n=1$ and $E=[a, b]$, thereby linking the possibility of a generalisation of the integral with that of a generalisation of measure and measurability.

It was in this measure-theoretic setting that aspiring young French mathematicians learned Riemann's theory of integration, for Jordan incorporated this approach into the second edition of his *Cours d'analyse* (*1893a*). Relatively little attention had previously been given by the French to the application of Cantorian set theory to analysis. Through his *Cours d'analyse* Jordan in effect put his stamp of approval upon the set-theoretic approach to analysis, and since he was one of France's prominent, established mathematicians, his tacit approval of that approach was not without effect. Within a few years of the publication of the second edition, the set-theoretic approach to analysis was extensively cultivated by three young French mathematicians: Emile Borel, René Baire and Henri Lebesgue.

4.6. *What is the measure of a countable set?*

Lebesgue very aptly described Jordan as a 'traditionalistic innovator' (*1926a*, lxi). Jordan's reformulation of the theory of integration was unquestionably innovative; he carried the set-theoretic approach to integration much farther than his predecessors. And yet he carried on the tradition they had established. He simply refined the notion of outer content by introducing the concepts of inner content and measurability. The resulting theory of measure is completely compatible with Riemann's approach to integration. Indeed, inner and outer content and measurability are the exact analogues of lower and upper Riemann integrals and Riemann integrability. Jordan's approach to integration became suggestive only after the possibility of a different, less traditional, approach to the measure of sets was recognised. Such a possibility was brought to light by Emile Borel. The radically different approach to the measure of sets proposed by Borel originated in his unorthodox answer to the question: What is the measure of a countable set?

The question had first been considered by Axel Harnack, who was one of the several German mathematicians involved with the development and exposition of Riemann's theory of integration and the associated theory of outer content. (German mathematicians spoke simply of 'content', by which they meant what Jordan termed 'outer content'.) In *1885a* Harnack observed that if in the definition of outer content the restriction to a finite number of covering intervals is dropped, there is a remarkable, paradoxical consequence: every countable set $E=\{e_n\}$ would have zero 'outer content'. The reason is that, for any $\epsilon>0$, each e_n can be enclosed in an interval I_n of length $\epsilon/2^n$, and the total length of these intervals is $\sum_{n=1}^{\infty} \epsilon/2^n = \epsilon$. Thus a countable set can always be enclosed within intervals of arbitrarily small total length *if* an infinite number of intervals is permitted.

To Harnack the implication of these observations was clear. They revealed the crucial importance of the restriction to a finite number of covering intervals in the definition of outer content. The idea that countable sets should have zero content appeared paradoxical to him because countable sets could be dense. For example, the set E of rational numbers in [0, 1] is countable and dense. Because it is dense $c_e(E)=1$, not 0. This seemed the appropriate measure of E by virtue of its ubiquitousness. It appeared absurd to regard a dense set as extensionless, as of negligible measure.

Cantor, who had introduced the notion of a countable set, certainly shared Harnack's viewpoint. He himself introduced the outer content of a set as its measure, and his formulation made it clear that in measuring a set E, the set of its limit points E' should be regarded as a part of E. That is, Cantor *defined* the outer content of a set E as $c_e(E \cup E')$. (Since $c_e(E \cup E') = c_e(E)$, the definition is the same.) Thus to measure the set E of rational points in [0, 1] is to measure $E \cup E' = [0, 1]$. At the same time, other consequences of Cantor's study of infinite point sets implied that, in some respects at least, countable sets are negligible.

One such result turned out to be especially important. It is contained in Cantor *1882a*, the third instalment of Cantor's 'Über unendliche, lineare Punktmannichfaltigkeiten'. In keeping with the title ('On infinite linear point sets'), the first two instalments had been devoted to subsets of the real line. Cantor's principal objective was a proof of the continuum hypothesis (see section 5.4). In the third instalment, however, he decided to indicate the relevance of his work to areas of mathematics currently of widespread interest in the mathematical community: the theory of functions of a complex variable and the foundations of geometry. To this end, he turned to consideration of point sets in n-dimensional space. After presenting a theorem which he felt was relevant to the theory of functions of a complex variable, he continued with some observations he felt had implications for the foundations of geometry.

Cantor made his observations in the context of n-dimensional space R_n, but I shall consider the special case of the plane R_2. He observed that if M is a countable dense subset of the plane R_2, then $U = R_2 - M$, the plane with the points of M deleted, is still 'continuously connected'. That is, if $u_1, u_2 \in U$, then there exists a continuous curve joining u_1 and u_2 which is entirely contained in U. The idea of the proof is as follows. Consider the segment $\overline{u_1 u_2}$ and its perpendicular bisector in figure 4.6.1.

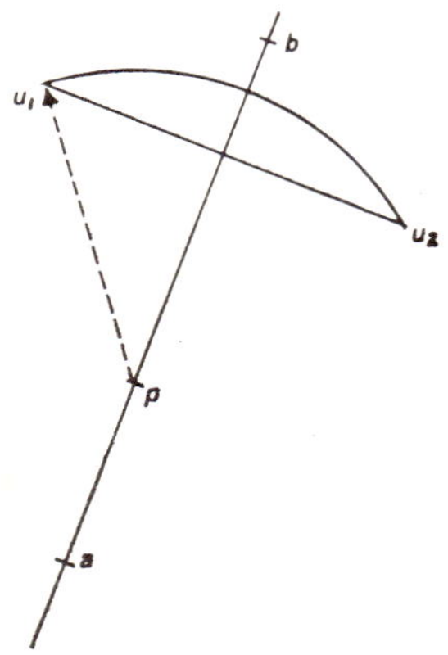

Figure 4.6.1.

Each point p on the perpendicular bisector determines a different circular arc joining u_1 and u_2. Since these arcs meet only in u_1 and u_2 they intersect M in distinct sets, and because M is countable, the number of circular arcs meeting M is at most countable. Suppose $p_1, p_2, p_3, \ldots$ are the centres of these arcs. As Cantor had proved in *1874a*, any interval $\overline{ab}$ on the bisector contains an uncountable number of points. Thus there are an uncountable number of points p on $\overline{ab}$ such that $p \neq p_n$ for any n, and hence the circular arc determined by p cannot meet M—it lies entirely in $R_2 - M$.

Cantor felt that the above observations had a bearing on the question of the geometrical nature of the physical world, a subject that had been popularised in the scientific community by Hermann von Helmholtz. Most mathematicians who were concerned with the foundations of geometry followed Riemann's lead in his *1854b* and identified space with the manifold of *all* triples of real numbers (x, y, z). Cantor felt that his observations showed that the identification of space with R_3 is *not* supported by the fact that continuous motion is possible in space. Continuous motion appeared just as possible in $R_3 - M$, where M is countable and dense. He therefore suggested the value of developing an approach to mechanics applicable to $R_3 - M$.

Cantor's remarks did not evoke a response from mathematicians interested in the foundations of geometry, but they made a great impression upon Emile Borel, who saw that they involved an idea that could be used in the theory of functions of a complex variable! Prior to Jordan, the only French mathematician who applied Cantorian set theory to analysis was Henri Poincaré. In *1883a* Poincaré considered the analytic expression

$$f(z) = \sum_{n=1}^{\infty} A_n/(z - a_n), \tag{4.6.1}$$

where A_n, z and a_n are complex numbers and $\sum_{n=1}^{\infty} |A_n| < \infty$. He observed that if C is a simple closed convex curve in the plane and if $\{a_n\}$ is a dense subset of C, then the expression $f(z)$ defines two distinct analytic *functions*, one defined inside C, the other outside of C. The reason is that the power-series expansion of $f(z)$ about any point z_0 inside C converges on a circular disc which touches C and does not go beyond it (see figure 4.6.2): the analytic function defined by $f(z)$ cannot be analytically continued across C in the sense of Weierstrass.[1]

[1] If two functions $f_1(z)$ and $f_2(z)$ of the complex variable z are analytically defined over regions D_1 and D_2 respectively, and if $f_1(z) = f_2(z)$ over $D_1 \cap D_2$, then $f_2(z)$ is an *analytic continuation* of $f_1(z)$ over $D_2 - D_1$ (and $f_1(z)$ is such a continuation of $f_2(z)$ over $D_1 - D_2$).

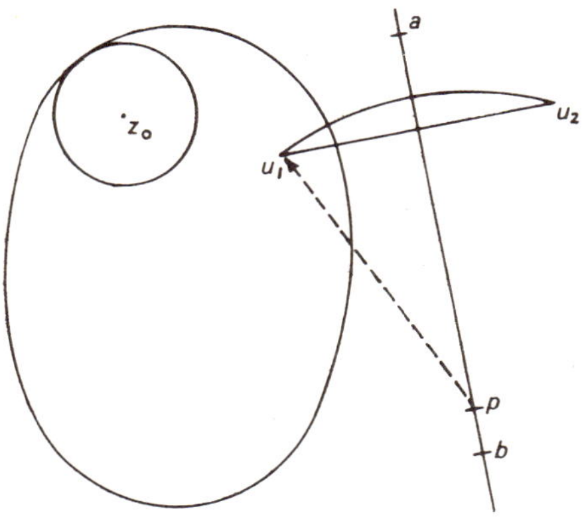

Figure 4.6.2.

In his doctoral thesis *1894a* Borel employed Cantor's idea to show that the two functions could be connected with each other so as to provide a generalised type of analytic continuation. By applying Cantor's observations to the countable set $M=\{a_n\}$, he concluded that there are at most a countable number of points $p_n \in \overline{ab}$ such that the circular arc from u_1 to u_2 meets C in one of the points a_n (see figure 4.6.2). Borel's problem, however, required showing that the series $f(z)$ converges absolutely and uniformly on circular arcs from u_1 to u_2. It led him to add a new twist to Cantor's observations which made them more relevant to the measure of a countable set.

Under the hypothesis that $\sum_{n=1}^{\infty} |A_n|^{1/2} < \infty$, Borel chose N such that $\sum_{n=1}^{\infty} |A_n|^{1/2}$ is less than half the length of $\overline{ab}$. He then applied Harnack's idea: enclose each p_n with $n \geqslant N$ in an interval I_n of length $2|A_n|^{1/2}$. Then the sum of these lengths is $2\sum_{n=N}^{\infty} |A_n|^{1/2}$, which is less than the length of $\overline{ab}$. Borel then deduced that there are uncountably many points p on $\overline{ab}$ such that $p \neq p_n$, $1 \leqslant n < N$, and $p \notin I_n$, $n \geqslant N$.[1] On the circular arcs corresponding to these points p the series $f(z)$ converges uniformly and possesses other properties that suggest, in a certain sense, that the functions inside and outside z are the same. Thus the singular points a_n, despite their dense distribution on C, do not prevent the

[1] The proof involved ' a theorem interesting in itself . . . : If one has an infinity of sub-intervals on a line [that is, a closed interval] such that every point of the line is interior to at least one of them, a finite number of intervals can effectively be determined having the same property '. Here we have the first statement of the so-called ' Heine–Borel theorem ' (see section 3.12).

series $\sum_{n=N}^{\infty} A_n/(z-a_n)$ from converging at uncountably many points on C, namely the points of intersection of C with the ' good ' arcs from u_1 to u_2. This result undoubtedly encouraged Borel to further investigate the nature of the set of points of convergence of these series and to develop in this connection the above idea on how to measure a countable set.

During the academic year 1896–1897 Borel was privileged to present a course of lectures on his new results at his *alma mater*, the *Ecole Normale Supérieure* in Paris. Probably Lebesgue attended these lectures, for he was a student there from 1894 until 1897. As a result of the enthusiastic response to the lectures, they were published in 1898 as a book, *Leçons sur la théorie des fonctions* (*1898a*).

In order to illustrate his methods in the simplest case, Borel began by considering the real-variable analogue of the absolute value of the series $f(z)$ in (4.6.1), namely $\sum_{n=1}^{\infty} A_n/|x-a_n|$, where $A_n>0$,

$$\sum_{n=1}^{\infty} A_n^{1/2} < \infty$$

and $\{a_n\}$ is a dense subset of [0, 1]. To study the points of convergence of the series, he proceeded much as he had in this thesis. Each point a_n is enclosed in an interval $I_n=(a_n-u_n, a_n+u_n)$, where $u_n=(1/2k)A_n^{1/2}$. If $B_k=\bigcup_{n=1}^{\infty} I_n$ then for $x\notin B_k$, $x\notin I_n$ for all n and hence $|x-a_n|\geqslant u_n$ or, equivalently,

$$A_n/|x-a_n| \leqslant \frac{A_n}{u_n} = 2kA_n^{1/2} \tag{4.6.2}$$

for all n. Therefore the series converges uniformly on the set $[0, 1]—B_k$. The set B_k consists of intervals of total length

$$\sum_{n=1}^{\infty} l(I_n) = \sum_{n=1}^{\infty} A_n^{1/2}/k = A/k, \quad \text{where} \quad A = \sum_{n=1}^{\infty} A_n^{1/2}. \tag{4.6.3}$$

Furthermore, if $B=\bigcap_{k=1}^{\infty} B_k$ and if D denotes the set of points of non-convergence of the series, then $D\subset B$, so that D can be enclosed in intervals $B_k=\bigcup_{n=1}^{\infty} I_n$ of arbitrarily small total length by making k suitably large.

These results, and many other fascinating ones on analogous complex-valued series, take up the second half of Borel's book. The first half is devoted to developing the set-theoretic and measure-theoretic notions needed to formulate the results of the second half in what he considered as a fitting manner. In this respect the theory of content is not helpful. Since $\{a_n\}$ is dense in [0, 1] and $\{a_n\} \subset D$, both D and the set of points of convergence, $[0, 1] - D$, have outer content 1 and inner content 0. They are indistinguishable in terms of their content and they are not measurable. Borel therefore deemed it appropriate to introduce a theory of measure that would distinguish between D and $([0, 1] - D)$ by assigning the former measure 0 and the latter measure 1.

Restricting his attention to subsets of [0, 1] Borel thus proposed the following definitions of measure and measurability (*1898a*, 46–48):

> When a set is formed of all the points in a denumerable infinity of intervals which do not overlap and have total length s, we shall say that the set *has measure s*. When two sets do not have common points and their measures are s and s', the set obtained by uniting them, that is to say their sum, has measure $s + s'$.
>
> More generally, if one has a denumerable infinity of sets which pairwise have no common point and having measures $s_1, s_2, \ldots, s_n \ldots$ their sum has measure $s_1 + s_2 + \ldots + s_n + \ldots$.
>
> All that is a consequence of the definition of measure. Now here are some new definitions: if a set E has measure s and contains all the points of a set E' of which the measure is s', *the set* $E - E'$ ... *will be said to have measure* $s - s'$ ... .
>
> *The sets for which measure can be defined by virtue of the preceding definitions will be termed measurable sets by us* ... .

No further elaboration or clarification was provided by Borel. His words must have appeared more like an enigma than a definition, although a reader familiar with modern real analysis will have no trouble detecting the notion of a Borel set.[1]

The nature of Borel's definition reflected a philosophical attitude that he later made explicit: his measurable sets are constructible from intervals by 'repeatedly' applying the operations of set union and difference. Jordan's measurable sets are not defined 'from the ground up' in this fashion, and Borel consequently regarded Jordan's definitions as more general than his own. The contrasting definitions, Borel suggested, reflect the completely different problems to which they are applied; he apparently did not envisage the applications of his ideas on measure to the theory of integration.

There was one slight difficulty that Borel had to face regarding his

[1] A Borel set was for Borel something like this: it is a Borel-measurable set constructed by performing a denumerable number of unions and intersections on intervals (or on Borel sets already constructed). Borel did not use the phrase 'Borel set', and some later definitions of the concept are rather broader than this one.

definition. The set D of points of non-convergence of his series is a subset of the Borel-measurable set B which has measure zero. But it does not follow from Borel's analysis that D is a Borel-measurable set. Consequently he had to adopt the following convention. If a set E is 'sandwiched between' Borel-measurable sets of measures a and b respectively, let us agree to say that the measure of E is $\geqslant a$ and $\leqslant b$ *without worrying about whether or not E is measurable.* The set D is, of course, 'sandwiched between' the empty set and the set B, both of which have measure zero. Thus by Borel's convention D is assigned the measure zero ; he is able to conclude that his series converges at all points of [0, 1] except a set of measure zero.

To someone with fewer philosophical scruples, a natural reaction to Borel's convention would be : Why not simply imitate the theory of content and define a set to have measure zero if it can be enclosed in a countable number of intervals of arbitrarily small total length ? Then D automatically has zero measure ! That is exactly what Lebesgue did to obtain his generalisation of Jordan's notions of measurability and measure. In fact, as Lebesgue pointed out in *1902a*, 241, the sets which are assigned a definite measure by Borel's convention—that is, the sets which are sandwiched between two Borel sets with the same measure—are precisely the Lebesgue-measurable sets. In other words, hidden away in Borel's convention, introduced more or less as a necessary evil, are the measurable sets of Lebesgue. What makes the matter all the more interesting is the fact that the set D of points of non-convergence actually is a Borel-measurable set. Borel's suggestive convention was really unnecessary !

Once Borel had introduced his novel ideas on the measure of sets, it was somewhat inevitable that other mathematicians would eventually combine them with Jordan's approach to produce the theory of Lebesgue measure. For subsets $E \subset [0, 1]$, Lebesgue proceeded as follows in his doctoral thesis *1902a*. Let $m_e(E)$ denote the greatest lower bound of all numbers $\sum_n l(I_n)$, where $E \subset \bigcup_n I_n$. This is the customary definition of $c_e(E)$ except that the number of intervals is allowed to be infinite. Since $c_i(E) = 1 - c_e(\tilde{E})$ (where $\tilde{E} = [0, 1] - E$), an analogous definition of inner measure is $m_i(E) = 1 - m_e(\tilde{E})$. From these definitions it follows easily that

$$c_i(E) \leqslant m_i(E) \leqslant m_e(E) \leqslant c_e(E). \tag{4.6.4}$$

The Lebesgue-measurable sets E are those such that $m_i(E) = m_e(E)$ and the above inequalities show that every Jordan-measurable set is Lebesgue measurable, and $c(E) = m(E)$ for such sets. Thus Lebesgue's theory of measure is a generalisation of Jordan's. Furthermore, the Lebesgue-measurable sets possess the properties posited by Borel in his definition.

For example, if $E = \bigcup_{n=1}^{\infty} E_n$ and the E_n are Lebesgue-measurable and satisfy $E_i \cap E_j = \varnothing$ for all $i \neq j$, then E is measurable and $m(E) = \sum_{n=1}^{\infty} m(E_n)$.

Equivalent definitions of measure and measurability were introduced independently by G. Vitali (*1904a*) and W. H. Young (*1905a*). Probably the new theory of measure did not initially appeal to many mathematicians, especially those who were young men in the 1870s and 1880s when Riemann's theory of integration seemed the ultimate in generality and the wave of the future. The ' generation gap ' to which I refer is evident in Arthur Schönflies's treatment of Borel's work. Schönflies was 22 in 1875, the year Vitali and Lebesgue were born. (Young was 12 years older but began his mathematical career at the age of 34.) Schönflies was commissioned by the German Mathematicians' Union to write a report on the theory of sets. The book-length report appeared as *1900a* ; it was the first treatise on the theory and application of point sets.

When Schönflies came to write the section on the measure of sets, he remarked that there was more than one theory that had been developed and that, as with all mathematical definitions, definitions of measure were somewhat subjective and would have to be judged by the degree to which their consequences were suited to the objectives behind their introduction. In this respect, Schönflies clearly felt that Borel's definition was not appropriate. For one thing, the content of a set E is identical with that of $E \cup E'$. Schönflies accepted the reasonableness of this property of content and pointed out that Borel's measure fails to have it. The fact that a dense set could have zero Borel measure appealed to Schönflies as little as it had to Harnack.

Schönflies's distaste for Borel's type of measure is reflected in his treatment of Borel's work on the series $\sum_{n=1}^{\infty} A_n/|x - a_n|$, discussed above around (4.6.2). He admired Borel's results but avoided Borel's measure-theoretic characterisation of them. Schönflies had earlier made the convention that ' content ' was to be used exclusively with reference to outer content since ' for applications it is always only a question of this outer content ' (*1900a*, 94). Accordingly, he observed that the series converges on a set of outer content 1. Instead of the result that the set D of points of non-convergence is contained in B, a Borel set of measure zero, Schönflies stressed the largeness of B :

$$[0, 1] - B = [0, 1] - \bigcap_{k=1}^{\infty} B_k = \bigcup_{k=1}^{\infty} ([0, 1] - B_k), \tag{4.6.5}$$

and each set $([0, 1] - B_k)$ is nowhere dense (because $\{a_n\}$ is dense).

Thus ([0, 1] − B) is a countable union of nowhere dense sets, a set of 'the first category' in the terminology introduced by Baire in *his* doctoral thesis *1899a* (see section 3.13). Baire proved that [0, 1] is not of the first category; hence B cannot be of the first category. Like [0, 1], it is of the second category.

4.7. *Conclusion*

By stressing the largeness of the set B, Schönflies seemed to be suggesting that B should not be regarded as negligible in measure, that a definition which implied such a conclusion was inappropriate. Others undoubtedly shared his sentiments. Indeed, we have seen that the idea that a dense set could have zero measure was contrary to the approach to the measure of sets adopted by Harnack, Cantor and many other mathematicians, and championed by Schönflies. Lebesgue's work really settled the issue over the most appropriate definition of measure, for he showed that a Borel-type measure is necessary—a necessary evil, perhaps, but nonetheless necessary. That is, the definition of the integral which accompanies Lebesgue's generalisation of Jordan's theory of measure (as explained at the beginning of section 4.5) is free from most of the defects of the Riemann integral, including those discussed in section 4.4. Thus if a uniformly bounded sequence of Lebesgue-integrable functions, $f_n(x)$, converges to a function $f(x)$ for each x in $[a, b]$, then f is Lebesgue-integrable and

$$\int_a^b f(x)\,dx = \lim_{n\to\infty} \int_a^b f_n(x)\,dx. \tag{4.7.1}$$

And if a function $f(x)$ has a bounded derivative $f'(x)$ on $[a, b]$, then f' is always integrable in Lebesgue's sense and

$$\int_a^b f'(x)\,dx = f(b) - f(a). \tag{4.7.2}$$

Lebesgue's signal achievement was the discovery that his generalisation of the integral possesses these and many other remarkable properties.[1] By creating his theory of integration Lebesgue had in effect confirmed Fourier's naïve belief that 'arbitrary functions' are not beyond the purview of mathematical analysis.

[1] For a more detailed historical analysis of Lebesgue's contributions see Hawkins *1970a*, chs. 5 and 6. An excellent exposition of Lebesgue's theory is given in Royden *1968a*, chs. 3–5.

Chapter 5

The Development of Cantorian Set Theory

Joseph W. Dauben

5.1. *Introduction*

This chapter explores the early development of set theory, in particular the contributions of the German mathematician Georg Cantor (1845–1918). Though he was joined by mathematicians in the 19th century like Riemann, Hankel, Harnack and du Bois Reymond (among others) in exploring the properties of point sets and their significance for mathematical analysis, Cantor's contributions were in many ways unique. His creation of transfinite numbers was controversial from the beginning, and his professional career was devoted to defending and to promoting his revolutionary work. Perhaps more than most branches of modern mathematics, set theory bears the special stamp of its originator's interests and personality. Thus the historical development of Cantorian set theory demonstrates how the abstract objectivity so often ascribed to scientific theory may be influenced by the character and interests of those who contribute most to its development. This is particularly true of so contentious a subject as the infinite in mathematics, for not only did Cantor have to face strong opposition from mathematicians, but also theologians and philosophers clung to traditions that refused to admit any ground to the actual infinite. In relentlessly supporting the validity of transfinite set theory, he promoted his research until its importance to virtually every branch of mathematics was recognised.

For convenience I shall usually cite particular passages in Cantor's writings from Zermelo's edition of his works (Cantor *Papers* in the bibliography). For more general studies of his life and work, see Fraenkel *1930a*, Meschkowski *1967a*, Grattan-Guinness *1971b* and Dauben *1977a* and *1979*.

5.2. *The trigonometric background : irrational numbers and derived sets*

Though Cantor's *Dissertation* of 1867, written at the University of Berlin under the auspices of Kummer and Kronecker, was devoted to a difficult problem in number theory (as was his *Habilitationsschrift*, published in 1869), this was not the area which first stimulated his interest in set theory. Having left Berlin early in 1869 to become a *Privatdozent* at the University of Halle, he found that one of his senior colleagues there, Eduard Heine, was working on problems dealing with the theory of trigonometric series (compare sections 3.11 and 3.12). Heine recognised in Cantor a young mathematician of great promise, and encouraged him to take up a very important question in analysis : If a given, arbitrary function could be represented by a trigonometric series, was the representation unique ? Heine had managed to solve a part of the problem in *1870a* by assuming that the given function was almost everywhere continuous, and that the trigonometric series in question was also uniformly convergent almost everywhere. But Cantor was anxious to do away with such restrictions, and to establish the uniqueness theorem in the most general terms possible. (For more details on this work, see Dauben *1971a.*)

This Cantor did, though for his first proof in 1870 he found it necessary to assume that the trigonometric series in question was convergent for all values of x (*Papers*, 80–83). In 1871 he published a short note indicating that it was in fact possible to establish the theorem even if, for certain values of x, either the representation of the function or the convergence of the series could be given up, so long as the total number of such exceptional points remained finite (*Papers*, 84–86). Cantor's greatest achievement (with respect to the uniqueness theorem) came in his *1872a*, when he succeeded in showing that even an infinite number of exceptions might be permitted, so long as they were distributed in a specified way.

Wanting to present this last proof in as simple and as rigorous a way as possible, Cantor found that he had to develop a satisfactory theory of the real numbers in order to deal precisely with the infinite sets of exceptional points which he now had in mind. Criticising earlier approaches for assuming the *existence* of the irrationals as limits of infinite sequences of rationals used to ' define ' them, he wanted to present a theory of the irrationals which in no way presupposed their existence. Beginning with the set of all rational numbers A, he introduced sequences of rationals : $a_1, a_2, \ldots, a_n, \ldots$. These sequences were further subject to the condition that, for whatever m, if n was taken large enough, $|a_{n+m} - a_n| < \epsilon$, for any rational number $\epsilon > 0$, however small. If the sequence satisfied this condition, Cantor called it a 'fundamental sequence' and said that it had a definite limit b (*Papers*, 93, 186). This was to be taken as a convention to express, not that the sequence $\{a_n\}$ actually had the limit b, or that the number b was presupposed as the limit, but merely that with each such sequence $\{a_n\}$ a number b was associated with it.

Cantor then denoted the collection of numbers b associated with such infinite sequences $\{a_n\}$ by B. Two numbers b and b' defined by two fundamental sequences $\{a_\nu\}$ and $\{a_\nu'\}$ were said to be equal, $b=b'$, if $a_\nu - a_\nu'$ became very small as ν increased without limit. He also noted that by virtue of the fact that any constant sequence $\{a\}$ was a fundamental sequence, then a must be an element of B. Consequently, $A \subset B$, though the converse was clearly false (*Papers*, 93–94).

In an analogous fashion, Cantor considered infinite sequences of elements from B: $b_1, b_2, \ldots, b_n, \ldots$. With each fundamental sequence $\{b_n\}$ there was associated a number c. All such sequences generated from B constituted the domain C. He went on in this way to define higher-order domains from C. Proceeding through λ such constructions, he reached the domain L. Given an element l in this domain L, he called it a number, value or limit (he took these to be the same for his theory of real numbers) of the λ-th kind (*Papers*, 95–96).

Though Cantor had built up the real numbers B from the domain of rationals A, and had then gone on in similar fashion, using infinite sequences to define higher-order domains, he was now faced with the problem of identifying the real numbers so constructed with points of the real line. It was clear that every such point could be associated with one of his real numbers, but it was by no means obvious that to each of his real numbers in B a unique point of the linear continuum must correspond. Therefore, he invoked the axiom: 'To every real number a definite point of the straight line corresponds, whose coordinate is equal to that number' (*Papers*, 97). This identification was to be especially important in terms of Cantor's definition of derived sets of the first and second species, which required the concept of limit-point: 'Given a point set P, if an infinite number of points of the set P lie within every neighbourhood, however small, of a point p, then p is said to be a "limit-point" of the set P' (*Papers*, 98; note that p may be a limit-point of P, and yet not belong to the set P itself).

Given any point set P, Cantor noted that every point was either a limit-point of P, or it was not. The set of all limit-points of P was denoted P', and called 'the first derived set' of P. Just as he was able to generate from B an entire system of λ-domains, he did the same with P'. If P' were an infinite point set, then it gave rise to a second derived point set P'', and so on, until after taking successively n such derived sets, it was possible to produce the $(n+1)$-th derived set of P, $P^{(n+1)}$.

The case important for Cantor's extension of his uniqueness theorem was the one in which, after n repetitions, the derived point set $P^{(n)}$ consisted only of a finite number of points, thus making the extension to further (non-empty) derived sets impossible. He designated such sets as point sets of the *first species*, for which the derived set $P^{(n)} = \varnothing$ for some finite value of n. Were $P^{(n)} \neq \varnothing$ for any finite value of n, then P was said to be a point set of the *second species*. It was from these point sets of the second species that he would eventually produce his transfinite numbers, but for his uniqueness theorem of 1872 he was concerned only with point sets of the first species.

Cantor showed that such point sets existed by appealing to a point on the line whose abscissa was determined by a number of the ν-th kind (hence the need for his axiom). Working backwards, he took the sequence of numbers of the $(\nu-1)$-th kind determining ν, then the numbers of the $(\nu-2)$-th kind determining each of these, and so on, until eventually he reached an infinite set of rational numbers in the domain A. By taking the point set corresponding to this set of rationals, he had clearly produced a point set of the ν-th kind. It was then possible to establish the most general of his theorems concerning the uniqueness of representations by means of trigonometric series (*Papers*, 99):

> If an equation is of the form
>
> $$0 = \tfrac{1}{2}d_0 + \sum_{n=1}^{\infty} c_n \sin nx + d_n \cos nx$$
>
> for all values of x with the exception of those which correspond to the points in a [closed] interval $(0 \ldots 2\pi)$ of a given point set P of the ν-th kind, where ν is any whole number, then $d_0 = 0$, $c_n = d_n = 0$.

Cantor's achievement was impressive. Following the success of his application of first species sets to establishing the uniqueness theorem, even for infinite sets of exceptional points, he must have been intrigued by the reasons which might account for the validity of the result. His proof had insisted that the points of exception be distributed in a carefully specified way. Nevertheless, there could be infinitely many such points, which raised the question : How could one characterise the important difference between the rationals and the reals ? The rationals were *dense* (between any two rationals there were always infinitely many others), but the set of all rationals was not continuous. It was perhaps natural to suspect that there were more irrationals than rationals, but what did that mean ? Try as he might, he could find no reason to establish or to deny the denumerability of the reals.

5.3. *Non-denumerability of the real numbers, and the problem of dimension*

On 29 November 1873, a year after their first meeting in Switzerland, Cantor wrote to his friend Richard Dedekind. In his letter, Cantor posed the problem which his analysis of irrational numbers had directly suggested. Was it possible to correspond uniquely in a one–one fashion the collection of all natural numbers N with the set of all real numbers X of the continuum ? He assumed that the answer was ' no ' ; but he had not been able to find a reason why this should be so, and he hoped that Dedekind might see a simple answer to the dilemma. But Dedekind replied that he could find no reason to prohibit any such correspondence (*Cantor/Dedekind*, 12–20).

Before the year was over, however, Cantor had discovered a valuable key to understanding the nature of continuity, and in 1874 he published an important theorem in Crelle's *Journal* : The set of all real numbers R cannot be corresponded in a unique, one–one fashion with the set of all natural numbers N. In other words, the set R is non-denumerable (*Papers*, 117). The proof ran as follows. Assuming that the real numbers ω were countable, it followed that they could be placed in a one–one correspondence with the natural numbers N :

$$\omega_1, \omega_2, \omega_3, \omega_4, \ldots, \omega_n, \ldots . \qquad (5.3.1)$$

Cantor then claimed that it was possible, given any closed interval $[a, b]$ in R, to find at least one real number $\eta \in R$ such that η failed to be listed as an element of (5.3.1). Assuming $a < b$, he picked the first two numbers from (5.3.1) which fell within the interval $[a, b]$. Denoted by a' and b' respectively, these were used to constitute another interval $[a', b']$. Proceeding analogously, he produced a sequence of nested intervals, reaching $[a^n, b^n]$, where a^n and b^n were the first two numbers from (5.3.1) lying within $[a^{n-1}, b^{n-1}]$. If the number of intervals thus constructed were finite, then at most only one more element from (5.3.1) could lie in $[a^n, b^n]$. It was easy in this case to conclude that a number η could be taken in this interval which was not listed in (5.3.1). Any real number in $[a^n, b^n]$ would suffice, so long as it was not the possible least number indexed in (5.3.1).

On the other hand, if the number of intervals $[a^n, b^n]$ were not finite, Cantor's argument shifted to alternatives in the limit. Since the sequence $a, a', \ldots, a^n, \ldots$ did not increase indefinitely, but was bounded within $[a, b]$, it had to assume an upper limit, which he denoted by a^∞. Similarly, the sequence $b, b', \ldots, b^n, \ldots$ was assigned the lower limit b^∞. Were $a^\infty < b^\infty$, then, as in the finite case, any real number $\eta \in (a^\infty, b^\infty)$ was sufficient to produce the necessary real number not listed in (5.3.1).

However, were $a^\infty = b^\infty$, he reasoned that $\eta = a^\infty = b^\infty$ could not be included as an element of (5.3.1). He designated η as ω_ρ. But ω_ρ, for sufficiently large index n, would be excluded from all intervals nested within $[a^n, b^n]$. Nevertheless, by virtue of the construction that he had given, η had to lie in every interval $[a^n, b^n]$, regardless of index. The contradiction established the proof : R was non-denumerable.

Cantor's proof, coupled with the fact that the set of all algebraic numbers was denumerable, provided an independent corroboration of Liouville's proof *1851a* that there were an infinite number of transcendental numbers in any given interval $[a, b]$ of reals. But this was hardly the most significant part of Cantor's conclusion. As he described it, without particular emphasis (*Papers*, 116) :

> This theorem shows why sets of real numbers (for example, the entirety of real numbers $\geqslant 0$ but $\leqslant 1$) cannot be uniquely corresponded with the set of all natural numbers N. Thus I have found the clear difference between a so-called continuum and a set of the nature of the entirety of all algebraic numbers.

But, as he was to discover, the features distinguishing continua from other kinds of sets were not completely described by the fact that they were non-denumerable. Nevertheless, with the idea of denumerability and the existence of non-denumerable sets established, Cantor was now able to make some of his earlier ideas more precise. For example, though he had the basic idea for the transfinite numbers in the sequence of derived sets P', P'', . . ., $P^{(n)}$, . . ., the basis for any articulate distinction between $P^{(n)}$ and $P^{(\infty)}$ was lacking. There was no precise basis for defining the first transfinite number ∞ following all finite natural numbers n until it was clear that in fact there were sets much larger than N, sets that could not be counted or enumerated by the indices of natural numbers.

Cantor's next subject of research produced surprising and unexpected results. (They are discussed in more detail in Dauben *1974a*.) Shortly after his discovery that the real numbers were non-denumerable, he must have begun to search for other distinct powers of infinity greater than the power of the real numbers. Early in 1874 he wrote to Dedekind, posing a new but clearly related problem : ' Might it be possible to correspond a surface (a square, perhaps, including its boundaries) with a straight line (perhaps an interval with the inclusion of its endpoints) so that to each point of the surface, one point of the line corresponds, and conversely ? ' (*Cantor/Dedekind*, 20).

Cantor cautioned that the solution was one of great difficulty, though one might be tempted to say that the answer was clearly ' no ', and, even more clearly, that a proof was superfluous. In fact, when he mentioned the same problem to friends while visiting Berlin in the spring of 1874, they were astonished at the seeming ridiculousness of the question (*Cantor/Dedekind*, 21).

More than three years passed before Cantor discovered a way to produce a one–one correspondence between lines and surfaces. Finally, in 1877, he wrote to Dedekind and explained that, contrary to prevailing mathematical opinion, the 'absurd' correspondence between lines and planes was not impossible. The discovery prompted one of his best-known remarks: 'I see it, but I don't believe it!' (*Cantor/Dedekind*, 34).

Although Cantor had originally constructed a one–one correspondence between the points of any ρ-dimensional space and the linear continuum, the basic idea of his proof can be expressed more easily. For the simplest case of the two-dimensional plane and the one-dimensional line, he took any point (x_1, x_2) in the plane and matched it with exactly one y of the line. He did so by considering the infinite decimal expansions

$$x_1 = \alpha_1, \alpha_2, \ldots, \alpha_\nu, \ldots, \tag{5.3.2}$$

$$x_2 = \beta_1, \beta_2, \ldots, \beta_\nu, \ldots . \tag{5.3.3}$$

The corresponding y under his mapping was then determined as follows:

$$y = \alpha_1, \beta_1, \alpha_2, \beta_2, \ldots, \alpha_\nu, \beta_\nu, \ldots . \tag{5.3.4}$$

Unfortunately, there was a difficulty which Dedekind explained in a letter to Cantor of 22 June 1877 (*Cantor/Dedekind*, 27–28). In order to avoid the representation of one and the same value x twice, the assumption had to be added that no representation be allowed which from a certain index on was always zero. Otherwise a number x would have two representations, for example: $x = 0{\cdot}3000 \ldots$ and $x = 0{\cdot}2999 \ldots$. The only exception to the above restriction would of course be the representation of zero itself. But under these conditions, Cantor's mapping was necessarily incomplete. Any y of the form

$$y = \alpha_1, \beta_1, \alpha_2, \beta_2, \ldots, \alpha_\nu, \beta_\nu, 0, \beta_{\nu+1}, 0, \beta_{\nu+2}, \ldots \tag{5.3.5}$$

was inadmissible under his assumptions, since it would have corresponded to the two points:

$$x_1 = \alpha_1, \alpha_2, \ldots, \alpha_\nu, 0, 0, 0, \ldots, \tag{5.3.6}$$

$$x_2 = \beta_1, \beta_2, \ldots, \beta_\nu, \ldots . \tag{5.3.7}$$

Fortunately, the damage was not irreparable, and he was soon able to find an alternative proof which, though more complex, nevertheless established the general validity of his theorem. It was possible to determine a one–one correspondence which mapped the points of the two-dimensional plane onto the one-dimensional line (*Papers*, 122–125).

Cantor's discovery was so startling because it was completely contrary to what mathematicians had believed for so long. He immediately criticised the work of others, particularly the work of Riemann and Helmholtz, who had assumed that the dimension of a space was uniquely determined by the number of coordinates needed to identify a point in that space. As Cantor had demonstrated, there was no such invariance. But, as Dedekind was quick to point out, Cantor's theorem involved a mapping that was *discontinuous*, and, as everyone had always assumed, dimension was invariant under continuous, one–one correspondences.[1] Above all, the value of Cantor's proof was the justification it gave him for narrowing his study of continuity to the linear continuum of real numbers. His next major publications presented a systematic study of linear point sets, and introduced his transfinite numbers as the keys to producing a general theory of infinite sets.

Though the paper published in 1878 was a triumph for Cantor, it was also the cause of some distress and unpleasantness. In fact, it was the first occasion for open hostilities between Cantor and one of his life-long opponents, his former teacher at the University of Berlin, Leopold Kronecker.

5.4. *First trouble with Kronecker*

The details of Kronecker's programme of arithmeticisation, basing all of mathematics on a finite number of operations involving only the integers, were outlined in his article ' Über den Zahlbegriff ' (' On the number-concept ' : Kronecker *1886a*). Cantor, who had written both his *Dissertation* and *Habilitationsschrift* under Kronecker, could not have been unaware of his extreme position. In the early 1870s Kronecker was building his forces in opposition to such basic concepts as the Bolzano–Weierstrass theorem, upper and lower limits, and the irrational numbers. At one point he had even tried to persuade Heine not to publish his *1872a*, but Heine was not deterred.

As an editor of Crelle's *Journal*, Kronecker was in a position to refuse any article for publication, and by 1879 he was so appalled at the direction Cantor's work was taking that he did just that. Though Cantor had sent his manuscript on the subject of dimension to the editors of Crelle's *Journal* on 12 July 1877, it did not appear immediately.

[1] This assumption was supported shortly after Cantor's startling article by a number of mathematicians who offered proofs in a number of forms and with varying degrees of generality, but the first completely satisfactory argument that dimension was invariant appeared only in Brouwer *1911a*. Cantor's alleged proof of the invariance of dimension is discussed in Dauben *1975a*.

Despite the editors' promise to accept it, and Weierstrass's efforts to promote its appearance, no steps were taken to prepare the paper for press. Cantor, suspecting Kronecker's intervention, became so agitated over the matter that he wrote a bitter letter to Dedekind complaining about the treatment of his work, and raising the possibility of withdrawing it from the journal and asking Vieweg to publish it separately (*Cantor/Dedekind*, 40). Dedekind, acting upon his own experience in such matters, was able to convince Cantor that he should wait. As it turned out, Dedekind was right: Cantor's paper finally appeared in the volume for 1878. But he was so offended by the apparent reluctance of the editors to give his paper speedy notice that he refused to publish again in Crelle's *Journal.*

The delay in publication represented the first major conflict that Cantor was to experience with regard to the acceptance of his work. No longer could he believe that his differences with Kronecker were purely academic. He recognised the extent to which Kronecker would go to prevent the spread of his work. In the years to come, he fought vigorously against all those who refused to allow the completed infinite, the transfinite numbers of the new set theory, into the bounds of accepted mathematics.

5.5. *Descriptive theory of point sets*

Beginning in 1879, Cantor published a series of four papers dealing specifically with the theory of linear point sets. He returned to the concept of derived set that had proven so useful in his research on trigonometric series, and stressed that the analysis of the properties of derived sets would eventually reveal the properties of the continuum.

Some preliminaries were required, starting with this definition: If P lies partially or entirely in the interval $[a, b]$, then it can happen that every interval $[c, d]$ in $[a, b]$ contains points of P. In such a case the set P is said to be *everywhere-dense in the interval* $[a, b]$ (*Papers*, 140–141). Cantor immediately connected the ideas of derived and everywhere-dense sets. A set P was everywhere-dense in an interval $[a, b]$ whenever the first derived set P' of P contained $[a, b]$ itself. Furthermore, everywhere-dense sets were necessarily sets of the second species, while first species sets could never be everywhere-dense.

The second definition is this: Two sets M and N are said to be of the *same power* if to every element of M one element of N corresponds, and conversely, to every element of N one element of M corresponds (*Papers*, 141). Cantor singled out two cases: denumerable sets, whose power was that of the natural numbers N; and continuous, or non-denumerable sets, whose power was that of the real numbers R. Countably infinite sets included the natural numbers, the rational and algebraic numbers. All sets of the first species were also of this first, denumerable, kind. But he noted that the rational and algebraic numbers showed that everywhere-dense sets, and hence sets of the second species, could also be denumerable.

Like the derived sets $P^{(n)}$ of a given set P, the power of P was given intrinsically with the set P. Cantor explained the mathematical importance of the concept of power as follows (*Papers*, 150, 152):

> The *concept of power*, which includes as a special case the concept of whole number, that foundation of the theory of number, and which ought to be considered as the most general genuine origin of sets [Moment bei Mannigfaltigkeiten], is by no means restricted to linear point sets, but can be regarded as an attribute of any *well-defined* collection, whatever may be the character of its elements . . . *Set theory* in the conception used here, if we only consider mathematics for now and forget other applications, includes the areas of arithmetic, function theory and geometry. It contains them in terms of the concept of power and brings them all together in a higher unity. *Discontinuity* and *continuity* are similarly considered from the same point of view and are thus measured with the same measure.

Cantor's next article of 1880 was short. It continued the same brick-laying work of the 1879 article, and sought to reformulate old ideas in the context of linear point sets. It also introduced for the first time his transfinite numbers. Given a set of the second species P, he explained how the first derived set P' of P could then be given the disjoint decomposition

$$P' = \{Q, R\}, \tag{5.5.1}$$

where Q was the set of all points belonging to first species sets of P', and R was the set of points contained in *every* derived set of P'. He defined this last property (intersection) by

$$R \underset{\text{Df}}{=} \mathscr{D}(P', P'', \ldots). \tag{5.5.2}$$

Since R was to consist of points belonging to *every* derived set of P', then it was equally true that

$$R = \mathscr{D}(P^{(2)}, P^{(3)}, \ldots), \ldots, \tag{5.5.3}$$

$$R = \mathscr{D}(P^{(n)}, P^{(n+1)}, \ldots). \tag{5.5.4}$$

Consequently, he felt justified in defining R, taken from P, as

$$R \underset{\text{Df}}{=} P^{(\infty)}. \tag{5.5.5}$$

$P^{(\infty)}$ was the derived set of P of order ∞ (*Papers*, 147). Assuming $P^{(\infty)} \neq \varnothing$, he denoted the first derived set of $P^{(\infty)}$ by $P^{(\infty+1)}$, the n-th by $P^{(\infty+n)}$. Continuing in this fashion, it was possible to generate derived sets of the following general form:

$$P^{(n_0\infty^{\nu}+n_1\infty^{\nu-1}+\ldots+n_{\nu})}.$$

By allowing ν to be taken as a variable, one could then produce 'an endless sequence of concepts', as he put it:

$$P^{(n\infty^{\infty})},\ P^{(\infty^{\infty}+1)},\ P^{(\infty^{\infty}+n)},\ P^{(\infty^{n\infty})},\ P^{(\infty^{\infty^{n}})},\ P^{(\infty^{\infty^{\infty}})},\ \ldots \tag{5.5.6}$$

Cantor summarised the entire procedure in decisive terms: 'we see here a dialectic generation of concepts', he said, 'which always continues further and thus is free of any arbitrariness' (*Papers*, 148). He also took care to add (in a footnote which Zermelo failed to include in Cantor's *Papers*) that he first had the idea of the second species sets and their corresponding transfinite symbols (soon to be named his transfinite ordinal numbers) a decade earlier (Cantor *1880a*, 358). This note was doubtless in response to an accusation made in 1879, in which du Bois Reymond claimed priority in the matter of designating 'everywhere-dense' sets with his own terminology 'pantachisch'.[1] Cantor was anxious to make clear that his work certainly had been done earlier than that of du Bois Reymond. Thus his footnote was intended to underscore the origins of his own work on the subject in his paper on trigonometric series published in 1872. But until he came to terms with the metaphysical nature of the transfinites, he referred to them only as 'infinite symbols' (*Papers*, 160). The derived sets remained the focus of his research for another few years, and the transfinite numbers, the infinite symbols themselves, were taken only for useful tags by means of which derived sets could be distinguished and identified.

The importance of the infinite symbols, however, was demonstrated without delay. The paper of 1879 had left unanswered the question of second species sets and whether they were necessarily everywhere-dense in any given interval. The new infinite symbols made it easy to describe a procedure by which second species sets could be identified

[1] du Bois Reymond had alluded to everywhere-dense sets in 1875; he had named them explicitly in 1879, soon followed with direct reference to Cantor's own paper of 1879. He even suggested in not-so-oblique terms that Cantor was misappropriating ideas not entirely his own. Though the concept of everywhere-dense sets was certainly of long standing, du Bois Reymond clearly felt some claim to priority in designating the property such sets exhibited. He must have hoped his terminology, 'pantachisch', would become standard. Even more clearly, he felt Cantor posed a threat for undisputed recognition. See du Bois Reymond *1880a*, 127–128.

which consisted of a single point (*Papers*, 148). These sets, which Cantor grouped with sets of the first species, were clearly losing candidates in the search for a complete explanation of the nature of the continuum.

In pressing his study of point sets of the second species further, Cantor offered a number of definitions and theorems dealing with various kinds of sets. For example, a set P was said to be *isolated* if it contained none of its limit-points ; in other words, when $\mathscr{D}(P, P')=\varnothing$ (*Papers*, 158). Then, given any set P, an isolated set Q resulted by simply removing $\mathscr{D}(P, P')$ from P. Thus

$$Q=P-\mathscr{D}(P, P'), \tag{5.5.7}$$

and consequently

$$P=Q+\mathscr{D}(P, P'). \tag{5.5.8}$$

(5.5.8) offered some immediate insights. Clearly, any set P could be considered as a disjoint combination of an isolated set Q and any set that was a divisor of P'. Since $P^{(n+1)}\subseteq P^{(n)}$, it followed that $P^{(n)}-P^{(n+1)}$ was *always* an isolated set. Two decompositions were of special importance, and led to some far-reaching conclusions :

$$P'=(P'-P'')+(P''-P''')+\ldots+(P^{(n-1)}-P^{(n)})+P^{(n)}\,; \tag{5.5.9}$$

$$P'=(P'-P'')+(P''-P''')+\ldots$$
$$+(P^{(n-1)}-P^{(n)})+(P^{(n)}-P^{(n+1)})+\ldots+P^{(\infty)}. \tag{5.5.10}$$

Following five theorems proving the denumerability of certain types of sets, Cantor offered a corollary which dealt with non-denumerable sets : ' If P is a non-denumerable point set, then $P^{(\alpha)}$ is also non-denumerable, whether α is a finite whole number, or if it is one of the infinite symbols ' (*Papers*, 160).

The transfinite numbers were still merely symbols, employed only as indices for the sake of adding precision to the distinction between first and second species sets. But it was only a matter of months before Cantor was to alter his goals dramatically, abandoning this older view of the ' infinite symbols ' to introduce the new transfinite numbers.

5.6. *The* Grundlagen : *transfinite ordinal numbers, their definitions and laws*

By the end of 1882, Cantor had finished a manuscript : *Grundlagen einer allgemeinen Mannichfaltigkeitslehre* (' Foundations of a general theory of sets ' : *1883a*), which outlined a defence of his new ideas in theological, philosophical and mathematical terms (*Papers*, 165–208). He was unusually anxious that its publication proceed rapidly. Towards the end of the year, he wrote nearly every day to Felix Klein, editor of *Mathematische Annalen*, and urged the greatest speed possible.[1] He even visited the press in Leipzig himself, hoping to expedite the appearance of his defence, both mathematical and philosophical, of an entirely new mathematics. The *Grundlagen* established his place as the founder of set theory, and he subsequently overshadowed even his closest rivals like du Bois Reymond and Harnack. It was the beginning of something new, quite startling and profoundly original.

The major achievement of the *Grundlagen* was its presentation of the transfinite numbers as an autonomous and systematic extension of the real numbers. Cantor had reached the point in his research where no progress in set theory, no advances in his study of continuity, were possible without recourse to the transfinite numbers. His own mathematical future hinged in large measure upon the acceptance of the actual infinite by mathematicians.

Cantor admitted that his new ideas might seem risky, but he argued their simplicity and necessity in a straightforward way. With concern for the introduction of ideas previously foreign to mathematics, he suggested (*Papers*, 165) that :

> So daring as this may seem, I can express not only the hope but the firm conviction, that this extension will, in time, have to be regarded as a thoroughly simple, appropriate, and natural one. But I in no way hide from myself the fact that with this undertaking I place myself in a certain opposition to widespread views about the mathematical infinite and to frequently advanced opinions on the nature of number.

Cantor's first concern was to counter mathematicians sympathetic to Kronecker's finitism who might easily have refused to read further than the *Grundlagen*'s first paragraph. He began by explaining the distinction that had long been recognised between the potential and actual infinite. The former was used in mathematics as the very roots of the calculus. It involved essentially the idea of variation, of growing beyond any ascertainable bound, but a state never actually considered as completed or final. He also referred to such infinities as improper infinities. In contrast with these were proper or actual infinities.

The realisation that his transfinite numbers were equally as real mathematically as the finite whole numbers had only recently come to

[1] The correspondence between Cantor and Klein, largely unpublished, is preserved in the archives of the Niedersächsische Staats- und Universitätsbibliothek, Göttingen. In particular, see Cantor's letter to Klein, No. 429, 18 December 1882, and his letter No. 430, 20 December 1882. This correspondence is referred to again in section 5.10 below.

consciousness in Cantor's mind. This recognition represented significant progress : ' I will define the infinite real whole numbers in the following, to which I have been led over the past few years without realising that they were concrete numbers of real meaning ' (*Papers*, 166).

In the *Grundlagen* Cantor explained how the sequence of natural numbers 1, 2, 3, . . . had its origin in the repeated addition of *units.* He called this process of defining finite ordinal numbers by the successive addition of units the *first principle of generation.* It was clear that the class of all finite whole numbers (I) had no largest element. Though it was incorrect to speak of a largest element for (I), he believed there was nothing improper in thinking of a new number ω which expressed the natural, regular order of the *entire* set (I). This new number ω, the first transfinite number, was the *first* number following the entire sequence of natural numbers ν. It was then possible to apply the first principle of generation to ω, and to produce additional transfinite ordinal numbers :

$$\omega, \omega+1, \omega+2, \ldots, \omega+\nu, \ldots . \qquad (5.6.1)$$

Again, since there was no largest element, one could imagine another number representing the entirety, in order, of numbers $\omega+\nu$. Denoting this entirety by 2ω, it was possible to continue further :

$$2\omega, 2\omega+1, 2\omega+2, \ldots, 2\omega+\nu, \ldots . \qquad (5.6.2)$$

(Later Cantor reversed the order of the terms in ordinal multiplication, so that, for example, ' 2ω ' became ' $\omega 2$ ' ; this latter is the modern notation.)

In attempting to characterise this mode of generation, Cantor allowed that ω could be regarded as a limit towards which the natural numbers N increased monotonically but never reached. Lest the analogy seem entirely mistaken, he added that by this he meant only to emphasise the character of ω taken as the first whole number following next after *all* the numbers $n \in N$. The idea of ω as a limit served to satisfy its role as an ordinal, the smallest integer larger than any integer $n \in N$.

This then was the *second principle of generation.* Whenever a sequence of numbers could be considered as limitless in extent, new transfinite numbers could always be generated by positing the existence of some least number larger than any in the given sequence. Cantor expressed the essential feature of this second principle of generation in terms of its logical function (*Papers*, 196) :

> I call it the *second principle of generation* of real whole numbers and define them more precisely : if any definite succession of defined whole real numbers exists, for which there is no largest, then a new number is created by means of this second principle of

generation which is thought of as the *limit* of those numbers, that is, it is defined as the next number larger than all of them.

By successive application of the two principles it was always possible to produce new numbers, and always in a completely determined succession. In their most general formulation, such numbers could be given in the following form :

$$\nu_0\omega^\mu + \nu_1\omega^{\mu-1} + \ldots + \nu_\mu.$$

But by proceeding apparently without constraint, there seemed to be no end to the numbers of this second number-class. Were this the case, what distinctions could be drawn between the first and second classes ? Cantor was able to add, however, a third principle which he called the *principle of limitation* (' Hemmungsprinzip '), and which was designed to produce natural breaks in the sequence of transfinite numbers. Consequently it was possible to place definite bounds upon the second number-class (II), and to distinguish it from the third and successively higher number-classes, with this definition (*Papers*, 197) :

> We define therefore the second number-class (II) as *the collection of all numbers* α *(increasing in definite succession) which can be formed by means of the two principles of generation :*
>
> $$\omega,\ \omega+1,\ \ldots,\ \nu_0\omega^\mu + \nu_1\omega^{\mu-1} + \ldots + \nu_\mu,\ \ldots,\ \omega^\omega,\ \ldots,\ \alpha,\ \ldots,$$
>
> *with the condition that all numbers preceding* α *(from 1 on) constitute a set of power equivalent to the first number-class (I).*

In the *Grundlagen* he went on to establish that not only were the powers of the two number-classes (I) and (II) distinct, but that in fact the power of the second number-class (II) was the next larger after that of the first number-class (I) (*Papers*, 197–201).

An important advance made possible by the new numbers was the distinction that Cantor made between ' Zahl ' and ' Anzahl '. Zahl, or Number, referred to the cardinal sense of the number of objects in a set without regard to the order in which elements occurred ; Anzahl, or Numbering, took into consideration the order of elements. The difference was fundamental. For example, all of the following sets have the same cardinal number, they are equal in power, and are all denumerable. Nevertheless, their Numberings, their ordinal numbers, are different :

$$\left.\begin{aligned} (a_1, a_2, \ldots, a_n, a_{n+1}, \ldots) &= \omega, \\ (a_2, a_3, \ldots, a_{n+1}, a_{n+2}, \ldots, a_1) &= \omega+1, \\ (a_3, a_4, \ldots, a_n, \ldots, a_1, a_2) &= \omega+2, \\ (a_1, a_3, a_5, \ldots;\ a_2, a_4, a_6, \ldots) &= \omega+\omega = 2\omega. \end{aligned}\right\} \qquad (5.6.3)$$

Once his transfinite numbers were defined, Cantor went on to describe their arithmetic and properties such as prime numbers among the transfinites (*Papers*, 201–204). Among the most significant characteristics of the transfinite ordinals was their non-commutativity. In general, $a+b \neq b+a$, nor did $ab=ba$ in all cases. For finite numbers, commutativity of operations was preserved, but not for transfinites. For example:

$$2+\omega=(1, 2, a_1, a_2, \ldots, a_n, \ldots) \neq (a_1, a_2, \ldots, a_n, \ldots, 1, 2)=\omega+2, \qquad (5.6.4)$$

$$2\omega=(a_1, a_2, \ldots; b_1, b_2, \ldots) \neq (a_1, b_1, a_2, b_2, \ldots, a_n, b_n, \ldots)=\omega 2. \qquad (5.6.5)$$

Two sets were defined to be of the same Numbering (that is, their corresponding ordinal numbers were equal) if they could be corresponded in a one–one fashion such that the order of elements was preserved in each case. In a similar way, the powers of two sets M and N were defined as equivalent if the elements of one set could be corresponded one–one with those of the other.[1]

The newly introduced distinction between Number and Numbering brought new insights to understanding the difference between finite and infinite sets. For finite sets, regardless of ordering, the Numbering of elements was always the same. Infinite sets were much more interesting because of the different Numberings one could find for sets of the same power. The Numbering of sets, therefore, was a concept totally dependent upon the order in which the elements of the set occurred. And there was a correlation between the Number of a set and the Numbering that its elements might produce, depending upon their arrangement: 'Every set of the power of the first class is denumerable by numbers of the second number-class and only by such numbers' (*Papers*, 169).

Though the difference between Number and Numbering was indistinct on finite sets, it helped to explain how the number concept functioned in a double sense, and why there had been confusion for centuries over potential and actual infinities. For finite sets, ordinal and cardinal numbers coincided. But because the two kinds of number were fundamentally different, Cantor could demonstrate the illegitimacy of trying to press properties of finite numbers onto infinite numbers. Once the ordinal/cardinal distinction had been recognised on transfinite

[1] Though Cantor did not identify the powers of infinite sets with transfinite cardinal numbers until after the *Grundlagen* had appeared, he later defined inequalities among cardinal numbers in full detail; see section 5.10 below, and *Papers*, 284–285. The somewhat more involved definitions for inequalities between transfinite ordinal numbers are discussed in section 5.11 below, and *Papers*, 320–325.

numbers, he was able to re-apply the same concepts to finite sets, and in the process find that it was another way to characterise the differences between finite and infinite domains : if a set were finite, then its cardinal and ordinal numbers were the same (*Papers*, 168–169).

5.7. *The continuum hypothesis and the topology of the real line*

One of the major goals of Cantor's transfinite set theory was the answer to a question that seemed quite simple, but one which to this day remains unanswered : What is the power of the continuum ? This question, with an answer always believed to be the only possible solution, has come to be known as ' Cantor's continuum hypothesis ' : The power of the continuum is equivalent to that of the second number-class (II) (*Papers*, 192).

Though Cantor was never able to establish the truth of this conjecture, his *Grundlagen* did manage to make some progress in refining the mathematical description of continuous sets. One such advance was his description of the general conditions necessary and sufficient to constitute a continuum, which involved the idea of perfect sets : A set P is said to be *perfect* if it equals its derived set ; in other words, if $P=P'$ (*Papers*, 193).

It was clear that continua had to be perfect sets, since the classic example, the set of all real numbers R, clearly equalled the set of its limit-points. But there was a further difficulty : perfect point sets were not necessarily everywhere-dense. As an example, Cantor offered his famous ternary set,[1] the set of all real numbers Z represented by

$$Z=\frac{c_1}{3}+\frac{c_2}{3^2}+\ldots+\frac{c_n}{3^n}+\ldots, \quad c_n=0 \text{ or } 2. \tag{5.7.1}$$

Though he could prove that perfect sets never had the power of the first number-class (I), his ternary set was everywhere-dense in no interval. Consequently, in addition to the fact that continua must be perfect sets, another definition was needed : ' T is a connected point set if, for any two points t and t' and for any arbitrarily small number ϵ, a finite number of point $t_1, t_2, t_3, t_4, \ldots, t_n$ can always be found in T, such that the distances $\overline{tt_1}, \overline{t_1t_2}, \overline{t_2t_3}, \ldots, \overline{t_nt'}$ are all less than ϵ' (*Papers*, 194). Thus the characteristic features of continua were identified : they were both perfect and connected. These, stated Cantor, were the necessary and *sufficient* conditions under which a point set could be

[1] Cantor introduced this set in note 11 of the *Grundlagen* (*Papers*, 207). In *Papers*, 235, he proved that it was of measure zero ; compare section 4.5.

considered continuous. But there was a serious lacuna in the *Grundlagen* : the question of the power of the continuum was still unanswered. He intimated that he was hopeful a proof would be forthcoming, establishing his conjecture that the power of the continuum was none other than that of the second number-class (II).

The corollaries to such a solution would be numerous. It would immediately follow that all infinite point sets were either of the power of the first or second number-class, something Cantor had long claimed but never proven. It would also establish that the set of all functions of one or more variables represented by infinite series was necessarily equal in power to the second number-class. Likewise, the set of all analytic functions, and that of all functions represented by trigonometric series, would also be shown to be equivalent in power to that of the second number-class (II).

In a continuation of the *Grundlagen* published in 1884 Cantor developed a number of ideas which were related directly to the continuum hypothesis. He began by establishing a number of theorems concerning perfect sets and various kinds of derived sets. Characteristic of these theorems is the following : If P is a point set such that its first derived set P' is of power greater than the first, then ' there are always points which belong to all derived sets $P^{(\alpha)}$, where α is any number of (I) or (II), and the set of all these points, nothing other than the derived set $P^{(\Omega)}$, is always a perfect set ' (*Papers*, 221).

Next, Cantor established a theorem (Theorem E) of interest because it corrected an error in the *Grundlagen* that had been discovered by the Swedish mathematician Ivar Bendixson. Originally, Cantor had claimed that if P' were equal in power to the second number-class, then P' could be uniquely decomposed into two sets, $P' = R \cup S$; the set R he took to be *reducible*, meaning there was always some γ of (I) or (II) such that $R^{(\gamma)} = \varnothing$ (*Papers*, 193 and 222–223). This, in fact, was not true, as Bendixson pointed out.[1] S was a *perfect* set, meaning that for any number γ of (I) or (II), $S^{(\gamma)} = S$. Thus the question arose as to what properties distinguished the denumerable set R from other denumerable sets. This question was answered in one of Bendixson's theorems, as Cantor acknowledged : If R is the set of first power mentioned in Theorem E, then there is always a smallest number α of (I) or (II) such that $\mathscr{D}(R, R^{\alpha}) = \varnothing$ (*Papers*, 224).

Cantor then turned his attention to another class of sets closely related to perfect sets : those he termed ' closed sets '. If a set con-

[1] Bendixson's letters to Cantor are kept in the archives of the Institut Mittag-Leffler, Djursholm, Sweden. These archives also contain the correspondence between Cantor and Mittag-Leffler, which is used occasionally in the rest of this chapter and which is mostly unpublished.

tained its first derived set, then it was said to be *closed*, namely :

$$\mathscr{D}(P, P')=P'. \tag{5.7.2}$$

Every set might be closed by simply adding its first derived set P' : thus $P \cup P'$ was a closed set. Any set P could be decomposed into the sum of two sets, one set Q which was isolated and therefore denumerable, the second set P' which was closed : $P=Q \cup P'$ (*Papers*, 226–227). He went on to discuss the properties of sets dense-in-themselves (for which $\mathscr{D}(P, P')=P$), perfect sets, everywhere-dense sets, and the like, as well as inter-connections between them (*Papers*, 225–229). But he did not come any nearer to answering the question of the power of the continuum itself.

5.8. *Cantor's mental breakdown and non-mathematical interests*

Nothing caused Cantor greater annoyance than did Kronecker and his persistent attacks upon transfinite set theory. Cantor was especially angered by the fact that Kronecker refused to be open, preferring to save his most critical remarks for lectures and informal discussions with students. Kronecker thus carried on his polemic privately, or semi-publically in university, but never openly in print.

In early September 1883, Cantor learned that Kronecker was writing to the French mathematician Hermite and criticising Cantor's work as ' Humbug ' (Cantor to Mittag-Leffler, 5 May 1883). Shortly before Christmas Cantor wrote to Gösta Mittag-Leffler, for a time one of Cantor's closest friends, who was responsible as the founding editor of *Acta mathematica* for having nearly all of Cantor's early work on set theory published in French translation (Cantor *1883b*). Cantor confided that he had written to the Ministry of Education, hoping to annoy Kronecker by applying for a position in Berlin available the following spring. This was a direct expression of Cantor's lifelong belief that he deserved the honour of a position at one of two German universities known for their great mathematicians : either Göttingen or Berlin. But on 30 December 1883 he admitted to Mittag-Leffler that the application in Berlin would come to nothing. He had heard from Weierstrass that the obstacles were largely financial, owing to Kronecker's large salary. The letter to Mittag-Leffler was another occasion for Cantor to re-iterate the bitterness which he felt towards his position at Halle. The entire episode underscored the frustration and hostility which he felt in realizing that there was little he could ever do in the face of powerful opposition to improve his position.

If Kronecker was annoyed at Cantor's move, he returned the challenge masterfully. Early in January 1884 he wrote to Mittag-Leffler asking to publish in the *Acta mathematica* a short paper in which he would show ' that the results of modern function theory and set theory are of no real significance ' (see Schönflies *1927a*, 5). At first Cantor was mildly receptive to the idea, believing that the article would at last bring Kronecker's opposition into the open, where it could be directly countered and presumably rejected. But Cantor began to have second thoughts. He feared that Kronecker might reduce his arguments to personal polemics. It all seemed as though Kronecker, by wanting to publish in the *Acta mathematica*, was trying to drive Cantor out of the one journal in which he had found a sympathetic editor, just as Kronecker years earlier had tried to prevent Cantor from publishing any work in Crelle's *Journal*.

Cantor threatened that Mittag-Leffler could expect him to withdraw his support for the journal in the years to follow should any polemical writings appear in the *Acta mathematica* under Kronecker's signature (Schönflies *1927a*, 5). Kronecker apparently never sent anything for the *Acta*, but the threats that Cantor was willing to make even to his friend Mittag-Leffler show how sensitive he could be to the conspiracy he felt was brewing against his work under the auspices of a single man : Leopold Kronecker.

While Cantor was becoming increasingly annoyed by the opposition to his work in Germany, there were other frustrations conspiring to upset his peace of mind in the early part of 1884. The continuum problem seemed as intransigent as ever, though he had reduced it to the problem of showing that perfect sets were equal in power to the second number-class (II).

Then, barely a month later, Cantor experienced his first mental breakdown (see Grattan-Guinness *1971b*, 356 ; and Peters *1961a*, 15, 27). It came upon him swiftly, unexpectedly, and apparently lasted somewhat more than a month. By the end of June he was sufficiently recovered to write to Mittag-Leffler, but complained that he lacked the energy and interest to return to rigorous mathematical thinking, and was content to take care of trifling administrative matters at the university. He felt capable of little more. But it was significant that he wrote to Mittag-Leffler saying that he was anxious to return to work, and would prefer his research to confining himself to the preparation of his lectures (see Schönflies *1927a*, 9). In fact, as soon as he had recovered sufficient strength, he set off for his favourite vacation resort in the Harz mountains and returned to his analysis of perfect sets. He also undertook the bold step of writing directly to Kronecker, and attempted to put their differences aside (see Meschkowski *1967a*, 237–241).

Less than a week after his letter of reconciliation to Kronecker,

Cantor wrote to Mittag-Leffler announcing at last an extraordinarily simple proof that the continuum was equal in power to the second number-class (II). The proof attempted to show that there were closed sets of the second power. Based upon straightforward decompositions and the fact that every perfect set was of power equal to that of the continuum, he was certain he had triumphed at last. He summarised the heart of his supposed proof in a single sentence : ' You see, therefore, everything comes down to defining a single closed set of the second power. When I have put everything in order, I will send you the details ' (Meschkowski *1967a*, 243).

But on 20 October Cantor sent a lengthy letter to Mittag-Leffler announcing the complete failure of the new proof. On 14 November he again wrote saying that he had just found a rigorous proof that the continuum did *not* have the power of the second number-class, or of any number-class. He consoled himself by saying that ' so fatal an error, which one has held for so long, makes it an even greater advance to overcome it ' (Schönflies *1927a*, 17). Nevertheless, within twenty-four hours he had decided that his latest proof was wrong, and that the continuum hypothesis was again an open question. It must have been embarrassing for him to have been compelled to reverse himself so often within so short a period of time. Even more discouraging must have been the realisation that the simplicity of the continuum hypothesis concealed difficulties of a high order that he was unable to resolve, despite his best efforts. However, he was not easily discouraged, and his continuing search for new methods and results marked the final and most devastating episode responsible for his disillusionment with mathematics and his discontent with colleagues both in Germany and abroad.

Cantor had developed a theory of simply ordered sets, sets for which given any two distinct elements a and b, either $a<b$ or $b<a$. He was confident that a systematic study of the types of simply ordered sets, in particular the rational and real numbers as given in their natural order, would make new advances possible (for details, see section 5.11 below). Consequently he prepared an article entitled ' Principien einer Theorie der Ordnungstypen ' (' Principles of a theory of order-types ' : *1885a*), which he sent to Mittag-Leffler for the *Acta mathematica*. The manuscript was partly set in type and dated 21 February 1885, but not published until it appeared in Grattan-Guinness *1970b*.

The *Principien* was a remarkable paper. To deal with simply ordered sets like the rationals or reals, Cantor introduced new concepts, including those of coherent and adherent sets and their order-types. Designating the type of simply ordered set represented by the rationals as η, he produced a general theorem establishing the necessary and sufficient conditions for any set to be of type η. He also introduced θ as the simply ordered type represented by the continuum of all real numbers taken in their natural order on [0, 1].

The *Principien*, in addition to such technical advances, contained Cantor's first explicit statement that pure mathematics was nothing other than pure set theory. By this Cantor meant that the principles and results of set theory were so general and so penetrating that all of mathematics could be understood in terms that were essentially set-theoretic in nature (Cantor *1885a*, 84).

Despite the importance of his theory of simply ordered types, to Cantor's dismay Mittag-Leffler wrote on 9 March 1885 to suggest that the *Principien* be withdrawn from press. Mittag-Leffler was convinced that the publication of Cantor's newest research, before he had been able to obtain any positive results from it, would harm his reputation rather than advance it. He added that if Cantor's set theory came into discredit because of the *Principien*, it would take much longer for any of Cantor's work to win general acceptance among mathematicians. He even added that though Cantor's ideas concerning simply ordered sets might never be appreciated in his lifetime, it would perhaps be re-discovered by someone a hundred years later, and then Cantor would receive the credit he deserved ! (Grattan-Guinness *1970b*, 102).

In recalling this episode more than a decade later, Cantor confided in Poincaré his real feelings about Mittag-Leffler's request that he not publish the *Principien* in the *Acta mathematica* : ' It was soon clear to me that he was doing this in the interest of his Acta Mathematica ' (*ibid.*, 105). Cantor was deeply hurt by Mittag-Leffler's rejection of his newest research. More than his polemic with Kronecker, more than his nervous breakdown or the trouble he was having in finding a proof that his continuum hypothesis was true, Mittag-Leffler's suggestion that he not print his work in the *Acta mathematica* seemed the cruellest blow of all. Though he never admitted that the incident affected in any way his personal regard and friendship for Mittag-Leffler, he wrote less frequently, and only seldom did he mention matters concerning his research. He felt as though the last mathematician at all sympathetic with his struggle to establish the transfinite numbers had abandoned him. He published only once more in the pages of the *Acta mathematica*, and that was a paper (*1885b*) which had already been accepted.

Instead, Cantor began to concentrate more and more upon problems of philosophy, theology, and the Bacon–Shakespeare controversy (see Grattan-Guinness *1971b*, esp. pp. 363–365 ; and Meschkowski *1967a*, 172, 264). Isolated and alone in Halle, he began to teach philosophy, and to correspond with theologians who provided a natural outlet for his need to communicate the importance and implications of his work. In turn, his contact with Catholic theologians may have made his own religious sympathies all the stronger. By the early part of 1884 he could write to Mittag-Leffler that he was not the creator of his new work, but merely a reporter (Schönflies *1927a*, 15–16). He was even more direct in a letter written to Hermite during the first month of 1894, in which he claimed that it was God's doing that had led him away from serious mathematics to concerns of theology and philosophy (Meschkowski *1965a*, 514–515) :

> But now I thank God, the all-wise and all-good, that He forever denied me the fulfilment of this wish [for a position at university in either Göttingen or Berlin], for He thereby constrained me, through a deeper penetration into theology, to serve Him and His Holy Roman Catholic Church better than I would have been able to with my probably weak mathematical powers through an exclusive occupation with mathematics.

At one stroke Cantor signalled the many disappointments and doubts accumulated over more than two decades. His remarks reflected the frustration that he must have felt at being unable to solve the continuum hypothesis, and the disastrous effects which both the relentless attacks from Kronecker and Mittag-Leffler's response to his work on order-types had occasioned. Realising that no positions were ever going to be offered him in either Göttingen or Berlin, Cantor turned to other interests less demanding and more positively reinforcing. By the end of his life, in the spirit of Pope Leo XIII's encyclical *Aeterni patris*, he saw himself as the servant of God, a messenger or reporter who could use the mathematics he had been given to serve the Roman Catholic Church (see Dauben *1977a*). He firmly believed that ' for the first time Christian philosophy will learn from me the true theory of the infinite ' (Meschkowski *1965a*, 513). Convinced that he had been inspired and helped by God, Cantor was sure that his work was of consequence, despite the failure of mathematicians to understand the importance of his discoveries.

5.9. *Cantor's method of diagonalisation and the concept of coverings*

One of Cantor's most important projects in the late 1880s was the formation of a professional society for the promotion of mathematics in Germany. He viewed the idea as an alternative to the tradition-bound universities and the poorly-organised *Gesellschaft Deutscher Naturforscher und Aerzte* (' Society of German scientists and physicians '). Above all, he felt his own career had been greatly damaged by the premature and prejudiced rejection of his work by the prevailing establishment, led by Kronecker, and he hoped that an independent organisation would provide an open forum, one that would give younger mathematicians encouragement and a fair hearing of new, even radical, ideas.

The new society, the *Deutsche Mathematiker-Vereinigung* (' German Mathematicians' Union '), held its first meeting in 1891 in Halle, and elected Cantor its first president. (On the history of the Union, see especially Gutzmer *1904a* and Gericke *1966a*.) As his own contribution to the Union's first proceedings, he presented a theorem using a powerful new method to establish the existence of non-denumerable sets (*Papers*, 278–280). He thereby reconfirmed his earlier (1874) proof of the non-denumerability of the real numbers (see section 5.3 above), but he was able to go much further. He had made a fundamental advance of major significance for the future of transfinite set theory : ' This proof seems remarkable not only because of its great simplicity, but above all because the principle which it follows can be extended immediately to the general theorem, that the powers of well-defined sets have no maximum, or, what is the same, that every given aggregate L can be replaced by another M which is of greater power than is L ' (*Papers*, 279).

The proofs that Cantor offered in his paper of 1891 hinged on his new method of diagonalisation. Relying upon only two elements, m and w, he considered the collection M of elements $E=(x_1, x_2, \ldots, x_\nu, \ldots)$, where each x_ν was either m or w. As examples, he suggested :

$$E^{\mathrm{I}}=(m, m, m, m, \ldots), \tag{5.9.1}$$

$$E^{\mathrm{II}}=(w, w, w, w, \ldots), \tag{5.9.2}$$

$$E^{\mathrm{III}}=(m, w, m, w, \ldots). \tag{5.9.3}$$

He then claimed that the collection of all such elements M was non-denumerable : ' If $E_1, E_2, E_3, \ldots, E_\nu, \ldots$ is any simply infinite sequence of elements of the set M, then there is always an element E_0 of M which corresponds to no E_ν ' (*Papers*, 278). In his proof he first listed the elements of M, assumed to be denumerable :

$$\left.\begin{array}{l} E_1=(a_{11}, a_{12}, \ldots, a_{1\nu}, \ldots), \\ E_2=(a_{21}, a_{22}, \ldots, a_{2\nu}, \ldots), \\ \quad\ldots\ldots\ldots\ldots \\ E_\mu=(a_{\mu 1}, a_{\mu 2}, \ldots, a_{\mu\nu}, \ldots), \\ \quad\ldots\ldots\ldots\ldots \end{array}\right\} \tag{5.9.4}$$

Each element $a_{\mu,\nu}$ of the array (5.9.4) was taken to be either m or w. He then defined a new sequence $b_1, b_2, \ldots, b_\nu, \ldots$. Again, each

b_ν was either m or w, but determined so that $b_\nu \neq a_{\nu\nu}$. Then $E_0 = (b_1, b_2, b_3, \ldots)$ was an element of M, but it was immediately clear that $E_0 \neq E_\nu$ for any value of the index ν. Using only two elements m and w, he had shown that from these alone a new set could be generated, and one of greater power. In fact, the method of diagonalisation provided him with an easy means of showing that the ascending sequence of powers of well-defined sets had no maximum. In other words, given any set L, it was always possible to produce from elements of L another set M which was necessarily of higher power than L itself.

Above all, Cantor had advanced significantly beyond the specific conclusion of his earlier paper of 1874, and had proved the existence of more than just one non-denumerable transfinite power without having to make any reference to irrational numbers or to the limits of infinite sequences. The comprehensiveness of his new method of diagonalisation made the paper of 1891 an important contribution to the development of set theory. As an example, he considered the linear continuum L, the set of all real numbers on $[0, 1]$, and the collection M of single-valued functions $f(x)$ which assumed only the values 0 and 1 for any value of x in $[0, 1]$, and proved that: The set M of single-valued functions $f(x)$ assuming only the values 0 and 1 on the interval $[0, 1]$ is greater in power than the set L of real numbers $x \in [0, 1]$ (*Papers*, 279–280; compare Fraenkel *1953a*, 64).

Similarly, the diagonalisation argument made it possible for Cantor to show that given any set, the set of all its sub-sets was always of a power greater than the parent set itself. Though he apparently did not realise it at the time, the means were now available whereby the continuum hypothesis itself could be given a direct algebraic formulation. But the discovery of how this could be done was not made until July 1895 (see section 5.10 below).

One last feature of Cantor's paper of 1891 deserves notice. Unlike the *Grundlagen*, where powers were never considered as numbers, he had come to see that powers actually represented the sole and necessary generalisation of the concept of cardinal number (*Papers*, 280). Thus regarded as powers, transfinite cardinal numbers should enjoy the same reality and definiteness as did the finite cardinals. The only difference involved the non-commutativity which distinguished transfinite numbers operationally from their finite counterparts. But in emphasising the fact that powers were to be regarded as necessary generalisations of the finite cardinal number concept, he was indicating a significant advance in his conceptualisation and representation of the basic principles of transfinite set theory.

5.10. *The* Beiträge : *transfinite alephs and simply ordered sets*

Cantor's last major publication was his *Beiträge zur Begründung der transfiniten Mengenlehre* (' Contributions to the founding of transfinite set theory '), issued in two parts as *1895a* and *1897a.*[1] In terms of new results, it made few startling advances. Where it was most innovative, it either improved the scope and presentation of procedures like multiplication or exponentiation of transfinite numbers, or it refined more successfully the details of conclusions already obtained. Above all, it introduced for the first time his special notation for the cardinal numbers, the alephs, where in particular $\aleph_0$ expressed the cardinality of denumerably infinite sets and was thus the smallest of the transfinite alephs.

Cantor opened the *Beiträge* with what has become a classic definition : ' By a " set " we understand any collection *M* of definite, distinct objects *m* of our perception or of our thought (which will be called the elements of *M*) into a whole ' (§ 1 ; *Papers*, 282). Powers, which he had come to regard since the *Grundlagen* as cardinal numbers, were defined in terms of the process of abstraction introduced in his *1887a* (*Papers*, 411–412). He now wrote : ' We call " power " or " cardinal number " of *M* that general concept which, with the help of our active thought-process, arises from the set *M*, abstracting from the character of its various elements *m* and from the order in which they occur ' (§ 1 ; *Papers*, 282).

Here Cantor's philosophical idealism was plain. Gottlob Frege later criticised him sharply, however, for depending upon such lax and unrigorous formulations. He particularly disliked Cantor's use of abstraction in defining both ordinal and cardinal numbers. Nevertheless, he believed that Cantor's theory could be salvaged from its uncritical presentation, and once opined that, basically, the results of transfinite set theory were sound, though its foundations required much more careful scrutiny.[2]

Equivalence between sets and their corresponding cardinal numbers was defined no differently in the *Beiträge* than in previous presentations

[1] Part I of the *Beiträge* was translated almost immediately into Italian as Cantor *1895b*. Both parts were translated into French as Cantor *1899a*. English-speaking readers had to wait until 1915 for a translation made by P. E. B. Jourdain (Cantor *1915a*). All references in this chapter to the *Beiträge* are made first to the appropriate section, and then to the corresponding pages of the *Beiträge* in Cantor's *Papers*.

[2] Frege criticized Cantor's idealism even before the *Beiträge* appeared (Frege *1892a*, 270). But he also made it clear that he thought the transfinite set theory could be given a rigorous foundation, one that would make it entirely acceptable mathematically (see Frege *1892a*, 272). For an even more critical evaluation of Cantor's use of abstractions, see Frege's unpublished draft version *1890a* of an earlier review of Cantor's work.

of Cantor's theory. One important question concerned the comparability of cardinals, for which he introduced this definition: Given $a = \overline{\overline{M}}$, $b = \overline{\overline{N}}$ (his way of denoting the double process of abstraction which produced cardinal numbers a and b), then if: (1) there is no proper subset $M' \subset M$ such that $M' \sim N$, and (2) there is a proper subset $N' \subset N$ such that $N' \sim M$, then it is to be said that $a < b$ or $b > a$ (§ 2; *Papers*, 284–285).

This was essentially the same definition of the order relation for cardinal numbers that Cantor had given as early as 1887. But he had gone further in 1887 and claimed that if M and N were two non-equivalent sets, then one always had to be equivalent to a proper subset of the other (*Papers*, 413). Moreover, he noted in consequence of his characterisation of order among the powers of sets that whenever two sets M and N could be mapped in a one–one fashion to proper subsets of each other, so that $M \sim N' \subset N$ and $N \sim M' \subset M$, then M and N were necessarily equivalent. This same theorem appeared as Theorem B of the *Beiträge*'s section 2, but it was one which he had never been able to prove directly himself. The theorem was later established independently by Felix Bernstein and E. Schröder, and has subsequently come to bear their names.[1] Cantor noted in the *Beiträge* that the Schröder–Bernstein Theorem, as well as three other theorems dealing with equivalence relations among sets, followed easily from a previously stated but *unproven* theorem, the trichotomy law: If a and b are any two cardinal numbers, then either $a = b$, $a > b$ or $b < a$ (§ 2; *Papers*, 285).

Cantor had already shown that, given any two cardinal numbers, only one of these order relations could hold, but he was unable to prove that *exactly* one always had to be true. The pernicious complication concerned the case of two cardinals represented by sets A and B, where A was assumed equivalent to no part of B and B was assumed equivalent to no part of A. He conjectured that this could only happen for finite sets where A and B were equivalent. But he was unable to show that the same could not occur on infinite sets. Consequently, there was no way he could establish the necessary comparability of all cardinal numbers, finite and infinite. The matter was critical, for if they were not comparable, it would be impossible to arrange all cardinal numbers in an ordered sequence, and in turn, it would be impossible to say whether one of two given cardinals was necessarily larger than the other. This was a grave matter for his continuum hypothesis, for if the power of the continuum were a non-comparable cardinal number, then he could never show that it was in fact equivalent to the power of the second number-class, which was a comparable cardinal by virtue

[1] For the papers by Bernstein and Schröder dealing with the equivalence of powers, see Schröder *1898a* and Bernstein *1898a*. See also Zermelo *1901a*, 34–38.

of its definition in terms of a well-ordered set. He assumed all along that every set could be well-ordered, and thus he could overlook a multitude of related difficulties while being able to assert, as a consequence, that all cardinal numbers were comparable. But he never published any proof of these claims, and chose not to raise the subject again in either part I of the *Beiträge* or its successor of 1897.

Like his definitions of order relations, rules for the addition and multiplication of any two cardinal numbers were first given in an earlier paper, Cantor's *Mitteilungen* (*1887a*), but now took this form: The union of two sets M and N, which have no elements in common, is denoted (M, N). The cardinal number of (M, N) depends only upon the cardinal numbers $a=\overline{\overline{M}}$ and $b=\overline{\overline{N}}$. This leads to the definition of the sum of a and b: $a+b=\overline{\overline{(M, N)}}$ (§ 3; *Papers*, 285–286). Since the power of the sets in question was independent of the order of the elements in either set or their union, the addition of cardinal numbers was commutative, and $a+b=b+a$. For multiplication, 'every element m of a set M may be joined with each element n of another set to form a new element (m, n); we denote the set of all these elements (m, n) by $(M . N)$, and call this the "union set of M and N". Thus $(M . N) = \{(m, n)\}$. If $a=\overline{\overline{M}}$ and $b=\overline{\overline{N}}$, then the product ab is defined as $ab=\overline{\overline{(M . N)}}$' (§ 3; *Papers*, 286).

There was a conspicuous difficulty with this definition. Products were defined for cardinal numbers in terms of their corresponding sets. These were combined to form a new set, a product set, but this was done for only two sets at a time. Thus the procedure could not be extended directly to include any more than finitely many products. But in 1895 Cantor explained how it was possible to represent the power of the continuum as an infinite exponentiation of the form $2^{\aleph_0}$. It was consequently of special importance to be able to define transfinite exponentiation. He was able to do so by a significant innovation: the idea of the covering of a set (§ 4; *Papers*, 287):

> By a 'covering of the set N with elements of the set M' or more simply, by a 'covering of N by M', we understand a rule by which, to each element n of N a definite element of M is corresponded, whereby one and the same element of M may be used repeatedly. The element corresponded with n from M is clearly a single-valued function of n and can therefore be designated $f(n)$; it is called the 'covering function of n', and the corresponding covering of N is called $f(N)$.

Drawing upon an example already used in the paper of 1891 to show that the set of single-valued functions was at least equivalent to the power of the continuum, Cantor suggested a covering by two elements of M, m_0 and m_1. If n_0 represented a particular element from N, then a covering function could be given such that $f(n_0)=m_0$, and $f(n)=m_1$ for all other values of n in N excepting the value n_0.

The set of all such covering functions served as the basis for defining the exponentiation of cardinal numbers. Specifically, the set of all different coverings of a set N by the set M produced a set with elements $f(N)$, which Cantor denoted $(N|M)$. This he called ' the covering set ', and $(N|M)=\{f(N)\}$. Since the definition depended only upon the cardinal numbers $a=\overline{\overline{M}}$ and $b=\overline{\overline{N}}$, then the cardinal number of $(N|M)$ served to define the exponentiation :

$$a^b \underset{\mathrm{Df}}{=} \overline{\overline{(N|M)}}. \tag{5.10.1}$$

Cantor was very much pleased with the new results he could suddenly obtain with merely ' a few strokes of the pen ' (*Papers*, 289), and he sent off word of his discovery on 19 July 1895 directly to Felix Klein, editor of *Mathematische Annalen*. By then the first part of the *Beiträge* was already in press, but he was determined to insert the new definition of exponentiation, and to explain its ramifications in a freshly written fourth section. The language of his letter to Klein appeared nearly word-for-word, without change, in the *Beiträge*, which still bore the date March 1895 though he had not by then included the very important conclusions made possible by the definition and rules of exponentiation for cardinal numbers. As he told Klein on 19 July 1895, ' from the following example one can see how fertile the simple formulae extended from the powers are '. The example offered was none other than an exact algebraic determination of the power of the continuum, expressed in terms of two other known cardinals of lesser degree. This was something that he had never managed to do previously.

Cantor recognised that the power of the linear continuum, denoted by $\mathfrak{o}$, could be represented as well by the set of all representations :

$$x=\frac{f(1)}{2}+\frac{f(2)}{2^2}+\ldots+\frac{f(\nu)}{2^\nu}+\ldots, \tag{5.10.2}$$

where $f(\nu)=0$ or 1 ; x represented numbers of the continuum $[0, 1]$ as given in the binary system (§ 4 ; *Papers*, 288–289). Though the numbers $x=(2\nu+1)/2^\mu<1$ were represented twice, they were the only numbers in $[0, 1]$ that failed to have a unique representation, and they could be discounted in the question of cardinality, as he showed, since they were only countably infinite in number. By denoting the set of elements with a double representation $\{s_\nu\}$, then

$$2^{\aleph_0}=\overline{\overline{(\{s_\nu\}, X)}}, \tag{5.10.3}$$

X representing the entire set of x given in (5.10.2). Removing from X any countable set $\{t_\nu\}$ and denoting the remainder by X_1, then

$$X = (\{t_\nu\}, X_1) = (\{t_{2\nu-1}\}, \{t_{2\nu}\}, X_1), \tag{5.10.4}$$

$$(\{s_\nu\}, X) = (\{s_\nu\}, \{t_\nu\}, X_1), \tag{5.10.5}$$

and since

$$\{t_{2\nu-1}\} \sim \{s_\nu\}, \{t_{2\nu}\} \sim \{t_\nu\}, X_1 \sim X_1, \tag{5.10.6}$$

then

$$X \sim (\{s_\nu\}, X), \text{ and thus } 2^{\aleph_0} = \overline{\overline{X}} = \mathfrak{o}. \tag{5.10.7}$$

For the first time, algebraically, Cantor had a firm grasp of what the power of the continuum must be. Even before he had introduced his new symbol ‘ $\aleph_0$ ’ for the smallest transfinite cardinal number, he was using it to show the light he hoped the exponentiation $2^{\aleph_0}$ would shed on his long-standing promise to establish the truth of his conjecture.

5.11. *Simply ordered sets and the continuum*

Cantor realised that, in order to describe completely the structure of the continuum, more refined means were necessary than appeals to well-ordered sets and their ordinal and cardinal numbers. Since no well-ordered set possessed the more interesting and essential properties of the continuum, in particular the property of everywhere-denseness, he turned to the study of simply ordered sets in order to advance his study of continuity. A set S for Cantor was *simply ordered* if there is some rule by which all its elements are ordered such that given any two, one can always be said to precede the other. Thus for any two elements m_1 and m_2 of a simply ordered set, either $m_1 \prec m_2$ or $m_1 \succ m_2$. If $m_1 \prec m_2$, $m_2 \prec m_3$, then it is always true that $m_1 \prec m_3$ (§ 7 ; *Papers*, 296). Different sets could be arranged with different ordinal properties. For example, the rationals in their natural order on the real line were everywhere-dense, though they could be arranged to form a denumerable sequence.

Cantor designated the order-type of any given set M by $\overline{M}$: ‘ Every ordered set M has a definite “ order-type ”, or more briefly a definite “ type ”, which we denote by $\overline{M}$; by this we understand the general concept which arises from M if we only abstract from the character of its elements, but retain the order in which the elements occur ’ (§ 7 ; *Papers*, 297). Two simply ordered sets were said to be *similar*, expressed $M \sim N$, if and only if there was a one–one correspondence between the two preserving the order in which elements occurred in both sets. Since two simply ordered sets could have the same order-type if and only if they were similar, it followed that $\overline{M}=\overline{N}$ and $M \sim N$, each implying the other. Moreover, since sets of equal type were always of equal cardinality, $M \sim N$ always implied that $\overline{\overline{M}}=\overline{\overline{N}}$, though the converse was clearly not true in general.

Cantor denoted ordinal numbers by lower case letters of the Greek alphabet. For example, the well-ordered set of natural numbers 1, 2, 3, . . . was denoted as type ω. The simply ordered set of rational numbers in their natural order on [0, 1] was denoted by the order-type η, and the simply ordered set of real numbers on [0, 1] in their natural order by θ.[1]

In reversing the order of elements,[2] a new set was produced which Cantor denoted $^*M$, with order-type $^*\alpha$, assuming that $\alpha=\overline{M}$. For finite types it was always true that $^*\alpha=\alpha$. To make possible the combination of various order-types, arithmetic operations were also introduced. These followed the definitions given earlier in the *Grundlagen*, and included the familiar caveats concerning the non-commutative character of transfinite operations whenever order-types were concerned.

The set R of rational numbers within (0, 1) was a particularly fascinating one as a simply ordered type. Though it was countable, and therefore of cardinality $\aleph_0$, it produced distinctly different order-types depending upon how the precedence of elements might be taken. In characterising the properties peculiar to R, Cantor noted that it was denumerable. that there was neither a lowest nor a highest element, and that between any two elements of R there were infinitely many others belonging to the set. Thus R was everywhere-dense. These properties, he claimed, were the necessary and sufficient properties determining the order-type η, exemplified by the rationals (§ 9 ; *Papers*, 304).

Characterising the ordinal properties of continuous sets, however, posed a special problem. The special character of limit-points, a familiar component of continuous sets, also had to be translated into the language of order-types. This Cantor did, first by defining fundamental sequences, and then by introducing the concept of limit elements (§ 10 ; *Papers*, 307–308). Drawing upon the elements of his theory of real numbers, he reminded his readers that every fundamental sequence $\{x_\nu\}$ in the linear continuum X had a limit element x_0 in X, and conversely, every element of X was the limit element of some fundamental

[1] Cantor's first substantial effort to study the types of simply ordered sets was the *Principien* (*1885a*). Cantor discussed the simply ordered type ω in the *Beiträge*, § 9 (*Papers*, 303–307), and the type θ in § 11 (*Papers*, 310–311).

[2] Cantor first introduced the concept of inverse order-types in the *Principien*, and later devoted part of section 7 of the *Beiträge* to the same material (*Papers*, 299).

sequence in X. Moreover, as the set of all real numbers in [0, 1], X contained as a subset the set of all rational numbers R of type η, and thus it was true that between any two elements x_0 and x_1 of X there were an infinite number of additional elements of X.

Collecting all of these properties together, Cantor claimed that quite apart from the specific example of X, such properties were both sufficient and necessary to characterise the ordinal type θ of any linearly continuous domain. He formulated the entire matter as follows (§ 11 ; *Papers*, 310) :

> If a set M is so constituted that
> 1) It is 'perfect',
> 2) That there is a set S contained in M with cardinality $\overline{\overline{S}} = \aleph_0$, which is so related to M that between any two elements m_0 and m_1 of M elements of S occur,
> then [the order of M is of type θ,] $\overline{M} = \theta$.

This theorem brought the first part of Cantor's *Beiträge* to a close. It was two years before the sequel to Part I appeared in 1897. Sometime in 1895 he discovered the first of the paradoxes of set theory, specifically those involving the largest ordinal and cardinal numbers (see section 6.6). He had probably come upon them in the course of trying to establish his comparability theorem for transfinite cardinal numbers, and in the attempt to deal with the related questions of whether every transfinite power was necessarily an aleph, and whether every set could in fact be well-ordered. These problems may account for his delay in forwarding the second half of the *Beiträge* to Felix Klein for printing. They may also explain why Part I made few references to well-ordered sets and went no further than to introduce the first transfinite aleph, $\aleph_0$.

Perhaps Cantor believed he might soon be able to resolve the complications which the paradoxes of set theory seemed to raise. It may be that he even hoped to apply the new ideas introduced in Part I of the *Beiträge* to produce a proof that every set could be well-ordered. In turn this would settle both the theorem concerning comparability of the powers of all transfinite sets, as well as the question of whether every cardinal number was also an aleph. He may even have thought that the new algebraic formulations of the relations between cardinal numbers he had most recently discovered would enable him to solve the continuum hypothesis itself and to prove conclusively that $2^{\aleph_0} = \aleph_1$. But this was not to be.

5.12. *Well-ordered sets and ordinal numbers*

Part II of the *Beiträge* presented the bulk of Cantor's important theory of the transfinite ordinal and cardinal numbers. These had not featured with any prominence at all in Part I, but they now appeared in a detailed study which would carry his readers beyond $\aleph_0$ to the first of the non-denumerable, transfinite alephs.

The first step was to define the concept of well-ordered set. This was essential, for Cantor had already shown that sets could be ordered in many different ways, with diverse properties. But in arithmetic, as he had indicated as early as the *Grundlagen*, well-ordered sets were intrinsic to the process of counting by the successive addition of units. Moreover, he found in the concept of well-ordered sets the rigorous foundation for his transfinite numbers, something that his earlier 'principles of generation' had failed to offer in the *Grundlagen* (§ 12; *Papers*, 312):

> A simply ordered set F is said to be well-ordered, if its elements f *increase* from a lowest f_1 *in a definite succession*, so that the following two conditions are fulfilled:
> I. '*There is in F a smallest element* f_1'.
> II. '*If* F' *is any subset of F and if F contains one or more elements larger than all elements of* F', *then there exists in F an element* f' *which follows next after the entirety* F', *so that there is no element of F which falls between* F' *and* f''.

Before Cantor could advance adequate definitions for his transfinite ordinal and cardinal numbers, he had to introduce the concept of sections or segments ('Abschnitte') of well-ordered sets: 'If f is any element of a well-ordered set F different from its first element f_1, then we call the set A of all elements of F which $\prec f$ a "section of F", and in fact the section of F determined by f. We call the set R of all other elements of F *including* f the "remainder of F" ...' (§ 13; *Papers*, 314). For example, given the well-ordered set $F=(a_1, a_2, \ldots; b_1, b_2, \ldots; c_1, c_2, c_3)$, a_3 determined the segment (a_1, a_2) and the remainder $(a_3, \ldots; b_1, \ldots; c_1, c_2, c_3)$; b_1 determined the segment $(a_1, \ldots)$ and the remainder $(b_1, \ldots; c_1, c_2, c_3)$; c_2 determined the segment $(a_1, \ldots; b_1, \ldots; c_1)$ and the remainder (c_2, c_3).

Inequalities were introduced between segments as follows. Given two segments A and A' determined respectively by two elements f and f' of F, A' was said to be a segment of A if $f' \prec f$. In such cases, A' was said to be the lesser segment, written $A' < A$. Clearly, for every segment A of F, $A < F$. In terms of such segments, Cantor was able to state clearly the relations that were possible under similar correspondences between any two well-ordered sets F and G in the following form: If F and G are any two well-ordered sets, then either (1) F and G are similar to each other, (2) there is a definite segment B_1 of G which is similar to F, or (3) there is a definite segment A_1 of F which is similar to G, and each of these three cases excludes the possibility of the other two (§ 13; *Papers*, 319).

The results of this theorem were translated directly into important conclusions concerning the order of any two ordinal numbers in general, once these had been defined as follows: ' Every simply ordered set M has a definite *order-type* $\bar{M}$; it is the general concept which arises from M if the character [but not the order] . . . is abstracted from the elements of M . . . . We call the order-type of a well-ordered set F the " ordinal number " corresponding to it ' (§ 14; *Papers*, 320–321).

Order-types, defined as the concept obtained from a well-ordered set by abstracting all individual properties of the elements while retaining their order, were represented as follows: $\alpha=\bar{F}$, $\beta=\bar{G}$. Given any two such sets F and G such that $\bar{F}=\alpha$ and $\bar{G}=\beta$, then the theorem above (concerning the relations possible between segments of similar sets) insured that only three mutually exclusive possibilities could occur. Either: (1) $F\sim G$, in which case $\alpha=\beta$; (2) G contained a definite segment B_1 such that $F\sim B_1$, then $\alpha<\beta$; or (3) There was a definite segment A_1 of F such that $G\sim A_1$: then $\alpha>\beta$.

Moreover, because of the comparability of segments, it followed immediately that if α and β were any two ordinal numbers, then exactly one of three possibilities was necessarily true: either $\alpha<\beta$, $\alpha=\beta$ or $\alpha>\beta$. The nature of segments also ensured that such relations were transitive: among any three ordinal numbers, if $\alpha<\beta$ and $\beta<\gamma$, then $\alpha<\gamma$, and it followed that the set of all ordinal numbers, taken in their order of magnitude, constituted a simply ordered set. Later Cantor showed that the set of all ordinal numbers was actually well-ordered, since every subset of the set of all ordinals had a least element and every element of the set had a definite and unique successor.

Cantor used the first number-class (I) of finite ordinals ν to determine the first transfinite cardinal number $\aleph_0$. However, to introduce the second transfinite cardinal $\aleph_1$, it was similarly necessary to establish securely the succession of transfinite ordinal numbers of the second number class. First he gave this definition: ' *The second number-class* $Z(\aleph_0)$ *is the entirety* $\{\alpha\}$ *of all order-types* α *of well-ordered sets of the cardinality* $\aleph_0$ ' (§ 15; *Papers*, 325). He went on to prove that $Z(\aleph_0)$ was well-ordered, from which he could then define its cardinal number $\aleph_1$, and establish that $\aleph_0<\aleph_1$.

Cantor ended the *Beiträge*, not with discussion of the cardinality of the continuum and the solution of his continuum hypothesis that $2^{\aleph_0}=\aleph_1$, but instead with a detailed study of the arithmetic character of the numbers of the second number-class. At first he paid special attention to the numbers of $Z(\aleph_0)$ which could be expressed as polynomials in ω. In fact, such numbers could always be uniquely expressed in the form :

$$\phi=\omega^{\mu}\nu_0+\omega^{\mu-1}\nu_1+\ldots+\nu_{\mu}. \qquad (5.12.1)$$

Later Cantor was able to generalise such representations for ordinals of the second number-class without restricting the degree μ to finite values. No transfinite ordinals of the form ω^{ω}, for example, could be included rigorously in his transfinite number theory until he had established a satisfactory means of introducing the product of transfinitely many ordinal numbers. To do so, he invented the process of transfinite induction which was similar, but by no means the same as, the familiar mathematical (or complete) induction on well-ordered sets of type ω (§ 18 ; *Papers*, 336–339).

In the final section of the *Beiträge*, Cantor investigated the properties of a special kind of number, the ' epsilon numbers ' ϵ of $Z(\aleph_0)$ for which $\omega^{\epsilon}=\epsilon$. These were central to his introduction of transfinite cardinals, since these were determined by those ' initial elements ' of each number-class which could not be reached or produced by any arithmetic or exponential combination of elements preceding them. Initial elements, like ω and Ω, were transfinite numbers which were preceded by no numbers of equal power. Moreover, to every transfinite power there was only one such number, and every such ' first number ' was necessarily an epsilon number. Thus there was a basic connection between the succession of transfinite alephs and the epsilon numbers which he introduced at the very end of the *Beiträge* (§ 20 ; *Papers*, 347–351).

Cantor's presentation of the principles of transfinite set theory in the *Beiträge* was elegant, but ultimately disappointing. One might have thought that, at long last, having given the extensive and rigorous foundations for the transfinite ordinal numbers of the second number-class, he would then have gone on to discuss the higher cardinal numbers in some detail. In particular, one might have expected him to fulfil his promise made in Part I to establish not only the entire succession of transfinite cardinal numbers $\aleph_0, \aleph_1, \ldots, \aleph_{\nu}, \ldots$, but to prove as well the existence of $\aleph_{\omega}$, and to show that in fact there was no end to the ever-increasing sequence of transfinite alephs. But instead, the final sections of the *Beiträge* were devoted to an analysis of the number-theoretic properties of transfinite ordinals. The entire manner of his handling of the transfinite cardinals in the *Beiträge* was fundamentally unsatisfying because it seemed so anti-climactic.

By the time that Cantor came to write the *Beiträge*, the continuum hypothesis seemed as elusive as ever, despite a tantalising hope that coverings, which led to the formulation $2^{\aleph_0} = \aleph_1$, might provide the key for which he had searched so long. But by 1897 he had discovered the paradoxes of set theory, he had failed to establish directly the comparability of all cardinal numbers, and he had not managed to find any proof that every set could be well-ordered (for details on all these matters, see sections 6.6 and 6.9). These obstacles seemed to leave no alternative: rather than produce the complete and absolutely certain solutions to the outstanding problems that his set theory had raised, he was forced to accept something less. Instead, he sought to present the elements and internal workings of his theory of transfinite sets as rigorously as his research completed since the appearance of the *Grundlagen* would allow. Abstract, and independent of point sets and physical examples, the *Beiträge* represented his last effort to present mathematicians with the basic features of his transfinite set theory. He hoped that at last the theory would speak for itself, and that its utility and interest would be acknowledged accordingly.

5.13. *Cantor's formalism and his rejection of infinitesimals*

Cantor always insisted that his transfinite numbers arose naturally and necessarily from the elements of *sets*, and thus he was convinced that his characterisation of the infinite was the *only* characterisation possible. This attitude was reflected in his eagerness to represent transfinite set theory as absolutely certain, complete, open neither to variant opinions nor to opposing interpretations. In this connection the work of the Italian mathematician G. Veronese was particularly unwelcome, because it advanced a theory of the infinite very different from Cantor's in a number of fundamental respects. Cantor devoted some of his most vituperative correspondence, as well as a portion of the *Beiträge*, to attacking what he described at one point at the 'infinitesimal Cholera bacillus of mathematics', which had spread from Germany through the work of Thomae, du Bois Reymond and Stolz, to infect Italian mathematics. Mostly at issue, but not exclusively, was the question of infinitesimals. (For more details, see Dauben *1977b*.)

Very early in his career Cantor had denied any role to infinitesimals in determining the nature of continuity, and by 1886 he had devised a proof that the existence of such entities was in his view impossible (*Papers*, 407–409). Thus any attempt to urge their legitimacy could be interpreted as a direct challenge to one of the most basic principles of Cantorian set theory, since it was in terms of the character of his transfinite numbers that he had argued the impossibility of infinitesimals. Moreover, any acceptance of infinitesimals necessarily meant that his own theory of number was incomplete. Thus to accept the work of Thomae, du Bois Reymond, Stolz and Veronese was to deny the perfection of Cantor's own creation. Understandably, Cantor launched a thorough campaign to discredit Veronese's work in every way possible. Veronese had just published a German translation of his *Fondamenti di geometria*, and Cantor felt it was timely to warn everyone of its manifold errors.[1] But the very nature of Cantor's own view of his set theory determined from the start the attack he would make, and the fate of its outcome.

Cantor underscored in the *Beiträge* that his concept of ' ordinal type ', together with that of ' cardinal number ' or ' power ', included *everything* capable of being numbered that was thinkable (§ 7 ; *Papers*, 300). To him this meant that no further generalisations were conceivable. Moreover, there was nothing in the least way arbitrary about his definitions of number ; cardinals and order-types were perfectly natural extensions of the number concept. Equally free from arbitrariness was his condition for the equality of two order-types, which was given in terms of their similarity. This condition followed with absolute necessity from the concept of order-type, and hence permitted no alteration. He claimed that Veronese's failure to understand this absolute character of the transfinite numbers was the major source of error in his misguided attempt to establish a different sort of infinite number in his *Fondamenti di geometria*. He could reject Veronese's definition of the equality of his ' numbers of ordered groups ' by pointing out that it was viciously circular and therefore meaningless. To employ the concept ' not equal ' in a definition of equality *presupposed* that one already knew what was meant by ' equality '. Thus the *petitio principii* rendered Veronese's entire approach suspect and mathematically unsound (§ 7 ; *Papers*, 301).

But there were more reasons for rejecting Veronese's theory of number, reasons which concerned the impossibility of results which he had obtained. Examining both Veronese's transfinite numbers and his infinitesimals, Cantor easily spotted ' erroneous ' conclusions. Not surprisingly, his criticisms were based on the incompatibility of Veronese's conclusions with his own.

[1] Veronese *1891a* ; the German translation by A. Schepp appeared three years later (Veronese *1894a*). Cantor told Killing in a letter of 3 June 1895, that it was appropriate to warn everyone of Veronese's errors ; see page 156 in the second of his letter-books. (These were volumes in which he would draft his letters before mailing a final copy. Three survive, and all are now part of the Cantor *Nachlass*, recently given to the Göttingen Academy of Sciences. In the following text, *Letter-book* I refers to the book used by him between 1884 and 1888, and *Letter-book* II denotes the one used between 1890 and 1895. For further details, see Grattan-Guinness *1971b*, 348–349.)

Concerning Veronese's infinitely large numbers, Cantor once commented that as soon as he saw the equation $2 \cdot \infty = \infty \cdot 2$, he knew that the entire theory which Veronese had developed was false (*Letter-book* II, 165). Assuming the absolute character of his own theory of transfinite numbers, Cantor concluded that any theory of the infinite would have to be comparable with his ; one major requirement was the non-commutativity of arithmetic operations for infinite ordinal numbers. Since Veronese's numbers clearly violated such necessary laws, they were inadmissable. To Cantor's way of thinking, it was as simple as that.

Infinitesimals, at the opposite extreme, were equally unwelcome, and were high on Cantor's list of mathematical ' ghosts and chimareas ' (*Letter-book* II, 30, 138). His proof of the self-contradictory nature of the idea of infinitesimals was based upon a property he regarded as common to all finite numbers, expressed by the Axiom of Archimedes : If a and b are any two positive numbers, then there exists a positive integer n such that $na > b$ (*Papers*, 408–409). Basically, he refused to regard this as an axiom at all, but argued that it followed directly from the concept of linear number (*Letter-book* I, 96 ; II, 16, 137 ; *Papers*, 409). Numbers were *linear* if finitely or infinitely many of them could be added producing yet another linear magnitude. But Cantor assumed all along what Hilbert in his *Grundlagen der Geometrie* (*1899a*) was to call the Axiom of Continuity, and consequently all of his assertions followed directly. In particular, since the Axiom of Continuity and the Axiom of Archimedes implied each other, one could be derived from the other, and from Cantor's view there was no difficulty in asserting that the Archimedean ' Axiom ' could be proven. In terms of his assumptions, it was as provable as the non-existence of infinitesimals. Moreover, had he agreed that the Archimedean property of the real numbers was merely axiomatic, then there was no reason to prevent the development of number systems by merely denying the axiom, so long as consistency was still preserved. But to have allowed this would have left him open to the challenge that, if infinitesimals could be produced without contradiction, then his own theory of number would have been contravened.

On the other hand, were the Axiom of Archimedes not an ' Axiom ' at all, but a theorem which could be proven from other accepted principles, then Cantor could rest assured that it was impossible simply to deny the proposition and produce a consistent theory of infinitesimals. He was so persuasive, in fact, with his disavowal of infinitesimals that he was able to convince Peano, who wrote an article *1892a* on the subject in his *Rivista di matematica.* Bertrand Russell went even further than Peano, and argued in *The principles of mathematics* that mathematicians, completely understanding the nature of real numbers, could safely conclude that the non-existence of infinitesimals was firmly established. He was wise to add, however, that if it were ever possible to speak of infinitesimal numbers, it would have to be in some radically new sense (Russell *1903a*, 334–337).

Finally, there was an additional argument, one that Cantor found equally persuasive in rejecting the attempts Thomae, du Bois Reymond, Stolz and Veronese had made to develop logically sound theories of the infinitesimal. Once more his reasons were based on his view of set as the ultimate origin of any concept of number. In writing to Veronese on the subject, Cantor accused the infinitesimalists of talking nonsense, since in the realm of the possible there were no infinitely small entities. He stressed that his transfinite numbers were linked with *real ideas* produced directly from *sets*, and he challenged Veronese to show any *real ideas* corresponding to the supposed infinitesimals (*Letter-book* II, 15). Until Veronese could do so, Cantor insisted that deviation from the ' Axiom ' of Archimedes, which he took as proven, was an error of the greatest seriousness.

5.14. *Conclusion*

Following publication of the *Beiträge* and its translation almost immediately into Italian and French, Cantor's ideas became widely known and were circulated among mathematicians of the highest rank the world over. The value of his transfinite set theory was recognised almost immediately, and soon his ideas were fuelling heated polemics between widely divided camps of mathematical opinion. Though he never seemed able to avoid controversy over the nature of his work, he was, after 1895, increasingly defended by younger and more energetic mathematicians. No longer was he left to face the opposition alone. Though Kronecker had died in 1891, he was replaced in the phalanx of dissenters by mathematicians like Poincaré ; but Cantor could begin to count an ever-increasing, always impressive array of those ready to join in support of set theory. For him, the crusade was nearly over, and though the difficulties were by no means satisfactorily resolved, it was widely recognised at last that Cantor had contributed something of lasting significance to the world.

Chapter 6

Developments in the Foundations of Mathematics, 1870–1910

R. Bunn

6.1. *Introduction*

This chapter deals with some of the most important work in foundations from the time of the investigations into the foundations of analysis in the early 1870s up to the publication in 1910–1913 of Whitehead and Russell's *Principia mathematica.* My treatment of the foundations of analysis is limited to the work of Richard Dedekind, for Cantor's contribution has already been covered in chapter 5 ; moreover, Dedekind is most suitable for my purposes because his motivation is most explicit. In general, I have restricted myself to positive contributions to the foundations of mathematics, and have not tried to include the criticism and polemic which is especially prominent in the writings of Frege, Poincaré and Brouwer. This in part explains why there is not a separate section on intuitionism. Brouwer's early writings (1907–1914) on intuitionism are largely critical ; the real development of intuitionistic mathematics does not occur until 1918. For readers who would like to supplement this chapter—which concentrates on the classical foundations of Dedekind, Frege, Russell and Zermelo—with an account of Brouwer's intuitionistic mathematics, I recommend chapter 5 of E. W. Beth's *Mathematical thought* (*1965a*).

The most prominent feature of the period under consideration here is the tendency, culminating in *Principia mathematica*, towards the logical systematisation of mathematics, and replacement of so-called ' intuitive ' explanations and arguments pertaining to the elements of mathematics by formal proofs based on logically precise definitions or systems of axioms. The first area of mathematics to be so reconstructed was mathematical analysis. As far back as 1858 Dedekind had been motivated to improve matters in the foundations of the calculus by his first experience with teaching differential calculus. This made him feel ' more keenly than ever before the lack of a really scientific foundation for arithmetic ' (*1872a*, preface). Although he had worked out his basic definitions already in 1858, his pamphlet *Stetigkeit und irrationalen Zahlen* (' Continuity and irrational numbers ') was not published until 1872. Even before this work appeared, he had a plan for a similar treatment of the foundations of the theory of natural numbers, which was eventually realised as *1888a* in his *Was sind und was sollen die Zahlen?* (known in English as ' The nature and meaning of numbers '). In *1889a*, art. 1 Giuseppe Peano presented independently a system of axioms for the theory of natural numbers which is very closely related to Dedekind's basic definition. Earlier, Gottlob Frege had published a penetrating investigation *1884a* of the general concept of cardinal number and of the finite cardinals.

Dedekind had intended to present the whole subject of natural, negative, rational, irrational and complex numbers ' in a systematic form ', but he never achieved it. A very comprehensive systematic presentation of logic and mathematics was, however, effected by Peano and his school, and an even more thorough-going exposition of a system of logic and arithmetic was given by Frege in his *Grundgesetze der Arithmetik* (' Foundations of arithmetic '), which was published in two volumes as *1893a* and *1903a*. Frege's work, though it dealt with a much less extensive part of mathematics, was executed with remarkable attention to detail and an exactness of expression which goes far beyond what was achieved by Peano or anyone else. For example, he was especially clear on the distinctions between the use and mention of expressions and between logical theses and rules of inference.

The foundations (in some cases only implicit) of the works of Dedekind, Cantor, Frege and Peano turned out to be inconsistent, as was shown by the antinomies (especially Russell's). These antinomies elicited a variety of responses from mathematicians and logicians interested in foundations. One reaction was the determination to find a system for avoiding the antinomies which would include as much as possible of the results obtained before their appearance. There are a number of methods of avoiding the antinomies, because there are a number of characteristics common to the paradoxical cases. Depending upon which characteristic is chosen as a basis for avoiding the antinomies and what supplementary assumptions (such as Russell's axiom of reducibility) are made, systems comprehending various portions of the classical results (perhaps in a somewhat modified form) are obtained. In some cases—especially Russell's—there has been an attempt to formulate a system which not only avoids the antinomies but also eliminates the precise fallacy to which they are due. It could, thus, count as a *philosophical* solution to the antinomies.

6.2. *Dedekind on continuity and the existence of limits*

I shall begin my more detailed discussions with the work of Dedekind. Like many other mathematicians of the mid-19th century, Dedekind was dissatisfied with the foundations which had so far been provided for the calculus and for the arithmetic of real numbers. The primary motivation behind his ' Continuity and irrational numbers ' was the desire to replace undefined concepts (more or less geometrical) and the so-called intuitive justifications based upon them with proofs from precisely formulated definitions. Above all, he wished to find definitions from which the basic theorems on the existence of limits could be proved. To accomplish this he needed to define a system having a certain sort of completeness or continuity. The term ' continuous ' is not an especially apt one for the characteristic involved, but it indicated the correlate in the old system—continuous magnitude. It had been the practice to present the calculus as being concerned with continuous magnitudes. But this property of continuity, which was attributed to such things as line-segments and motions, was not defined—at least not in a way which could serve as a basis for proofs. As Dedekind put it : ' By vague remarks upon the unbroken connection in the smallest parts obviously nothing is gained ; the problem is to indicate a precise characteristic that can serve as the basis for valid deductions ' (*1872a*, pt. 3 ; *1901a*, 10).

The sort of complete system Dedekind needed to define had to be a densely ordered system for which arithmetical operations could be defined and in which could be proved propositions such as that every element of the system has a square root. This completeness of arithmetical operations may be described by saying that the system must be closed under arithmetical operations satisfying the laws of elementary algebra. Furthermore, the system must be complete with respect to limits ; that is, every convergent sequence of its elements must have a limit.

The property of completeness or continuity which Dedekind found satisfactory for his purposes is a characteristic of ordered systems. A *cut* in an ordered system M is a pair of classes M_1, M_2, called the ' lower ' and ' upper ' sections of the cut, which together exhaust M and are such that every element of M_1 precedes every element of M_2. Then a densely ordered system is *complete* (*continuous*) in Dedekind's sense if every cut in the system is produced by exactly one element of the system, that is, if there is an element of the system which is either the maximum of the lower section or the minimum of the upper section.

Dedekind seems to have thought of this definition by reflecting on lines, and contrasting the system of points on a line with the system of rational numbers. But he was not willing to settle for an axiomatisation of the concept of magnitude, which would feature an axiom of continuity. Geometry was to serve only as the source of the idea for constructing an arithmetical foundation. The continuous system which was to be Dedekind's foundation would be arithmetical in the sense that its operations would ultimately be defined in terms of operations on natural numbers, and no mention would be made of any geometrical objects. Thus, besides a basis for proofs, Dedekind also sought a purely arithmetical foundation for the calculus.

A demonstrably complete (continuous) domain was defined by Dedekind in terms of the cuts in the rationals. This domain, called ' the system of real numbers ', is to contain all rational numbers, and, in addition, for each cut in the rationals which is not produced by a rational, exactly one new object, which is called an irrational number. By reference to the system of cuts, Dedekind defined among the real numbers a relation which could be proved to have the properties of a dense ordering. Furthermore, and this is the main point, it can be proved from the definition of the real numbers and the definition of ' $<$ ' that the system of real numbers is complete : every cut in it is produced by one of its elements, which is to say that for every cut in the system of real numbers, either its lower section has a maximum or its upper a minimum. The fundamental theorems on limits follow from this property of completeness. In particular, he proved (1) that every bounded increasing sequence of real numbers has a limit, and (2) that a function f whose arguments and values are real numbers has a limit when $x \to \infty$, if for every positive δ there is an x_0 such that $|f(x) - f(x_0)| < \delta$ for all $x > x_0$.

Arithmetical operations (addition, multiplication, and so on) may be defined in terms of the cuts in the system of rationals and the corresponding arithmetical operations defined for the rationals. The operations on rationals can in turn be defined in terms of the operations between natural numbers. From the definitions of the operations on the real numbers such algebraic laws as

$$a + b = b + a \quad \text{and} \quad a(b + c) = ab + ac \qquad (6.2.1)$$

can be proved. It also becomes possible to demonstrate equations like $\sqrt{2} \cdot \sqrt{3} = \sqrt{6}$, which so far as Dedekind knew had ' never been established before '. Although the difficulties concerning existence theorems for limits were fairly well recognised, some mathematicians were reluctant to admit that the rule for multiplying square roots had never really been proved. For instance, R. Lipschitz maintained in correspondence with Dedekind that the basis for the proof of the theorem in question was already in Euclid's *Elements*. Dedekind's reply was that, even setting aside the desire to avoid a geometrical foundation of arithmetic, no axiom of completeness for the domain of magnitudes is to be found in Euclid. But no generally valid definition of, for example, multiplication is possible for an incomplete domain, because for any two quantities of such a domain it may be that *no* quantity of the domain is the product of the two quantities. ' If, to be sure, general definitions of addition, subtraction, multiplication, and division are relinquished, one only needs to say : I understand by the product $\sqrt{2} . \sqrt{3}$ the number $\sqrt{6}$, consequently $\sqrt{2} . \sqrt{3} = \sqrt{6}$, which was to be proved ' (see Dedekind *Works*, vol. 3, 474).

Many authors who adopted Dedekind's basic ideas preferred not to follow him in defining the real numbers as creations of the mind corresponding to cuts in the system of rational numbers. In *The principles of mathematics* (*1903a*) Bertrand Russell emphasised the advantage of defining the real numbers simply as lower sections of cuts whose upper sections have a minimum or, as he said, segments of the rationals. A *segment* of the rationals is a non-empty proper subclass of the rationals having the property of being identical ' with the class of rationals x such that there is a rational y of the said class such that x is less than y ' (Russell *1903a*, 271). Russell needed to use a definition like this because it was his aim to define all mathematical concepts in terms of the vocabulary of logic and the theories of classes and relations. But Dedekind had his reasons (though not necessarily good ones) for defining the real numbers as he did. When Heinrich Weber expressed his opinion in a letter to Dedekind that an irrational number should be taken to be the cut, instead of something new which is created by the mind and supposed to correspond to the cut, Dedekind replied (*Works*, vol. 3, 489) :

> We have the right to grant to ourselves such a creative power, and besides it is much more appropriate to proceed thus because of the similarity of all numbers. The rational numbers surely also produce cuts, but I will certainly not give out the rational number as identical with the cut generated by it ; and also by introduction of the irrational numbers, one will often speak of cut-phenomena with such expressions, granting to them such attributes, which applied to the numbers themselves would sound quite strange.

But of course Dedekind's method of introducing the irrationals was only a matter of preference, and in a letter to Lipschitz he remarked parenthetically that if one does not wish to introduce numbers in his sense, ' I have nothing against it ; the theorem I prove [on completeness] then reads : the system of all cuts in the discontinuous domain of rational numbers forms a continuous manifold ' (*ibid.*, 471).

Although there were often different preferences and philosophical convictions among classical mathematicians, they were not such as to lead to any essential differences in the mathematics produced. By contrast, the philosophical differences between classical and constructive mathematicians result in completely different mathematical theories. Let me illustrate the point. Some of the classical mathematicians were realists, and regarded their work as formulating truths describing the objective facts concerning such abstract entities as functions and sets which were taken as existing independently of thought. Others were idealists and considered mathematical systems to be created by the mind. Many constructivists were also idealists. But the idealism of a classical mathematician like Dedekind was much more similar to realism than to the idealism of a constructivist like Poincaré or Brouwer. In fact, the idealism sometimes advocated by classical mathematicians could very well be called ' quasi-realism '. For, although a mathematical structure is said to be created in the sense of being thought up, rather than discovered, it is conceived of as, so to speak, a system of simultaneously existing inter-related entities. But the constructive idealist conceives of his mathematical entities as things coming into being one after another, individually created. Thus the infinity which enters into constructive mathematics is only potential infinity, whereas the infinity treated in classical mathematics is always actual infinity.

While Dedekind's definition or ones serving the same purposes were quickly adopted by many mathematicians interested in greater rigour in foundations, Leopold Kronecker found such definitions completely unacceptable. He disapproved of definitions for which it is not decidable in the case of any entity whether it satisfies the definition or not (see his *1882a*, art. 4). But Dedekind and Cantor considered knowledge of a decision procedure for membership in a particular set to be of no account for their purposes ; as Dedekind explained, ' the general laws to be developed in no way depend upon it ; they hold under all circumstances ' (*1888a*, art. 2, note ; compare Cantor *Papers*, 150–151, and Frege *1884a*, art. 80). Kronecker also had the conviction that the infinite should not be introduced except in cases in which it could be eliminated (see *1886a*, 334–336 ; *Works*, vol. 3, 155–156), and maintained that the various concepts of irrational numbers are of the sort which ' must be avoided in arithmetic-algebraic theories '. Dedekind's reply was simply that the restrictions on concept formation advocated by Kronecker did not seem justified, but that there was no point in going into the matter further until Kronecker published his reasons (Dedekind *1888a*, art. 2, note). Unfortunately, Kronecker seems never to have published the desired reasons. While he had carried on vigorous debates with his colleague Weierstrass concerning the foundations of analysis, his writings contain only a few brief expressions of his opinions on the foundations of mathematics (compare sections 5.4 and 5.8 above).

It is interesting to consider the account given by Cantor of the point of view of those who disapprove of the method of introducing the irrational numbers by means of infinite sets or sequences. In his 1883 *Grundlagen* Cantor explained that those who reject this approach hold that ' a merely *formal* significance should belong to irrational numbers in pure mathematics, in that they serve as it were only as marks of computation for fixing and describing in a simple, uniform way properties of groups of whole numbers ' (*Papers*, 172). The advantages of reducing the content of analysis to relations among finite integers are a greater security and completeness in its foundations as well as an improvement in its methodology. Hence (*Papers*, 173) :

> In this way, a definite, even if rather prosaic and obvious principle is assumed, which is recommended to all as a guiding principle ; it is supposed to serve for indicating the true limits for the flight of the desire for mathematical speculations and conceptions, where it runs no danger of falling into the abyss of the transcendent, where, as it is said with fear and holy dread, ' everything is possible '. Setting this aside, who knows if it has not been just the point of view of expediency alone which caused the authors of this opinion to recommend it to the powers, so easily endangered by enthusiasm and extravagance, as an effective regulative [principle], as a protection against all errors, although a fruitful principle cannot be found in it . . .

Cantor did not think that those who recommended the restrictive principles had made their discoveries by adhering to them. No true advances are due to the observation of such principles, and if science actually proceeded in accordance with them, it ' would be held back or still would be bound within the narrowest limits ' (*ibid.*). He himself recognised only two restrictions on concept formation : the concepts must be consistent, and the new concepts must be related by definitions to concepts already recognised. He did not think that there was much danger to science in his point of view, because poor ideas would fade away ; on the other hand, he did see a real threat to science in unnecessary restrictions (*Papers*, 182).

6.3. *Dedekind and Frege on natural numbers*

The desire to base arithmetic on a system of precisely formulated definitions motivated both Dedekind's and Frege's work on the foundations of the theory of natural numbers. Regarding what was perhaps the most significant point, the basis of inference by mathematical induction, their systems are the same. But Dedekind, unlike Frege, did not define a particular set of objects as the natural numbers. Rather, he defined a class of structures, which he called ' simply infinite systems ', any one of which could serve as the subject of the arithmetic of natural numbers. Indeed, he understood the primary concern of arithmetic to be the ' relations or laws ' derivable from the essential characteristics of simply infinite systems (*1888a*, art. 73). A class N is *simply infinite* if there is a one–one function ϕ mapping N into itself and an object b such that b is not a value of ϕ for an argument in N, and N is the intersection of all classes containing b and also $\phi(y)$ whenever they contain y. Thus the essential characteristics of a simply infinite system N are the following (Peano's axioms in *1889a* coincide with or are immediate consequences of them) :[1]

$$(\forall y)(y\in N \rightarrow \phi(y)\in N), \tag{6.3.1}$$

$$N = \bigcap\{Z \mid b\in Z \,.\, \& \,.\, (\forall y)(y\in Z \rightarrow \phi(y)\in Z)\}, \tag{6.3.2}$$

$$(\forall y)(y\in N \rightarrow b \neq \phi(y)), \tag{6.3.3}$$

$$\phi \text{ is one–one.} \tag{6.3.4}$$

Dedekind said that the elements of a simply infinite system may be called ' numbers ' if ' we entirely neglect the special character of the elements ; simply retaining their distinguishability and taking into account only the relations to one another in which they are placed by the order-setting transformation ϕ ' (*1888a*, art. 73). Thus it may be said that any simply infinite set could be defined to be the set of (natural) numbers.

The appropriateness of Dedekind's definition is easily perceived. A number series must have a first element, $b = 1$, from which the succession of numbers proceeds without end. That 1 is the first term of the succession of numbers means that 1 is a number which is not the successor of any number. The series proceeds without end in that every number has a successor. Moreover, each number has only one successor, that is, a successor relation is a function ; and different natural numbers have different successors, so that the successor function is one–one. Thus we have the basic requisites concerning 1 and a successor function ϕ. But the most significant part of the definition of a natural number series (simply infinite system) N is the specification that N be the intersection of all classes Z which contain the ' base-element ' 1 and which contain $\phi(x)$ whenever $x\in Z$ and ϕ has a value for x. The idea for the definition of a set of natural numbers N is that it should contain a base-element 1, the successor $\phi(1)$ of 1, and so on, and *no*

[1] Attention is recalled to the list of notations in section 0.5.

other elements, that is,

$$N = \{1, \phi(1), \phi(\phi(1)), \ldots\}. \quad (6.3.5)$$

The means of defining N so as to exclude undesirable elements was, Dedekind says, ' one of the most difficult points of my analysis and its mastery required lengthy reflection ' (letter to Keferstein, in van Heijenoort *1967a*, 100). For it cannot be said without ' the most pernicious and obvious kind of vicious circle ' that n belongs to N if n is 1 or a value of ϕ after a *finite number* of iterations of ϕ beginning from 1. The method which Dedekind found sufficient for his purposes was to define N as the class which contains exactly those things which belong to *every* class Z which contains 1 and also all values of ϕ for arguments in Z.

This definition of the finite numbers has the following principle of mathematical induction as a consequence :

$$1 \in M \,.\, \& \,.\, (\forall y)(y \in N \cap M \rightarrow \phi(y) \in M) \,.\, \rightarrow \,.\, N \subseteq M. \quad (6.3.6)$$

In words, any class M containing 1 as well as the successor of any number belonging to it contains every number. Dedekind also proved a theorem (*1888a*, art. 126) justifying inductive or recursive definitions such as the following definition of addition :

$$x + 1 = \phi(x), \quad (6.3.7)$$

$$x + \phi(y) = \phi(x + y). \quad (6.3.8)$$

His theorem justified this definition in that it asserts that there exists one and only one function of two arguments which satisfies the conditions (6.3.7) and (6.3.8). In general, the theorem on definition by induction states that for any functions g and h there is one and only one function f such that

$$f(x, 1) = h(x) \quad (6.3.9)$$

$$f(x, \phi(y)) = g(f(x, y)) \quad (6.3.10)$$

(*1888a*, arts. 126 and 135).

We have seen that Dedekind did not specify any particular set of things as the natural numbers, but he did prove the existence of simply infinite systems in order to show the consistency of his definition. Instead of the example of a system N, ϕ, b, which he actually used, we shall consider a more mathematical example suggested by his *1888a*, arts. 66 and 72. Let $\phi(x) = \{x\}$ and $b = \varnothing$. From these definitions it follows that ϕ is one–one and that b is not a value of ϕ. The class

$$N = \{\varnothing, \{\varnothing\}, \{\{\varnothing\}\}, \{\{\{\varnothing\}\}\}, \ldots\}, \quad (6.3.11)$$

which is the intersection of all classes containing $\varnothing$ as well as the unit class of anything belonging to them, is simply infinite. This method of proving the existence of a simply infinite system is, however, affected by the antinomies (see the end of section 6.8 below).

Dedekind's treatment of the concept of the cardinal number of a set or the number (*Anzahl*) of things in a set was restricted to finite sets. He proved that for each finite set M there is a unique natural number n such that there is a one–one correspondence between M and the initial segment

$$Z_n \underset{\text{Df}}{=} \{x \mid 1 \leqslant x \leqslant n\} \tag{6.3.12}$$

of the set of natural numbers (*1888a*, art. 160). This justified his definition of the number of elements belonging to a finite set M, or the cardinal of M, as the natural number n such that M is in one–one correspondence with Z_n.

I turn now to Frege's definition, in which the cardinal number of a class A is the class of all classes in one–one correspondence to A (*1884a*, art. 72).[1] This definition is quite appropriate, if a cardinal number is conceived as something belonging to a class of things which is common to different classes when they are 'equal in number'. Now it is possible to say, without using numerical terms, when two classes are equal in number: they are *numerically equal* or *equipollent* when there is a one–one correspondence between them. By saying a relation R is 'one–one' it is meant that for every x, y and z,

$$zRy\ .\ \&\ .\ zRx\ .\ \rightarrow .\ x=y, \tag{6.3.13}$$

$$xRz\ .\ \&\ .\ yRz\ .\ \rightarrow .\ x=y. \tag{6.3.14}$$

Thus classes are numerically equal if and only if there is a relation having the properties (6.3.13) and (6.3.14) between their elements. Frege specified something which is common to different classes when they are equipollent in the sense just defined. Moreover, in his system it could be proved that each class has a unique cardinal number and that the cardinals belonging to classes in one–one correspondence with each other are identical. The same treatment was later given independently by Russell, first in his *1901b*, arts. 1–2.

Frege defined the number 0 as the number belonging to the empty class; since there is only one empty class, $0=\{\varnothing\}$ (*1884a*, art. 74). The number 1 may now be defined as the class of all classes equipollent to 0, that is, to $\{\varnothing\}$, the class whose only element is the empty class.

[1] In this chapter I use both the words 'class' and 'set'. Certain historical and philosophical considerations have guided the choice of those terms, but the reader can take them as synonymous if he so wishes.

Any class equipollent to $\{\varnothing\}$ could have been used to define 1, but Frege wished to use only classes definable in the vocabulary of logic. The number 1 could also be defined thus :

$$1 \underset{\mathrm{Df}}{=} \{Z|(\exists y)[y \in Z\ .\ \&\ .\ (\forall x)(x \in Z \rightarrow x = y)]\}. \quad (6.3.15)$$

In either case, we get 1 as the class of all unit classes. In similar ways, the cardinals 2, 3, and so on could be defined (*1884a*, arts. 77–83).

The relation S of immediate succession between finite numbers is also easily defined in the vocabulary of logic (*1884a*, art. 76). Let $C(M)$ be the cardinal number of the class M. Then the relation S may be defined as follows :

$$nSm \leftrightarrow (\exists x, A)\{x \in A\ .\ \&\ .\ C(A) = n\ .\ \&\ .\ C(A - \{x\}) = m\}. \quad (6.3.16)$$

Using this definition and the definitions of 0 and 1, it can be proved that 1 immediately succeeds 0, that is, $1S\,0$: just let $x = \varnothing$ and $A = \{\varnothing\}$. It can also be demonstrated from Frege's definitions that the relation S is one–one and that 0 is not the successor of any number.

Using the same method which Dedekind employed in his definition of a simply infinite system, Frege defined the set N of natural numbers (finite cardinals) as the intersection of all classes Z such that $0 \in Z$ and whenever $x \in Z$ and ySx, then also $y \in Z$. The principle of mathematical induction is a consequence of this definition, and it can be proved that every natural number n has a successor $\neq n$. Moreover, from this proposition and his definitions and logical assumptions, Frege could prove that there exists an infinite cardinal number. The cardinal number of the class N is infinite, for N is equipollent to, for example, the class $N \cup \{N\}$. Therefore, the cardinal number of N succeeds itself and, hence, is not a finite number. This also shows why Frege's S is only suitable as the relation of immediate succession among finite numbers.

Not all mathematicians were satisfied with such foundations of arithmetic as were formulated by Dedekind and Frege. Some preferred to define the natural numbers as the results of a process of construction which proceeds according to a rule (see, for example, Kronecker *1887a*, art. 1 ; Poincaré *1913a*, 469 ; and Brouwer *1907a* in *Works*, vol. 1, 15), and to take the methods of proof and definition by induction as fundamental. The principle of induction involved in ' constructive ' number theory has a totally different content from the proposition which occurs in the systems of Dedekind and Frege, and it has a different justification. The latter is a principle concerning sets, while the former brings in the concept of possibility. According to the principle of induction belonging to constructive mathematics, if a property can be verified for 1 and a method is known for turning a verification of the property for n into a verification of the property for $n+1$, then the property can be verified for an arbitrary number, which is what is meant, from the constructive point of view, by saying that ' all ' numbers have the property. The justification of this method of induction was explained by Poincaré thus : ' the mind . . . knows it can conceive of the indefinite repetition of the same act, when the act is once possible. The mind has a direct intuition of this power, and experiment can only be for it an opportunity of using it, and thereby becoming conscious of it ' (*1913a*, 39).

Of course, such a justification would appear quite unsatisfactory to those seeking a logical foundation of arithmetic ; indeed, it was the very thing they wished to avoid. Thus Russell preferred the justification of inductive reasoning as a consequence of the definition of the finite integers rather than ' in virtue of any mysterious intuition ' (*1919a*, 27), and both Frege and Russell found the assertions of possibility which are involved in justifications like Poincaré's doubtful (Frege *1903a*, arts. 125 ff. ; Russell *1904a*). On the other hand, Poincaré had his reasons for rejecting the methods used to define the finite numbers and prove the principle of induction used in the logical foundation (*1913a*, 481).

6.4. *Logical foundations of mathematics*

Both Dedekind and Frege intended to reconstruct arithmetic, understood in the broad sense, on the basis of a system of precisely formulated definitions, from which the theorems were to be rigorously deduced by means of logical principles. But Frege went further than Dedekind by systematising the logical principles used in these deductions. He gave the following description of his ' ideal of a strictly scientific method in mathematics ' (*1893a*, introduction) :

> It cannot be demanded that everything be proved . . . but we can require that all propositions used without proof be expressly declared as such, so that we can see distinctly what the whole structure rests upon. After that we must try to diminish the number of these primitive laws as far as possible by proving everything that can be proved. Furthermore, I demand—and in this I go beyond Euclid—that all methods of inference employed be specified in advance . . .

Other mathematicians and logicians such as Peano and the American C. S. Peirce were engaged in formulating a logic of mathematical arguments, but no one carried it out with Frege's thoroughness and rigour.[1]

[1] In this chapter I am not describing the development of Boolean algebra by Boole, Peirce, Schröder and others ; consult, for example, Kneale and Kneale *1962a*, ch. 6.

It is not possible to go into the details and peculiarities of Frege's system of logic here, but the basic ideas may be presented.

The principles of inference used in mathematical proofs were intended to be such that when applied to truths of any subject as premises, the conclusions derived would also be truths. Thus, by means of basic logical principles whose correctness or validity could be recognised, and certain axioms whose truth could be apprehended, further principles and truths could be derived.

The rules of Frege's logic were formulated solely on the basis of the meaning intended for the logical symbols ' $\neg$ ', ' $\rightarrow$ ', ' & ' ' $\vee$ ', ' $\leftrightarrow$ ', ' $(\forall x)$ ', ' $(\exists x)$ ' (see, for example, the account in Frege *1923a*, 40). The first five of these are used to form sentences from sentences in order to have compound sentences whose truth or falsehood depends solely on the truth or falsehood of the sentences from which they are constructed. Sentential connectives used in this way are called ' truth-functional '. The intended meaning is given in the following table (where the letters ' A ' and ' B ' represent sentences) :

A	B	$\neg A$	$A \vee B$	$A \& B$	$A \rightarrow B$	$A \leftrightarrow B$
T	T	F	T	T	T	T
T	F		T	F	F	F
F	T	T	T	F	T	F
F	F		F	F	T	T

Some of these connectives can be defined in terms of others. For example, ' $A \rightarrow B$ ' could be defined as ' $\neg A \vee B$ ', and ' $A \vee B$ ' could be introduced, alternatively, as an abbreviation for ' $\neg A \rightarrow B$ '.

When the sentential connectives are used in the way just explained, the basic logical rules and formulas of classical logic are obviously valid. Thus, given the intended meaning (from the table) of the conditional connective ' $\rightarrow$ ', the correctness of such rules as the following is evident :

$$\frac{A,\ A \rightarrow B}{B} \quad \frac{A}{B \rightarrow A} \quad \frac{\neg A}{A \rightarrow B}. \tag{6.4.1}$$

The first of these rules, the rule of detachment, may be read : Given the sentences ' A ' and ' $A \rightarrow B$ ', the sentence ' B ' may be inferred. This rule is sometimes called the ' *modus ponens* '.

The general theory of deduction constituting Frege's logic comprises, besides the principles concerning arbitrary sentences, also principles relating to universal and existential statements. For example, it follows from the meaning of ' for all ' and ' there is ' that it is not the case that everything lacks the property P if and only if there is something having P.

The obvious rule of inference for existential propositions is this: From the statement 'Pa' infer '$(\exists x)Px$'. Thus, the existence of a thing having a certain property may be proved by giving an example of a thing having that property. But Frege's logic also includes another way of proving existence; for if a contradiction can be derived from the supposition '$\neg(\exists x)Px$', then, by means of the rule:

$$\frac{\neg(\neg A)}{A}, \tag{6.4.2}$$

it can be concluded that $(\exists x)Px$. (Inferences of this type are not valid in the logic of intuitionistic mathematics, where a different meaning is given to the logical symbols.)

Frege's programme of deriving mathematical theorems from definitions using only a small list of logical principles of the sort just described involved only arithmetic and did not extend to geometry. For Frege considered the axioms of Euclidean geometry to be intuitively apprehended truths, and therefore did not intend to reduce them to logic by the definition of basic geometrical concepts (see his *1906a*). By contrast, Russell aimed at a logical reconstruction of all pure mathematics, including geometry insofar as it belonged to pure mathematics as Russell conceived it: 'As a branch of pure mathematics, Geometry is strictly deductive, indifferent to the choice of its premisses and to the question whether there exist (in the strict sense) such entities as its premisses define' (*1903a*, 372). Geometry throws light on actual space only indirectly through 'increased analysis and knowledge of possibilities' (*1903a*, 374).

According to Russell's explication of the concept *proposition of pure mathematics*, the axioms of a particular mathematical theory (for example, Euclidean geometry) are *not* propositions of pure mathematics. What any branch of pure mathematics asserts is that 'such and such consequences follow from such and such premisses . . .' (*1903a*, 373; compare p. 458). The axioms of a mathematical theory are not actually propositions asserted in pure mathematics: 'The so-called axioms of Geometry, for example, when Geometry is considered a branch of pure mathematics, are merely the protasis [conditional clause] in the hypotheticals which constitute the science. They would be primitive propositions if, as in applied mathematics, they were themselves asserted . . .' (*1903a*, 430). Russell construed these 'so-called axioms' as 'parts of a definition' of a class of relations (*1903a*, 397). The asserted propositions of the theory, in the form 'Axioms $\rightarrow T$' (where T is a theorem), would then be those derivable by logic from the definition of a class of relations. Thus, the only genuine axioms used in the deduction of the asserted propositions of a mathematical theory would be logical axioms.

Besides the derivation of mathematical propositions from definitions, logic (in Russell's sense) had another very important function. It provided the existence theorems justifying the definitions formed from the axioms of a theory. An *existence theorem* for a theory asserts that there is at least one relation satisfying the conditions of the definition formed by its axioms. The demonstration of the existence theorem was also regarded as a proof of consistency.

The appearance of the antimonies showed that existence theorems proved by logical constructions did not give complete assurance of consistency. Russell consequently did not claim absolute certainty for his method, and he pointed out that such certainty is not had in any science. Since the antinomies have shown that logical common sense is not infallible, ' an element of uncertainty must always remain, just as it remains in astronomy. It may with time be immensely diminished ; but infallibility is not granted to mortals . . . ' (*1906b*, 631 ; *1973a*, 194). Moreover, he thought that ' it follows from the very nature ' of an attempt to base mathematics on a system of undefined concepts and primitive propositions that the ' results may be disproved ' by the discovery of a contradiction, ' but can never be proved . . . All depends, in the end, upon immediate perception ' (*1903a*, 129). As Whitehead put it, ' the only rigid proofs of existence theorems are those which are deductions from the premises of formal logic. Thus there can be no formal proofs of the consistency of the logical premises themselves ' (*1907a*, 3).

6.5. *Direct consistency proofs*

Logicians and mathematicians like Frege, Russell and Whitehead had been inclined to think that the only way the consistency of a definition or axiomatic theory could be established was by an existence theorem. Russell, for example, asserted that ' freedom from contradiction can never be proved except by first proving existence : it is impossible to perform *all* the deductions from a given hypothesis, and show that none of them involve a contradiction ' (*1910b*, 438). He apparently did not recognise the possibility of proving general theorems about the proofs which are possible in a theory. The first one to seek consistency proofs along such lines was David Hilbert. His consistency proofs for arithmetic theories were to be ' direct ' ; it was to be shown by purely formal considerations that from the axioms of a particular theory a statement and its negation cannot be derived by means of a finite number of logical inferences.

Hilbert's ' Über die Grundlagen der Logik und Arithmetik ' (' On the foundations of logic and arithmetic ' : *1904a*) contains the first presentation of his ideas for direct consistency proofs. The first step in his method involves the specification of symbols which are to be used, the axioms, and the means of inference ; thus logic and mathematics are formulated concurrently. Actually his procedure in this first attempt was not nearly so rigorous as it was to become in another twenty years. In *1904a* he did not make a clear distinction between the formal axiomatic system which is to be investigated for consistency and the metatheory in which the object theory (that is, the axiomatic system) is studied ; and, what is of particular importance, there is no characterisation of the means of proof to be used in demonstrating the consistency of logico-arithmetic object theories. Indeed, it was on issues related to this point that critics concentrated.

Since Hilbert needed to prove a general proposition relating to a denumerable infinity of possible formal derivations, it seemed that a form of inductive argument would be necessary ; but one of the systems whose consistency he aimed to establish was the theory of finite integers, which contains an axiom of induction (see Poincaré *1913a*, 455 ; and Brouwer *1907a* in *Works*, vol. 1, 93). Hilbert's *1904a* has nothing to satisfy people who wonder about that situation, but later he did clarify this matter. Proofs about the formal axiomatic theory of numbers were to be based solely on the construction and decomposition of numerical expressions, and this method was, in his view, essentially different from that principle of induction which needs and is capable of proof (*1922a*, 164 ; compare what was said in section 6.3 above concerning the constructive principle of induction). His clarifications, however, were not entirely satisfactory to everyone. (See van Heijenoort *1967a*, 480–482, where the editor gives an excellent account of this issue.)

The system that Hilbert dealt with in ' On the foundations of logic and arithmetic ' contained two axioms of identity and three arithmetical axioms. Because this system is rather peculiar, I shall discuss his treatment of a very similar, but simpler, system which is found in his ' Neubegründung der Mathematik ' (' New foundations of mathematics ' : *1922a*) and contains the following five axioms :

$$a = a, \tag{6.5.1}$$

$$a = b \rightarrow a + 1 = b + 1, \tag{6.5.2}$$

$$a + 1 = b + 1 \rightarrow a = b, \tag{6.5.3}$$

$$a = c \rightarrow (b = c \rightarrow a = b), \tag{6.5.4}$$

$$a + 1 \neq 1, \tag{6.5.5}$$

together with the rule of detachment (6.4.1). A *proof* with respect to these axioms is defined to be a list of formulas having a last entry, such that each formula in the list is either one of the axioms or the result of substituting numerals or other variables for the variables in an earlier entry, or a formula B with the formulas A and $A \rightarrow B$ occurring among the preceding entries (that is, a formula inferred in accordance with the rule of detachment). For example, the following list of formulas is a proof of the symmetrical property of equality :

$$a=c \rightarrow (b=c \rightarrow a=b), \tag{6.5.6}$$

$$a=a \rightarrow (b=a \rightarrow a=b), \tag{6.5.7}$$

$$a=a, \tag{6.5.8}$$

$$b=a \rightarrow a=b. \tag{6.5.9}$$

The last line of a proof, the formula proved, is called the ' end formula '.

The axiom system (6.5.1)–(6.5.5) is consistent if an equation $\alpha=\beta$ and its negation $\alpha \neq \beta$ are never both provable. In order to prove that the system is consistent, Hilbert established the lemma : A provable formula can contain the sign ' $\rightarrow$ ' at most twice. Suppose that a proof contained a formula in which there were more than two occurrences of ' $\rightarrow$ '. If a proof contains such a formula at all, it must have a first such. But the first formula B in a proof containing ' $\rightarrow$ ' more than twice could neither be derived from earlier lines by substitution nor by means of the rule

$$\frac{A,\ A \rightarrow B}{B},$$

for the premise $A \rightarrow B$ would already have to contain more than two occurrences of ' $\rightarrow$ '.

Hilbert also proved a second lemma stating that an equation $\alpha=\beta$ is provable only if α is the same symbol as β. It should be noted here that the symbols ' α ' and ' β ' do not belong to the vocabulary of the object theory under investigation ; they are variables belonging to the meta-theory, and their range of values consists of the numerals and variables of the object theory together with the expressions made up from them by using ' + '. Thus, according to his second lemma, an equation written in the language of the object theory is only provable from axioms (6.5.1)–(6.5.5) if the same symbol occurs on both sides of the sign ' = '.

By means of the two lemmas, Hilbert demonstrated the consistency of his axiom system as follows. Since only equations having the same sign on both sides of ' = ' are provable, if the system were inconsistent, some formula of the form

$$\alpha \neq \alpha \tag{6.5.10}$$

would have to be provable. But only inequalities of the form

$$\alpha+1\neq 1 \qquad (6.5.11)$$

are provable by direct substitution in the axioms (for only axiom (6.5.5) contains the sign ' $\neq$ '). If a formula $\alpha\neq\alpha$ were provable by means of the detachment rule, then a premise of the form $C\rightarrow\alpha\neq\alpha$ would have to be used ; but since it could not be derived directly by substitution, it would require a premise $B\rightarrow(C\rightarrow\alpha\neq\alpha)$, which in turn would depend on a premise $A\rightarrow(B\rightarrow(C\rightarrow\alpha\neq\alpha))$. But according to the first lemma, no provable formula has more than two occurrences of ' $\rightarrow$ '.

The axiom system which Hilbert used to illustrate direct proofs of consistency was a rather elementary part of the arithmetic of natural numbers and did not contain an axiom of mathematical induction. But he was confident that essentially the same methods of reasoning would yield consistency proofs for much more advanced systems, such as the full arithmetic of integers, the theory of real numbers, and also the theories of Cantor's higher number-classes. However, in *1931a* Kurt Gödel published a theorem, called ' Gödel's theorem on consistency proofs ', which showed that consistency proofs for (say) Peano's axiomatisation of arithmetic, using only the very elementary (finite combinatorial) methods of the sort which Hilbert had in mind, are impossible. Nevertheless, interesting results (in particular, Gentzen's theorem on the consistency of elementary number theory) have been obtained by modifying Hilbert's programme. The interested reader may consult Andrzej Mostowski's *Thirty years of foundational studies* (*1966a*) for further information on the work of Gödel and Gentzen, and on other advances in foundations occurring after the period covered in this chapter.

6.6. *Russell's antinomy*

Cesare Burali-Forti was the first mathematician to publish an antinomy of set theory. In his *1897a* he considered the class of all Cantorian ordinals and showed that its ordinal Ω satisfied the contradictory properties

$$\Omega+1>\Omega \quad \text{and} \quad \Omega+1\leqslant\Omega. \qquad (6.6.1)$$

However, his result does not seem to have made much of an impression. People were perhaps inclined to believe that some error had been made in reasoning in a new and unfamiliar subject. The same applied to the next antinomies that were reported in print—of the set of all alephs and the set of all powers—which were similar to Burali-Forti's. For example, when Hilbert mentioned them to Frege in 1900, Frege seems to have been uninterested (see Frege *1971a*, 12). While he was a firm supporter of the new theory of the transfinite, he considered Cantor's formulations to be imprecise ; perhaps he thought that some error had been made on that account.

The case was entirely different with the antinomy of the class of all classes which do not belong to themselves, which was discovered independently by Russell and by Zermelo and first published in 1903 in the second volume of Frege's *Grundgesetze* (*1903a*, appendix) and in Russell's *The principles of mathematics* (*1903a*, ch. 10). This antinomy involves only the concepts of class and membership, and it follows almost immediately from the axiom which had been the implicit basis of set theory. Yet it is an interesting fact that Russell was led to discover it by reflecting on the implications of Cantor's theorem which asserts that for every set there is another set, its power-set, of greater cardinal number (see section 5.9).

Russell noticed that although Cantor had proved that there is no greatest cardinal number, there must nevertheless be such a number. In fact, the greatest number should be the number of the class of all entities, for there cannot possibly be a larger class. This antinomy, which was already known to Cantor, is usually called ' Cantor's paradox '.[1] Russell became aware of it in January of 1901 and gave it its first mention in print in an article *1901a*, in which he said that if Cantor's proof that there is no greatest number were valid, ' the contradictions of infinity would reappear in a sublimated form. But in this one point, the master has been guilty of a very subtle fallacy, which I hope to explain in some future work ' (*1917a*, 89).

Thus Russell's reaction to the first antinomy which he discovered was that there must be some subtle mistake in Cantor's argument ; he did not conclude that there is something fundamentally wrong in set theory. Russell minutely examined the proof of Cantor's theorem, expecting to find some error ; in cases like that of the class of all entities, the class of all classes, or the class of all propositions, it seemed to him ' as though Cantor's proof must contain some assumption which is not verified ' (*1903a*, 362).

Expecting an error of this sort, Russell examined the results of applying the method of Cantor's proof to such classes. The result was his discovery of the antinomy which has come to be called ' Russell's para-

[1] I avoid using the word ' paradox ' to describe such results, for ' paradox ' is used very loosely in common speech to cover antinomies, correct arguments with puzzling or counter-intuitive though non-contradictory conclusions, and also only apparently valid arguments, such as the old-fashioned ' paradoxes of the infinite '. The antinomies of set theory are real contradictions derivable by means of specified logical rules (correctly applied) from an apparently true thesis concerning the existence of sets.

dox '. In his *Introduction to mathematical philosophy* he says that when he first came upon the contradiction of the greatest number in 1901, ' I attempted to discover some flaw in Cantor's proof that there is no greatest cardinal . . . Applying this proof to the supposed class of all imaginable objects, I was led to a new and simpler contradiction . . . ' (*1919a*, 136).

In order to see how Russell discovered the antinomy of the class of all classes which do not belong to themselves, recall from section 5.7 the method of showing in the proof of Cantor's theorem that any set A is not equipollent to the power-set $P(A)$ of all its subsets. For any one–one correspondence f whose domain of arguments is $A' \subseteq A$ and whose range of values is a subset of $P(A)$, the class

$$K \underset{\text{Df}}{=} \{x \mid x \in A' \;.\; \& \;.\; x \notin f(x)\} \tag{6.6.2}$$

belongs to $P(A)$ but not to the range of values of f. If it is supposed that K is an f-correlate of some element y of A', that is, if

$$y \in A' \;.\; \& \;.\; f(y) = K, \tag{6.6.3}$$

it follows that

$$y \in K \leftrightarrow y \notin K, \tag{6.6.4}$$

which implies the contradiction $y \in K \;.\; \& \;.\; y \notin K$. Now let us see what happens when A is the class U of all things, A' is the subclass C of U containing all classes, and for each $c \in C$, $f(c) = c$. In this case, K is the class R of all classes which do not belong to themselves, and it *follows* that

$$R \in C \;.\; \& \;.\; f(R) = R \tag{6.6.5}$$

(compare (6.6.3), which was only a supposition), and consequently that

$$R \in R \leftrightarrow R \notin R. \tag{6.6.6}$$

It is in this or some very similar way that Russell discovered his antinomy ; I would guess that the same is true in the case of Zermelo.

Once R is thought of, it is evident that its existence and the antinomy (6.6.6) follow immediately from the principle that for any sentential form ' ϕx ',

$$(\exists M)(\forall x)\{x \in M \leftrightarrow \phi x\}, \tag{6.6.7}$$

and hence that this principle is false. Thus, while Russell set out to correct a supposed error in Cantor's reasoning, he obtained another antinomy which made it entirely clear that something was wrong with the very first principles of set theory, or, as Russell would have said, with the principles of logic, since he regarded the general theories of classes and relations as branches of logic.

The principle (6.6.7) usually referred to as ' the naive principle of comprehension ' (or abstraction), asserts that every propositional function ' ϕx ', or every property, determines a class. This was taken to be an evident truth by Russell (see *1903a*, 102). He also pointed out that the naive axiom of abstraction is unofficially present in Peano's system : ' Peano holds (though he does not lay it down as an axiom) that every proposition containing only one variable is reducible to the form " x is an a " ' (*1903a*, 28 ; compare pp. 19 and 103). A principle having the same effect as (6.6.7) was also advanced by Frege.

Perhaps no one was so upset by Russell's antinomy as Frege. At first he made some attempt to resolve this antinomy (*1903a*, appendix), but apparently he became dissatisfied with any of the methods which were introduced to avoid the antinomies. Near the end of his life, he said that set theory had been ' destroyed ' by the antinomies (*1969a*, 289). He gave up attempting a logical reconstruction of arithmetical theories and settled for a geometrical foundation of arithmetic (*1969a*, 298–302). In contrast, Russell's attitude was that the antinomies ' can all be removed by patience in distinguishing and defining ' (*1910a*, 373).

6.7. *The foundations of* Principia mathematica

When Russell began to deal with the problem of the antinomies, he ' hoped the matter was trivial and could be easily cleared up ' (*1944a*, 13). But it took five years of effort before he produced the system for avoiding the antinomies which is used in *Principia mathematica.* Indeed, he never succeeded in formulating a system which completely satisfied him. One thing that made the problem of the antinomies so hard for him was that he sought not merely a way of avoiding them ; he also desired an independent explanation for the necessity of making the particular restrictions. In *My philosophical development* he says that while he was ' looking for a solution ' he considered it a requisite of a ' wholly satisfying ' solution that it ' should, on reflection, appeal to what may be called " logical common sense "—i.e. that it should seem, in the end, just what one ought to have expected all along ' (*1959a*, 79). And in *1906b* he said that satisfactory principles should ' recommend themselves to intuition ' and ' show exactly how we formerly fell into error ' (*1906b*, 631 ; *1973a*, 195). Although he wanted to find a way of avoiding the antinomies which appeals to logical common sense, in his *The principles of mathematics* he had said the antinomy of the class of all classes which do not belong to themselves ' springs directly from common sense, and can only be solved by abandoning some common-sense assumption ' (*1903a*, 105).

Russell's main idea for a solution of the antinomies is already contained in *Principles* : the doctrine that each propositional function[1] has a ' range of significance ', and is *meaningless* with arguments outside this range. The formulas which give rise to the antinomies are, of course, the meaningless ones : in particular, the expression ' x belongs to x ' and its negation were declared to be meaningless. But in order to have a ' solution ' of the antinomies, Russell needed an explanation of why certain apparently meaningful statements about classes are really meaningless ; he did not wish merely to propose that certain combinations of symbols be declared not well-formed as one way of avoiding the antinomies. Now Russell did not believe that (for example) ' $x \in x$ ' and its negation would be meaningless if there are such things as classes : ' That it is meaningless . . . to regard a class as being or not being a member of itself, must be assumed for the avoidance of a . . . mathematical contradiction ; but I cannot see that this could be meaningless if there were such things as classes ' (*1910a*, 376).

Consequently, another feature of Russell's efforts with the antinomies was his tendency to employ some form of ' no-class theory ', that is, a theory in which it is not assumed that many entities ever ' collectively form a single entity which is the class composed of them ' (*1906a*, 46 ; *1973a*, 155). Russell formulated two such theories. The first one, the substitutional theory, was only briefly described in Russell's publications, but he wrote a long exposition of it which has now been published (*1973a*, 165–189 ; for commentary, see Grattan-Guinness *1974b*, 389–401). The second no-class theory appears in Russell's ' Mathematical logic as based on the theory of types ' (*1908a*) and in *Principia mathematica.* Both of Russell's no-class theories provided a meaning for some statements purporting to be about classes, but neither provided a meaning for such statements as ' $x \in x$ ' or ' $x \notin x$ '.

Russell eventually became convinced that the basic fallacy underlying the antinomies was some sort of vicious circle. This idea had appeared in Poincaré *1906a*, where it was explained by reference to the following antinomy formulated in Jules Richard *1905a*.[2] Let E be the set of all

[1] The term ' propositional function ' is not clearly used with a single definite meaning in Russell's writings. Sometimes it seems to mean a function whose values are propositions in the sense of objective truths and falsehoods, while at other times it means a sentential form, that is, an expression like a sentence except that it contains one or more variables and becomes a sentence when constants are substituted for its variables. Unfortunately, it is sometimes impossible to be sure what Russell intended.

[2] Although the idea that a sort of vicious circle was involved in the use of bound functional variables had occurred to Russell as early as 1904, he did not (or was at least uncertain whether to) regard the antinomies as ' vicious circle fallacies ' until 1906. The paper on the substitutional theory written before Poincaré's article does not claim that the antinomies are due to vicious circles. Rather the suggestion was that the antinomies are due to ' false abstraction ' (see *1973a*, 165). In a letter of January 1906

decimals which can be defined in a finite number of words ; the set is obviously denumerable and, hence, can be arranged in a sequence. But, by reference to a sequence of the elements of *E*, it is possible to define in a finite number of words a decimal *N* which does not belong to *E*. *N* is the decimal containing the digit $p+1$ in its *n*-th place if *p* is in the *n*-th place of the *n*-th element of *E* and is not 8 or 9, but containing the digit 1 if *p* is 8 or 9. Now, according to Poincaré, ' the true solution ' of Richard's antinomy is this : ' *E* is the aggregate of *all* the numbers definable by a finite number of words *without introducing the notion of the aggregate E itself*. Else the definition of *E* would contain a vicious circle ; we must not define *E* by the aggregate *E* itself ' (*1913a*, 480). Since *N* is defined ' with the aid of the notion of the aggregate *E* ', it does not belong to *E*. The other antinomies are supposed to be explicable in a similar manner.

But no definition given in the statement of Richard's (or any other) antinomy is circular in the ordinary sense : the term defined or a synonymous term does not occur in the defining expression, and the definitions are not defective for the reason that the definitions usually called ' circular ' are. Thus Peano asserted that ' the definitions of Richard do not contain a vicious circle ' (*1906a*, art. 4 ; *1973a*, 214), and pointed out that the usual definition of the least common multiple of two integers is of the kind which Poincaré alleges to contain a vicious circle (*1973a*, 215). Zermelo also emphasised in *1908a* the fact that definitions of the form called ' viciously circular ' by Poincaré have been very frequently used in mathematics ' and up to now it has not occurred to anyone to regard this as something illogical ' (van Heijenoort *1967a*, 190–191). As a matter of fact, Russell himself had denied, in an appendix to *The principles of mathematics*, that there is any vicious circle involved in the method of Frege's definition of the natural numbers (*1903a*, 522), though that definition is a prime example of the sort classified as viciously circular by Russell and Poincaré in 1906.

In *Principia mathematica* Russell tried to explain why statements purporting to assert something about absolutely *all* propositions, propositional functions, or classes must be considered meaningless. The reason why such general statements are supposed to be meaningless is that the totalities to which they refer cannot be definite. A proposition such as : all propositions are true or false, ' could not be legitimate unless " all propositions " referred to some already definite collection, which it cannot do if new propositions are created by statements about " all propositions " ' (*1903a*, 37 ; see also *1959a*, 82). Many years

---

Russell wrote to Philip Jourdain that ' The error seems to me to lie in supposing that many entities ever combine to form one new entity, the class composed of them ' (see Grattan-Guinness *1977a*, 68).

later he said in his book *My philosophical development* : ' I must confess that this doctrine has not won wide acceptance, but I have seen no argument against it which seemed to me cogent ' (*1959a*, 83). It may be remarked, however, that some have found his explanations hard to understand (see Chihara *1973a*).

The scheme for avoiding the antinomies in *Principia* is called the ' theory of types ', or the ' ramified theory of types ' (to distinguish it from the ' simple theory of types ' which was developed later by L. Chwistek, F. P. Ramsey and others). The principle guiding its formulation is the *vicious circle principle* (VCP), which may be stated as follows : A totality *T* may not contain elements which are only definable by means of an expression containing a bound variable (such as ' *x* ' in ' for all *x*, . . . *x* . . . ' and in ' there is an *x* such that . . . *x* . . . ') whose range of values contains *all* elements of *T*. For example, the property *P* of having all properties *Q* of the class *T* must not be a member of *T*.

In accordance with the VCP, propositional functions which have a value for an object *a* as argument (*a*-functions or functions ' significant ' for *a*) are classified into orders. The VCP rules out a totality of all *a*-functions, for there are *a*-functions which can only be defined by means of a bound variable ranging over some totality of *a*-functions. Such functions must lie outside of the totalities in terms of which they are defined : at least, they must according to the VCP. But note that, strictly speaking, we cannot say that any ' legitimate ' totality of *a*-functions does not include *all a*-functions, or that for any legitimate totality of *a*-functions *there is* some *a*-function not belonging to it : such statements are supposed to be meaningless (see *PM*, vol. 1, 55).

Let us say that a function presupposes or involves a totality *T* if it is definable only by means of a bound variable whose range includes all of *T* ; it also presupposes whatever is presupposed by the members of such a totality, and so on. The *order* of a function depends on what totalities it presupposes. The functions significant for a particular entity *a* as argument are of infinitely many different orders above the order of *a*. The objects of the absolutely lowest order are the individuals (concrete objects). First-order functions are those which presuppose only the totality of individuals, while second-order functions of individuals presuppose only a totality of first-order functions, in addtion to the totality of individuals. There are also functions of individuals of arbitrarily many higher orders, and functions of order *m* can be arguments to functions of arbitrarily many orders above *m*. A function is said to be *predicative* if its order is next above the order of its highest-order argument ; in other words, ' if it is of the lowest order compatible with its having the arguments it has ' (*PM*, vol. 1, 53). Alternatively explained, ' a predicative function of a variable argument is one which involves no totality except that of the possible argument, and those that are presupposed by any one of the possible arguments ' (*ibid.*, 54). It should be noted that the concept of a predicative function is a primitive idea in *Principia mathematica*.

In addition to orders, there is also a hierarchy of types. The *type* of a function depends not only upon its order but also on the number and kind of the arguments that it takes ; thus it is a sub-classification of orders. For example, second-order functions of two individual arguments constitute a type.

Russell defined finite cardinal numbers as classes of equipollent classes independently of, though in broadly the same manner as, Frege's definitions in section 6.3 above ; but because of type theory numbers have to be defined for each type. (Russell also extended his definitions to include transfinite cardinal and ordinal arithmetic.) Now adherence to the system of orders is sufficient to develop these definitions and also to avoid the antinomies, but it makes it impossible to define the finite numbers by means of the property of having absolutely all hereditary properties of 0 (a property is *hereditary* if whenever a number has it, so does its successor). According to Russell's doctrine, expressions containing the phrase ' all properties ' or ' all functions ' are meaningless, and the property of having all of a certain totality of properties is of higher order than any of the properties belonging to that totality. If the integers are defined as the things having all hereditary properties of a particular order m which belong to 0, then the principle of induction is not a consequence of the definition. Since, by the original definition, the property N of being a number is the property of having *all* hereditary properties of 0, it follows that if P is a hereditary property of 0, then whatever has N has P. But if N is only the property of having all hereditary properties of order m which belong to 0, and P is a hereditary property of 0 whose order is higher than m, it does not follow from the definition of N that every number has P. To take a simple example from Russell *1908a*, art. 5, if the finite numbers are defined as those having all first-order hereditary properties of 0, then ' we shall be unable to prove that if m, n are finite numbers, then $m+n$ is a finite number. For, with the above definition, " m is a finite number " is a second-order property of m . . . ' (van Heijenoort *1967a*, 167).

The theory of real numbers is also affected. It will be recalled from section 6.1 that the virtue of Dedekind's definition was that it had the property of completeness (continuity) as a consequence : that is, using Russell's modification of the definition of the system of real numbers, it follows that every segment of the real numbers has an upper limit. A segment of reals is a certain class of classes of rationals, and its upper limit is its union. But, owing to the bound variable occurring in its definition, the union of a class A of classes will generally be a class of higher order than the elements of A and consequently cannot, according to the VCP, belong to a class containing A as a subclass. Thus, according to the VCP, the upper limit of a class of real numbers cannot generally be a member of the system of real numbers, and the system will not be complete.

In order to compensate for the much too negative effect of the vicious circle principle, Whitehead and Russell postulated the axiom of reducibility : ' The axiom of reducibility is introduced in order to legitimate a great mass of reasoning, in which, prima facie, we are concerned with such notions as " all properties of a " or " all a-functions ", and in which, nevertheless, it seems scarcely possible to suspect any substantial error ' (*PM*, vol. 1, 56). The axiom of reducibility is the statement that any propositional function satisfied by an object a is formally equivalent to a predicative function (or predicate) of a. Two propositional functions are formally equivalent if they are satisfied by exactly the same arguments, that is, if they are co-extensive. Thus, the axiom of reducibility means that for a propositional function of *any order* whatsoever which is satisfied by the object a there exists a co-extensive function whose order is next above that of a.

Assuming the axiom of reducibility, if the finite numbers are defined as those having all hereditary predicates of 0, then it will be possible to prove that a higher-order hereditary property P of 0 belongs to all finite numbers. For it follows from the definition that a hereditary predicate of 0 belongs to all finite numbers, and the axiom of reducibility asserts that there is a predicate Q of numbers which is coextensive with the property P. Since Q is coextensive with P, Q is a hereditary property of 0, and P belongs to all finite numbers because Q does. Similarly, the axiom of reducibility saves the theory of real numbers in spite of the VCP.

6.8. *Axiomatic set theory*

After Cantor discovered that a contradiction is sometimes implied by the supposition that there exists a set of all things having a certain property, he began to distinguish two kinds of multiplicities, which he called ' consistent ' or ' sets ' and ' inconsistent ' or ' absolutely infinite '. Though this distinction was not satisfactory, his basic idea for avoiding the contradictions has—after being extended and improved—become the most widely accepted reformulation of set theory.

Cantor explained an inconsistent multiplicity, in a letter of 1899 to Dedekind, as one for which ' the assumption that all of its elements " are together " leads to a contradiction, so that it is impossible to conceive of the multiplicity as a unity, as " one finished thing " ' (*Papers*,

443 ; van Heijenoort *1967a*, 114). But this is practically as much as to say that there are no such things as inconsistent multiplicities ; as Cantor himself said to Jourdain, ' inconsistent multiplicities . . . can never be conceived *complete* and *actually existing* ' (Grattan-Guinness *1971a*, 119). Thus Cantor does not really have ' two kinds ' of multiplicities. It is no wonder that Dedekind found Cantor's purported distinction ' unclear ' and did not know what Cantor meant by ' Zusammensein aller Elemente einer Veilheit ' (' togetherness of all elements of a multiplicity '; Grattan-Guinness *1974a*, 129). The fact is, Cantor spoke of an inconsistent multiplicity when he might better have spoken of a property such that the supposition that there is a set of all things having that property leads to a contradiction.

Later, John von Neumann formulated a system which avoids the antinomies by not assuming that every class belongs to further classes (see especially his *1925a*). The term ' set ' is reserved for those classes which are elements of other classes. Here two kinds of classes really are distinguished. It is usually stated that Cantor's inconsistent multiplicities are von Neumann's classes which are not elements ; but in the case of the former a contradiction results from the supposition of their existence, while the latter are such that the supposition that they are elements leads to a contradiction. What is common to the systems of Cantor and von Neumann is not the distinction between ' sets ' and ' proper classes ' (Cantor having no satisfactory distinction), but the supposition that it is the very large totalities which are involved in the antinomies.

Which multiplicities Cantor thought to be inconsistent is, it seems to me, evident from his alternate description of them as ' absolutely infinite '. He had always characterised the series of transfinite ordinals as absolutely infinite. His idea seems to have been something like this : each transfinite number is surpassed by other transfinite numbers, but what is absolutely infinite has the property of being essentially incapable of enlargement (see *Papers*, 167, 175 and 375). Thus I take his identification of the inconsistent multiplicities with those which are absolutely infinite as an expression of his belief that the *only* inconsistent multiplicites are the absolute totalities such as the totality of all things, all sets, or all ordinals, or totalities which would have to be as large, that is, which would have to contain a part equipollent to one of these absolutes. As Gerhard Hessenberg put it in Grelling and Nelson *1907a*, 330 (Nelson *1959a*, 82) :

> Certainly every set containing a part equivalent to *W* [the set of all ordinals] is infected with the same contradiction as *W* itself, and its power is greater than every aleph. Now Herr Cantor conjec-

tures that conversely this condition is sufficient, that therefore every set, whose power is an aleph, is consistently conceivable, so that one has to designate the type W so to speak as the ' first ' or ' smallest ' paradoxical type.

Cantor's method of dealing with the antinomies is of the sort which Russell discussed (but did not adopt) in his *1906a* under the title ' theory of limitation of size '. ' This theory ', Russell said, ' is naturally suggested by the consideration of Burali-Forti's contradiction, as well as by certain general arguments tending to show that there is not . . . such a thing as the class of all entities ' (*1906a*, 43 ; *1973a*, 152). He also stated that the theory of limitation of size ' naturally becomes particularised into the theory ' that a propositional function determines a class if there is a one–one relation between the things satisfying it and some initial segment of the ordinals. He even thought certain considerations made it seem likely that if an antinomy can be derived from the supposition of a class of all things having the property ϕ, then it will be possible to define a one–one relation between all ordinals and some (or all) of the things having ϕ (*1906c*, 36 ; *1973a*, 144). In Cantor's terminology : If A is an inconsistent multiplicity, then some sub-multiplicity of A is equivalent to W. This is just a particularisation of the conviction, maintained by Cantor, that *only* absolutely infinite multiplicities are inconsistent.

Russell himself showed that there are at least as many classes which do not belong to themselves as there are ordinal numbers. The following argument is suggested by his considerations (*1906a*, 35 ; *1973a*, 143), but is not exactly the one he gave. If x is a set of sets which do not belong to themselves, then neither x nor $x \cup \{x\}$ belongs to itself. The union of a set of sets whose members do not belong to themselves is a set of sets not belonging to themselves ; hence it does not belong to itself. Consequently, if x is any set not a member of itself, the series :

$$\left.\begin{aligned} S_0 &\underset{\text{Df}}{=} x, \\ S_{\alpha+1} &\underset{\text{Df}}{=} S_\alpha \cup \{S_\alpha\}, \\ S_\lambda &\underset{\text{Df}}{=} \bigcup_{\alpha<\lambda} S_\alpha, \text{ for limit numbers } \lambda, \end{aligned}\right\} \qquad (6.8.1)$$

is a series of sets which do not belong to themselves, and the series is isomorphic to the series of all ordinals.

In order to reconstruct set theory on the basis of Cantor's analysis of the antinomies, it is necessary to formulate a system of axioms from which the theorems can be proved but from which the existence of the large 'sets' involved in the antinomies does not follow. In correspondence with Dedekind, Cantor formulated several of the most important axioms of set theory. Let us first consider these axioms formulated in Cantor's terminology (writing '$S(A)$' for 'A is a set'):

$$S(A) \,.\, \& \,.\, B \sim A \,.\, \rightarrow .\, S(B), \tag{6.8.2}$$

$$S(A) \,.\, \& \,.\, A' \subseteq A \,.\, \rightarrow .\, S(A'), \tag{6.8.3}$$

$$S(A) \rightarrow S(\cup A). \tag{6.8.4}$$

But what is the range of the variables 'A', 'A'', and 'B'? It surely cannot be multiplicities in general comprising both consistent and inconsistent multiplicities, for the latter do not exist. This defect in the formulation of Cantor's system is, however, easily removed. Instead of inconsistent multiplicities, we could speak, as Russell did, of properties which do not determine a set. (A property does not determine a set if there is no set having as its elements exactly the things possessing the property.) Thus, Cantor's axioms could be formulated as follows: (1) If ϕ is a property of sets and there is a one–one relation between the things having ϕ and the set M, then ϕ determines a set, that is:

$$(\exists A)(\forall x)\{x \in A \leftrightarrow \phi(x)\}. \tag{6.8.5}$$

(2) If M is a set, then the property of having ϕ and belonging to M determines a set. (3) The property of being a member of a member of the set M of sets determines a set.

The first published system of axioms based on the theory of limitation of size was formulated in 1908 by Ernst Zermelo.[1] But his system was designed to avoid the antinomies of 'finite definability', such as Richard's antinomy, as well as those concerning very large totalities. Russell made no distinction between the various antinomies, but considered them all as 'vicious circle fallacies'. By contrast, Zermelo, following Peano and Hessenberg, distinguished between antinomies which could be formulated in terms of the primitive concepts of set theory and antinomies involving definability.

In order to avoid the antinomies of finite definability, Zermelo introduced the concept of a 'definite' assertion. An assertion about sets is definite if the relations of set membership and identity 'by means of the axioms and the universally valid laws of logic, determine without arbitrariness whether it holds or not' (*1908b*, art. 1; van Heijenoort *1967a*, 201). A sentential form 'ϕx' is definite, if every assertion resulting from an assignment of a value to 'x' is definite. Now, instead of axiom (2) above, Zermelo's system contains the following

[1] Some similar ideas are in Harward *1905a*, which seems not to have been influential.

axiom of 'separation' (axiom III): A definite sentential form determines a subset of the set M; that is, if 'ϕx' is definite, then

$$(\exists A)(\forall x)\{x{\in}A \,.\leftrightarrow .\, x{\in}M \,.\,\&\,.\, \phi x\}. \tag{6.8.6}$$

This is to avoid including such a set as the set of all decimals which are definable in a finite number of words.

While Zermelo's definition of 'definite' is hardly satisfactory, it is not in the least doubtful which assertions he had in mind. The definite assertions were to be all those which can be expressed using only variables for sets, '$=$', '$\in$', and logical symbols such as '$\neg$', '&', '$\rightarrow$', '$(\forall x)$' and '$(\exists x)$'. The first precise definition of the concept of a definite assertion was given in *1910a* by Hermann Weyl (*Papers*, vol. 1, 304).

In addition to the axiom of separation, Zermelo's system included the axiom of extensionality

$$(\forall x)(x{\in}A{\leftrightarrow}x{\in}B) \,.\rightarrow .\, A=B, \tag{6.8.7}$$

and axioms asserting the existence of: an empty set, a set $\{a\}$ for any a, a set $\{a, b\}$ for any a and b, the union of any set, the set of all subsets of any set. It also included the following axiom of choice: The union of any set T of pairwise disjoint non-empty sets contains at least one subset whose intersection with each member of T is a unit set (see section 6.9 below). He also formulated an axiom of infinity suggested by Dedekind's proof (*1888a*, art. 66) that a simply infinite system exists. According to this axiom, there exists a set which contains the empty set, and whenever it contains x, also contains $\{x\}$. A set whose elements may be called 'natural numbers' is defined as follows: Let Z be a set whose existence is postulated by the axiom of infinity. The set Z_0 of natural numbers is the intersection of the set of all subsets of Z which contain $\varnothing$ and the unit set of each of their elements:

$$Z_0 \underset{\mathrm{Df}}{=} \bigcap\{X \mid X\subseteq Z \,.\,\&\,.\, \varnothing{\in}X \,.\,\&\,.\, (\forall x)(x{\in}X{\rightarrow}\{x\}{\in}X)\}. \tag{6.8.8}$$

Z_0 cannot simply be defined as the intersection of the set A of all sets containing the empty set as well as the unit set of any element they contain, because such a set A would be 'too big', and its existence could not be proved from Zermelo's axioms.

In *Principia mathematica* axioms of choice and infinity are formulated, but not assumed to be true; they are taken as hypotheses, and many conditional theorems are established having one of these propositions as antecedent. There is a significant difference between Zermelo's and Russell's axioms of infinity: the former asserts the existence of an infinite set of sets, while the latter asserts the existence of an infinite set of individuals or concrete objects. Russell had to formulate such an axiom of infinity because of his theory of types (see especially his *1908a*, art. 10).

6.9. *The axiom of choice*

Before Zermelo's *1904a* focussed attention on the axiom of choice, various mathematicians had without realising it formulated proofs whose validity depended upon that axiom. A good example is provided by Dedekind's argument for the proposition that every set S containing for each natural number n a subset equivalent to Z_n (which is the set (6.3.12) of natural numbers between 1 and n) is equipollent to some proper subset of itself (*1888a*, art. 159). According to the hypothesis of the theorem, for each n there is a one–one function mapping Z_n into S; that is, for each n, the set A_n of one–one mappings from Z_n into S is not the empty set. Dedekind went beyond the hypothesis of the theorem when he assumed that there is a sequence of functions α_n such that α_n belongs to A_n.

Russell also unwittingly used an argument (due to Cantor) which needs the axiom of choice as a premise (*1903a*, 122–123). But he eventually recognised as a separate assumption the form of the axiom of choice which he called the ' multiplicative axiom ' : For every class A of mutually exclusive non-empty sets, there exists at least one set comprising exactly one element from each member of A. In a letter to Jourdain written in 1906, he relates how this came about (Grattan-Guinness *1972a*, 107 ; *1977a*, 80) :

> As for the multiplicative axiom, I came on it so to speak by chance. Whitehead and I make alternate recensions of the various parts of our book, each correcting the last recension made by the other. In going over his recensions, which contained a proof of the axiom, I found that the previous proposition used in the proof had surreptitiously assumed the axiom. This happened in the summer of 1904. At first I thought probably a proof could easily be found ; but gradually I saw that, if there is a proof, it must be very recondite.

The problems which eventually made the axiom of choice prominent were those concerning the comparability of powers and well-ordering. Cantor had always been convinced that any two transfinite powers are comparable, and so fully deserve the name ' cardinal number '. Two powers are comparable if for any sets having those powers, one is equipollent to a subset of the other. Now he had succeeded in showing that any two well-ordered sets are comparable, and hence that any two alephs are comparable, and that a transfinite power which is less than some aleph is itself an aleph. It only remained to show that every power is an aleph, or that every set has at least one well-ordering. Now my guess is that he thought that any power must at least be comparable with the alephs (compare Hardy *1904a*, 88), and, therefore, that two incomparable powers must both be greater than any aleph. I speculate further that it was in pursuing this line of thought that he came upon the antinomies and subsequently made his distinction between consistent and inconsistent multiplicities. He then formulated the following argument, which he communicated to Dedekind in 1899, for the proposition that every transfinite power is an aleph. Suppose that the multiplicity V does not have an aleph as its power. In that case, for every ordinal α, V is not equivalent to W_α (the initial segment of the ordinals determined by α). Cantor assumes then that W is ' projectible into ' V, which means that ' there must exist a sub-multiplicity V' of V that is equivalent to the system ' W (*Papers*, 447 ; van Heijenoort *1967a*, 117). By the axiom (6.8.2) V' is inconsistent because W is, and therefore V is inconsistent by (6.8.3). Thus, if V does not have an aleph for its power it is an inconsistent multiplicity ; consequently, if V is a set, then its power is an aleph.

In an editorial note to this argument in his edition of Cantor's works, Zermelo explained objections which led him to formulate his own proof. To explain why Cantor asserts that W is ' projectible ' into a multiplicity V whose power is not an aleph, Zermelo supposes Cantor to have thought in terms of a procedure of successive assignments of members of V to ordinals. Zermelo's objection is that ' the intuition of time is applied here to a process that goes beyond all intuition . . . ' (Cantor *Papers*, 451 ; van Heijenoort *1967a*, 117). But Cantor may not have intended any such thing ; it is possible that his reasoning is based on the following proposition :

$$(\forall\alpha)\{\neg(V \sim W_\alpha)\} \rightarrow (\exists V')\{V' \subseteq V \;.\; \& \;.\; V' \sim W\} \qquad (6.9.1)$$

taken as an axiom.

Zermelo remarks that the theorem which Cantor wished to prove could only be established by means of the axiom of choice ' which postulates the possibility of a *simultaneous* choice . . . '. But how it would be possible to make so many arbitrary choices ' simultaneously ' is not evident. Actually Zermelo did not understand his axiom to assert anything about the possibility of choices. Indeed, Sierpinski quotes Zermelo as having said in a letter that the formulation of the axiom in terms of choice ' concerns only the psychological method of presentation, while the axiom, as its wording by the way makes sufficiently clear, should be regarded as a pure axiom of existence ' (*1965a*, 96). Cantor might well have attempted to defend his argument in some such terms.

Had he done this, Zermelo could have brought into play his much more serious objection to Cantor's argument—the objection to the employment of inconsistent multiplicities. Surely the purported mention of inconsistent multiplicities cannot occur in an axiom.

Zermelo's own proof in *1904a* that every set has at least one well-ordering, which implies the comparability of powers, was constructed in accordance with the following requisites. The proof was to avoid ' not only all notions that were in any way dubious [such as that of an inconsistent multiplicity] but also the use of ordinals in general ' ; also, only ' principles and devices that have not yet by themselves given rise to any antinomy ' were used (Zermelo *1908a*, art. 2, sect. c ; van Heijenoort *1967a*, 192). The idea of using the axiom of choice to prove the well-ordering theorem was due to Erhard Schmidt. The form of the axiom in Zermelo *1904a* is : for any set M, there is at least one mapping γ such that for each non-empty subset M' of M, $\gamma(M') \in M'$.

The axiom of choice has many important consequences in set theory. It is used in the proof that every infinite set has a denumerable subset, and in the proof that every set has at least one well-ordering. From the latter, it follows that the power of every set is an aleph. Since any two alephs are comparable, so are any two transfinite powers of sets. The axiom of choice is also essential in the arithmetic of transfinite numbers. For example, it is needed to prove that the cardinal of the union of α disjoint sets each having β elements is $\alpha \times \beta$.

The axiom also plays a role in various parts of Weierstrassian analysis (see Sierpinski *1918a*, and Grattan-Guinness *1977a*, *passim*), whose development was described in sections 3.11–3.14. Here are some of its uses : to prove that a limit-point of a set is an accumulation point ; to prove that every field has an algebraic closure which is unique (up to isomorphism) ; to construct non-measurable sets ; and to prove the Bolzano–Weierstrass theorem, if it states that an infinite (in the non-inductive sense) set has a limit-point (as opposed to an accumulation point).

After Zermelo's proof of the well-ordering theorem in *1904a*, the proof and the axiom on which it was based became the subject of a considerable amount of controversy (see Zlot *1960a* ; and Fraenkel, Bar-Hillel and Levy *1973a*). In *1908a* Zermelo published a new proof of the well-ordering theorem and answered the criticisms directed against the first proof and the axiom of choice. One reason for the disagreements was the fact that not everyone understood the axiom in the same way. Thus someone to whom it seems an evident truth might well grant that it is quite doubtful when interpreted in a different way. But metaphysical convictions determine what a given author considers to be the possible interpretations.

Let us first consider Peano's remarks on the axiom of choice, which he understood to mean 'that we may arbitrarily choose an infinite number of elements' (*1906a*, art. 1; *1973a*, 207). He pointed out that he had already rejected this as a principle of inference in 1890. The only objection which he mentioned to the axiom of choice was that it is not provable from the axioms of his system of logic, which he apparently considered as definitive of the concept of proof: 'In some cases we do not know how to eliminate the postulate of Zermelo. Then these proofs are not reduced to the ordinary forms of argument, and the proofs are not valid, according to the ordinary meaning of the word "proof"' (*1973a*, 210). In a letter written to Russell in 1906, Peano expressed his point as follows: 'this form of reasoning is not reducible to the usual forms (for example, to those contained in pages 1–14 of the *Formulaire*, vol. 5 [*1908a*]); and to prove a proposition means to deduce it from known propositions by the usual forms of reasoning, without adding new principles' (Kennedy *1975a*, 209). Peano considered the question of the truth or falsehood of the axiom of choice to be of no consequence (*1973a*, 210). Thus he did not pass any judgment on the truth-value of the axiom of choice, but only on its legitimacy as a principle of demonstration.

What did Zermelo say to this? He had stated in *1904a* that the principle 'cannot, to be sure, be reduced to a still simpler one' (van Heijenoort *1967a*, 141). But he did not consider, as Peano apparently did, that the system of mathematical principles was already complete. Moreover, the axiom of choice emerged in the same way as the principles included in Peano's system must once have done, by analysis of 'the modes of inference that in the course of history have come to be recognised as valid'. It is also justified in the same way, namely, 'by pointing out that the principles are intuitively evident and necessary for science' (*1908a*, art. 1; van Heijenoort *1967a*, 187). The very extensive implicit use of the axiom of choice by many mathematicians could, Zermelo said, 'be explained only by its *self-evidence* . . . No matter if this self-evidence is to a certain degree subjective—it is surely a necessary source of mathematical principles, even if it is not a tool of mathematical proofs, and Peano's assertion [*1973a*, 210] that it has nothing to do with mathematics fails to do justice to manifest facts' (*ibid.*).

Russell was one of those who had implicitly used the axiom of choice (or the multiplicative axiom) in arguments; as we have already seen, when he first became aware of the latter as a proposition which had not yet been proved, he thought it must be provable. Perhaps he had thought this because the proposition seemed evident and, as Zermelo would say, 'necessary for science'. But, unlike Zermelo, Russell, after realising that the axiom was probably independent of the system of assumptions he had made so far, became sceptical about the axiom of choice and its equivalents. Fortunately, he explained quite clearly the source of his doubt.

Russell's first publication dealing with the multiplicative axiom is his *1906a*. Although he opens his discussion of the axiom by presenting the difficulty as one about the possibility of making an infinite number of arbitrary choices, the real point at issue is the *existence* of a selection set for each class *k* of mutually exclusive, non-empty sets : 'What is required is not that we should actually be able to pick out one term from each class which is a member of *k*, but that there should be (whether we can specify it or not) at least one class composed of one term from each member of *k*' (Russell *1906a*, 48 ; *1973a*, 158). Now because of what he meant by a class it seemed doubtful to Russell that there always is a selection set. He conceived of a class as something determined by a property or propositional function : If no property, then no class. Thus, from this point of view, 'what we are primarily in doubt about is the existence of a norm or property such as will pick out one term from each of our aggregates ; the doubt as to the existence of a *class* which will make this selection is derivative from the doubt as to the existence of a norm' (Russell *1906a*, 52 ; *1973a*, 162–163).

Now the multiplicative axiom does seem to be as evident as any of the other axioms of set theory on the pure extensional concept of set. Moreover, as Gödel has said (with this concept in mind), 'nothing can express better the meaning of the term "class" than the axiom of classes and the axiom of choice' (*1944a*, 151). But with Russell's concept of a class as something dependent on a property, the multiplicative axiom really is doubtful. For then it amounts to the assertion that for any class *k* of pairwise disjoint, non-empty classes, there is at least one property possessed by exactly one element from each member of *k* and by no other things. It would be quite possible to agree with Russell's opinion that 'this is not at all obvious' (*1911a*, 33) and yet think the multiplicative axiom an evident truth, by taking sets in the purely extensional sense to be the objects of set theory. His metaphysical convictions prevented him from doing this, but it is interesting to note that later he was persuaded for a while by Frank Ramsey and Henry Sheffer to assert the truth of the multiplicative axiom (Russell *1927a*, 299 ; compare Ramsey *1931a*, 58).

What was Zermelo's concept of set ? Unfortunately, he made no positive statement, but that he did not conceive sets as extensions of properties (as Russell did) is suggested by a couple of passages in his writings (van Heijenoort *1967a*, 189, last para. ; Cantor *Papers*, editorial note on p. 442). It remains possible that Zermelo intended his system to concern the purely extensional concept of set.

6.10. *Some concluding remarks*

What kinds of conclusion can we draw from such a miscellany of studies and techniques ? Perhaps two main points will suffice. Firstly, the introduction of set theory into mathematics and propositional functions into logic brought these two topics into newly intimate contact. Russell and Frege saw the connection as so close that they espoused a doctrine of ' logicism '—that mathematics (for Frege, only arithmetic) was a branch of logic. These forms of logicism are not normally asserted today, but the location of the dividing line between logic and mathematics is still a controversial matter. Secondly, the development of meta-mathematics by Hilbert and the distinction between use and mention by Frege (though, unfortunately, not by Russell) led mathematicians and philosophers to see the profound importance of the distinction between theory and meta-theory in the study of the foundations of logic and mathematics.

These remarks largely refer to the later developments in foundational studies. They lie outside the time-period of this book, which now draws to its close. A fitting conclusion to the book is provided by a return to the 17th century, where the final words of Descartes's *La géométrie* (*1637a*) may apply here also :

> But it is not my purpose to write a large book. I am trying rather to include much in a few words, as will perhaps be inferred from what I have done . . .
>
> I hope that posterity will judge me kindly, not only as to the things of which I have explained, but also as to those which I have intentionally omitted so as to leave to others the pleasure of discovery.

Bibliography

I. Grattan-Guinness

The works of each author are presented in the order : editions of collected or selected works (to which the catchword ' *Works* ', ' *Writings* ' or ' *Papers* ' is assigned) ; dated references, in chronological order (usually the nominal date of publication, but the date of composition when that is significantly different) ; and other works cited by catchword, in alphabetical order (these being items either published in several different editions, or never published by the author and of uncertain date of composition). Translations into English, unless extremely obscure or in the source-books Birkhoff *1973a* or Struik *1969a*, are also listed.

To save space, the number(s) of the edition(s) of a work are indicated by the appropriate sub-script number(s), photographic reprints of works are indicated by ' = ', English translations are prefaced by ' E ', and the titles of journals have been abbreviated in a manner similar to international catalogues of serials. Editorial comments are enclosed in square brackets.

Abel, N. H.

1826a. ' Untersuchungen über die Reihe . . . ', *J. rei. ang. Math.*, 1(1826), 311–339 ; *Oeuvres complètes*$_2$ (2 vols., ed. L. Sylow and S. Lie : 1881, Christiania = 1965, New York), vol. 1, 219–250. [Also other editions.]

d'Alembert, J. le R.

1743a. *Traité de dynamique* (1743, Paris).

1764a. ' Différentiel ', d'Alembert and Diderot *Encyclopaedia*, vol. 4 (1764), 985–989.

1765a. ' Limite ', d'Alembert and Diderot *Encyclopaedia*, vol. 9 (1765), 542.

1768a. ' Réflexions sur les suites et sur les racines imaginaires ', *Opuscules mathématiques* (8 vols., 1761–1780, Paris), vol. 5 (1768), 171–215.

d'Alembert, J. le R., and Diderot, D.

Encyclopaedia. [Eds.] *Encyclopédie ou dictionnaire raisonné* . . . (28 vols., 1751–1765, Paris, Neuchâtel and Amsterdam).

Ampère, A.-M.

1806a. ' Recherches sur quelques points de la théorie des fonctions . . . ', *J. Ecole Polyt.*, cah. 13, 6(1806), 148–181.

1826a. ' Démonstration du théorème de Taylor, . . . ', *Ann. math. pures appl.*, 17(1826–27), 317–329.

Arbogast, L.-F. A.

1789a. ' Essai sur de nouveaux principes du calcul différentiel . . . ' (probably 1789, manuscript, library of the *Ecole Nationale des Ponts et Chaussées*, Paris).

1791a. *Mémoire sur la nature des fonctions arbitraires* . . . (1791, St. Petersburg).

Archimedes

Works$_1$. *Archimedes opera omnia cum commentariis Eutocii* (3 vols., ed. J. L. Heiberg : $_2$1910–1915, Leipzig).

Works$_2$. *Les oeuvres complètes d'Archimède* (2 vols., ed. P. Ver Eecke : $_2$1960, Paris).

Works$_3$. *The works of Archimedes with the method of Archimedes* (ed. T. L. Heath : 1897, Cambridge = 1953, New York).

The method. Manuscript ; *Hermes*, 42(1907), 243–297 ; *Works*$_1$, vol. 2, 426–507 ; *Works*$_2$, vol. 2, 477–519 ; *Works*$_3$, supplement. [First published in a German translation by J. L. Heiberg and H. G. Zeuthen, ' Eine neue Schrift des Archimedes ', *Bibl. math.*, (3)7(1906–07), 321–363.]

Arzela, C.

1885a. ' Sulla integrazione per serie ', *Mem. Accad. Sci. Ist. Bologna*, (4)1 (1885), 532–537, 566–569.

Baire, R. L.

1899a. ' Sur les fonctions des variables réelles ', *Ann. math. pura appl.*, (3)3(1899), 1–123 = (1899, Milan).

Bar-Hillel, Y. *see* Fraenkel, Bar-Hillel and Levy.

Baron, M. E.

1969a. *The origins of the infinitesimal calculus* (1969, Oxford).

Baron, M. E., and Bos, H. J. M.

1976a. *Origins and development of the calculus* (1976, Milton Keynes). [A five-unit Open University course, each unit published separately : 1, *Greek mathematics* (Baron) ; 2, *Indivisibles and infinitesimals* (Baron) ; 3, *Newton and Leibniz* (Baron and Bos) ; and 4 and 5, *The calculus in the eighteenth century* (Bos).]

Barrow, I.

1670a. *Lectiones geometricae* (1670, London).

Berkeley, G.

1734a. *The analyst* (1734, London) ; *Works* (9 vols., ed. A. A. Luce and T. E. Jessop : 1948–1957, London and Edinburgh), vol. 4, 53–102. [Also other editions.]

Bernoulli, D.

1738a. *Hydrodynamica* (1738, Strassburg).

1755a. ' Réflexions et éclaircissemens sur les nouvelles vibrations des cordes . . . ', *Mém. Acad. Roy. Sci. Berlin*, 9(1753 : publ. 1755), 147–172.

Bernoulli, Jakob

Works. *Opera omnia* (2 vols. [consecutively paginated], ed. N. Bernoulli : 1744, Lausanne and Geneva).

1690a. ' Analysis problematis antehac propositi . . . ', *Acta erud.*, (1690), 217–219 : *Works*, 421–426.

1694a. ' Curvatura laminae elasticae . . . ', *Acta erud.*, (1694), 262–276 ; *Works*, 576–600.

1697a. ' Solutio problematum fraternorum . . . ', *Acta erud.*, (1697), 211–217 ; *Works*, 768–778.

1713a. *Ars conjectandi* (1713, Basel). [Various later editions.]

Bernoulli, Johann

Works. *Opera omnia* . . . (4 vols., ed. N. Cramer : 1742, Lausanne and Geneva = 1968, Hildesheim).

1691a. ' Lectiones mathematicae de methodo integralium ' (1690–1691, manuscript) ; *Works*, vol. 3, 385–558.

1691b. ' Solutio problematis funicularii . . . ', *Acta erud.*, (1691), 274–276 : *Works*, vol. 1, 48–51.

1696a. ' Supplementum defectus geometriae Cartesianae . . . ', *Acta erud.*, (1696), 264–269 ; *Works*, vol. 1, 155–161.

1697a. ' Curvatura radii . . . ', *Acta erud.*, (1697), 206–211 ; *Works*, vol. 1, 187–193.

1743a. *Hydraulica* (1743, Lausanne and Geneva).

1924a. *Die Differentialrechnung* (ed. P. Schafheitlin : 1924, Leipzig (Ostwald's Klassiker, no. 211)). [German translation of Bernoulli manuscript.]

Correspondence. *Der Briefwechsel von Johann Bernoulli*, vol. 1 (ed. O. Spiess : 1955, Basel).

Bernstein, F.

1898a. [Proof, as reported in] Borel *1898a*, 103–104.

Beth, E. W.

1965a. *Mathematical thought* . . . (1965, Dordrecht).

Biermann, K.-R.

1966a. ' Karl Weierstrass. Ausgewählte Aspekte seiner Biographie ', *J. rei. ang. Math.*, 223(1966), 191–220.

Birkhoff, G.

1973a. [Ed.] *A source book in classical analysis* (1973, Cambridge, Mass.). [Contains translations of bits of several works discussed in this book.]

Boas, P. van E.

1969a. ' Nowhere differentiable continuous functions ', *Mathematisch Centrum, Amsterdam*, report ZW–012(1969).

Bois Reymond, P. D. G. du

1870a. *Antrittsprogramm, enthaltend neue Lehrsätze* . . . (1870, Berlin); Ostwald's Klassiker, no. 185, 3–42.

1874a. ' Über die sprungweise Wertveränderungen analytischer Funktionen ', *Math. Ann.*, 7(1874), 241–261.

1875a. ' Versuch einer Classification der willkürlichen Functionen . . . ', *J. rei. ang. Math.*, 79(1875), 21–37.

1876a. ' Untersuchungen über die Konvergenz und Divergenz der Fourierschen Darstellungsformeln ', *Abh. Bayer. Akad. Wiss., II Kl.*, 12(1876), pt. 2, i–xxiv, 1–102; Ostwald's Klassiker, no. 186.

1880a. *Zur Geschichte der trigonometrische Reihen. Eine Entgegnung* (1880, Tübingen).

1880b. ' Der Beweis des Fundamentalsatzes der Integralrechnung . . . ', *Math. Ann.*, 16(1880), 115–128.

1882a. *Die allgemeine Funktionentheorie* . . . (1882, Tübingen).

1883a. ' Über das Doppelintegral ', *J. rei. ang. Math.*, 94(1883), 273–290.

Bolzano, B. P. J. N.

1817a. ' Rein analytischer Beweis des Lehrsatzes . . . ', *Abh. Böhm. Gesell. Wiss.*, (3)5(1814–17), 60 pp.; Ostwald's Klassiker, no. 153, 3–43. [Various later editions.]

1851a. *Paradoxien des Unendlichen* (ed. F. Prihonsky: 1851, Leipzig). E: *Paradoxes of the infinite* (trans. and ed. D. Steele: 1950, London).

Borel, E. F. E. J.

1894a. ' Sur quelques points de la théorie des fonctions ', *Ann. Ecole Norm. Sup.*, (3)12(1895), 9–55; (1894, Paris). [Also other editions.]

1898a. *Leçons sur la théorie des fonctions*$_1$ (1898, Paris).

Bos, H. J. M.

1974a. ' Differentials, higher-order differentials and the derivative in the Leibnizian calculus ', *Arch. hist. exact sci.*, 14(1974), 1–90.

See also Baron and Bos

Boyer, C. B.

1939a. *The history of the calculus and its conceptual development* (1939 = 1949 = 1959, New York [with this title]).

1968a. *A history of mathematics* (1968, New York).

Brill, A. von, and Noether, M.

1894a. ' Die Entwicklung der Theorie der algebraischen Funktionen in älterer und neuerer Zeit ', *Jber. Dtsch. Math.-Ver.*, 3, pt. 2 (1894).

Brodén, T.

1896a. ' Ueber das Weierstrass-Cantor'sche Condensationsverfahren ', *Ofversigt Vetens.-Akad. Förhandlingar*, 53(1896), 583–602.

Brouwer, L. E. J.

Works. *Collected works* (2 vols., ed. A. Heyting and H. Freudenthal : 1975–1976, Amsterdam).

1907a. *Over de grondslagen der wiskunde* (1907, Amsterdam). E : *Works*, vol. 1, 11–101.

1911a. ' Beweis der Invarianz der Dimensionenzahl ', *Math. Ann.*, 70 (1911), 161–165 = *Works*, vol. 2, 430–434.

Burali-Forti, C.

1897a. ' Una questione sui numeri trasfiniti ', *Rend. Circ. Mat. Palermo*, 11(1897), 154–164, 260 [a correction]. E : van Heijenoort *1967a*, 104–112.

Burkhardt, H. F. K. L.

1908a. ' Entwicklung nach oscillierenden Funktionen . . . ', *Jber. Dtsch. Math.-Ver.*, 10, pt. 2 (1908).

1910a. ' Über den Gebrauch divergenter Reihen in der Zeit von 1740–1860 ', *Gratulationsschrift zum 60. Geburtstag von A. Pringsheim* (1910, Leipzig), 41–78 ; *Math. Ann.*, 70(1910–11), 169–205.

Cajori, F.

1919a. *A history of the conceptions of limits and fluxions in Great Britain from Newton to Woodhouse* (1919, Chicago and London).

1929a. *History of mathematical notations*, vol. 2 (1929, Chicago).

Cantor, G. F. L. P.

Papers. *Gesammelte Abhandlungen* . . . (ed. E. Zermelo : 1932, Berlin = 1962, Hildesheim).

1872a. ' Über die Ausdehnung eines Satzes der Theorie der trigonometrischen Reihen ', *Math. Ann.*, 5(1872), 122–132 ; *Papers*, 92–102.

1874a. ' Über eine Eigenschaft des Inbegriffes aller reellen algebraischen Zahlen ', *J. rei. ang. Math.*, 77(1874), 258–262 ; *Papers*, 115–118.

1878a. ' Ein Beitrag zur Mannichfaltigkeitslehre ', *J. rei. ang. Math.*, 84(1878), 242–258 ; *Papers*, 119–133.

1880a. ' Über unendliche, lineare Punktmannichfaltigkeiten [pt. 2] ', *Math. Ann.*, 17(1880), 355–358 ; *Papers*, 145–148.

1882a. [*1880a*, pt. 3], *Math. Ann.*, 20 (1882), 113–121 ; *Papers*, 149–157.

1883a. [*1880a*, pt. 5], *Math. Ann.*, 21(1883), 545–591 ; *Grundlagen einer allgemeinen Mannichfaltigkeitslehre* . . . (1883, Leipzig [with new preface]) ; *Papers*, 165–209.

1883b. [French translations of various of Cantor's papers], *Acta math.*, 2(1883), 305–414.

1884a. [*1880a*, pt. 6], *Math. Ann.*, 23(1884), 453–488 ; *Papers*, 210–246.

1885a. ‘ Principien einer Theorie der Ordnungstypen ’ (1885, manuscript) ; Grattan-Guinness *1970a*, 83–101.

1885b. ‘ Über verschiedene Theoreme der Punctmengen . . . ’, *Acta math.*, 7(1885), 105–124 ; *Papers*, 261–277.

1887a. ‘ Mitteilungen zur Lehre von Transfiniten ’, *Ztschr. Phil. philos. Kritik*, 91(1887), 81–125, 252–270, 92(1888), 240–265 ; *Papers*, 378–439.

1895a. ‘ Beiträge zur Begründung der transfiniten Mengenlehre [pt. 1] ’, *Math. Ann.*, 36(1895), 481–512 ; *Papers*, 282–311. E : *1915a*, 85–136.

1895b. ‘ Contribuzione al fondamento della teoria degli insiemi trasfiniti ’ *Riv. di mat.*, 5(1895), 129–162. [Italian translation of *1895a* by F. Gerbaldi.]

1897a. [*1895a*, pt. 2], *Math. Ann.*, 49(1897), 207–246 ; *Papers*, 312–356. E : *1915a*, 137–201.

1899a. *Sur les fondements de la théorie des ensembles transfinis* (1899, Paris). [French translation of *1895a* and *1897a* by F. Marotte.]

1915a. *Contributions to the founding of the theory of transfinite numbers* (1915, Chicago = 1955, New York). [E of *1895a* and *1897a* by P. E. B. Jourdain.]

Cantor/Dedekind. *Briefwechsel Cantor–Dedekind* (ed. E. Noether and J. Cavaillès : 1937, Paris). [French edition : J. Cavaillès, *Philosophie mathématique* (1962, Paris), 179–251.]

Carnot, L. N. M.

1785a. ‘ Dissertation sur la théorie de l’infini mathématique ’ (1785, manu script) ; Gillispie *1971a*, 171–262.

Reflections. *Réflexions sur la métaphysique du calcul infinitésimal* ($_1$1797, Paris : $_2$1813 ; 1921 ; 1970 ; Paris). E_1 : *Phil. mag.*, (1)8(1800), 222–240, 335–352, (1)9(1801), 39–56. E_2 : *Reflexions on the metaphysical principles of the infinitesimal calculus* (trans. W. R. Bowell : 1832, Oxford).

Carslaw, H. S.

1925a. ‘ A historical note on Gibbs’ phenomenon in Fourier’s series and integrals ’, *Bull. Amer. Math. Soc.*, (2)31(1925), 420–424.

1930a. *Introduction to the theory of Fourier’s series and integrals*$_3$ (1930, Cambridge = n.d., New York).

Cassina, U.

1964a. ‘ Storia del concetto di limite ’, *Per. di mat.*, (4)16(1936), 1–19, 82–103, 144–167 ; *Dalla geometria egiziana alla matematica moderna* (1961, Rome), 142–214.

Cauchy, A.-L.

Works. *Oeuvres complètes* (12 + 15 vols., ed. Académie des Sciences : 1882–1974, Paris).

1814a. ‘ Mémoire sur les intégrales définies ’, *Mém. prés. div. sav. Acad. Roy. Sci.*, (2)1(1827), 601–799 ; *Works*, ser. 1, vol. 1, 319–506. [Written in 1814.]

1817a. ' Sur une loi de réciprocité qui existe entre certaines fonctions ', *Bull. sci. Soc. Philom. Paris,* (1817), 121–124 ; *Works,* ser. 2, vol. 2, 223–227.

1821a. *Cours d'analyse* . . . (1821, Paris) ; *Works,* ser. 2, vol. 3.

1822a. ' Sur le développement des fonctions en série . . . ', *Bull. sci. Soc. Philom. Paris,* (1822), 49–54 ; *Works.* ser. 2, vol. 2, 276–282.

1823a. *Résumé des leçons . . . sur le calcul infinitésimal* . . . (1823, Paris) ; *Works,* ser. 2, vol. 4, 5–261.

1826a. ' Mémoire sur les développements des fonctions en séries périodiques ', *Mém. Acad. Roy. Sci.,* 6(1823 : publ. 1827), 603–612 ; *Works,* ser. 1, vol. 2, 12–19. [Written in 1826.]

1827a. ' Sur les residues des fontions exprimées par des intégrales définies ', *Exercices des mathématiques,* vol. 2 (1827, Paris), 341–376 ; *Works,* ser. 2, vol. 7, 393–430.

1833a. *Résumés analytiques* (1833, Turin) ; *Works,* ser. 2, vol. 10, 5–184.

1849a. ' Mémoire sur les fonctions discontinues ', *C. r. Acad. Roy. Sci.,* 28(1849), 277–282 ; *Works,* ser. 1, vol. 11, 120–126.

1853a. ' Note sur les séries convergentes . . . ', *C. r. Acad. Roy. Sci.,* 36(1853), 454–459 ; *Works,* ser. 1, vol. 12, 30–36.

Cavalieri, B.

1635a. *Geometria indivisibilibus continuorum nova quadem ratione promota* ($_1$1635, $_2$1653, Bologna).

1647a. *Exercitationes geometricae sex* (1647, Bologna).

Chihara, C. S.

1973a. *Ontology and the vicious-circle principle* (1973, Ithaca and London).

Child, J. M.

1920a. *The early mathematical manuscripts of Leibniz* . . . (1920, Chicago and London).

Cleave, J. P.

1971a. ' Cauchy, convergence and continuity ', *British j. phil. sci.,* 22(1971), 27–37.

d'Alembert, J. le R. *see* Alembert

Darboux, J. G.

1875a. ' Sur les fonctions discontinues ', *Ann. sci. Ecole Norm. Sup.,* (2)4(1875), 57–112.

Dauben, J. W.

1971a. ' The trigonometric background to Georg Cantor's theory of sets', *Arch. hist. exact sci.,* 7(1971), 181–216.

1974a. ' Denumerability and dimension : the origins of Georg Cantor's theory of sets ', *Rete,* 2(1974), 105–134.

1975a. ' The invariance of dimension : problems in the early development of set theory and topology ', *Hist. math.*, 2(1975), 273–288.

1977a. ' Georg Cantor and Pope Leo XIII : mathematics and theology of the infinite ', *J. hist. ideas*, 38(1977), 85–108.

1977b. ' " Hypotheses non fingo " : theological dimensions of Cantorian set theory ', *Proc. XIVth Int. Cong. Hist. Sci. Tokyo, 1974* (to appear).

1979a. *Georg Cantor . . .* (1979), Cambridge, Mass.),

Dedekind, J. W. R.

Works. *Gesammelte mathematische Werke* (3 vols., ed. R. Fricke, E. Noether and O. Ore : 1930–1932, Braunschweig=[in part] 1969, New York).

1872a. *Stetigkeit und irrationale Zahlen* ($_1$1872, Braunschweig) ; $_5$*Works*, vol. 3, 315–334. E : *1901a*, 1–27.

1888a. *Was sind und was sollen die Zahlen ?* ($_1$1888, Braunschweig) ; $_3$*Works*, vol. 3, 335–391. E : *1901a*, 29–115.

1901a. *Essays on the theory of numbers* (1901, Chicago=1963, New York). [E of *1872a* and *1888a* by W. W. Beman.]

See also Cantor, *Cantor/Dedekind*

De Morgan, A.

1835a. *Elements of algebra* (1835, London).

1842a. *The differential and integral calculus* (1842, London).

Descartes, R.

Works. *Oeuvres* (13 vols., ed. C. Adam and P. Tannery : $_1$18 97–1913, $_2$1964– Paris).

1637a. *Discours de la methode . . . la dioptrique, les meteores et la géométrie* (1637, Leyden) ; *Works*, vol. 9. E : *1925a.* [Also other editions.]

1925a. *The geometry of René Descartes* (1925, Chicago=1954, New York). [E of ' La géométrie ' in *1637a* by D. E. Smith and M. L. Latham, with facsimile of original.]

Dickstein, S.

1899a. ' Zur Geschichte der Prinzipien der Infinitesimalrechnung . . . ', *Abh. Gesch. Math.*, 9(1899), 65–79.

Diderot, D. *see* d'Alembert and Diderot

Dijksterhuis, E. J.

1956a. *Archimedes* (1956, Copenhagen).

Dini, U.

1878a. *Fondamenti per la teorica delle funzioni di variabili reali* (1878, Pisa). [German translation : *1892a.*]

1892a. *Grundlagen für eine Theorie der Functionen einer veränderlichen reellen Grösse* (1892, Leipzig). [German translation of *1878a* by J. Lüroth and A. Schepp.]

Dirichlet, J. P. G. Lejeune-

Works. *Gesammelte Werke* (2 vols., ed. L. Fuchs and L. Kronecker : 1889–1897, Berlin = 1969, New York.)

1829a. ' Sur la convergence des séries trigonométriques . . . ', *J. rei. ang. Math.*, 4(1829), 157–169 ; *Works*, vol. 1, 117–132.

1837a. ' Über die Darstellung ganz willkürlicher Funktionen . . . ', *Rep. Physik*, 1(1837), 152–174 ; *Works*, vol. 1, 133–160 ; Ostwald's Klassiker, no. 116, 3–34.

1837b. ' Sur les séries . . . qui servent à exprimer des fonctions . . . ', *J. rei. ang. Math.*, 17(1837), 35–56 ; *Works*, vol. 1, 283–306.

1837c. ' Beweis eines Satzes . . . ', *Abh. Preuss. Akad. Wiss. Berlin*, (1837), math.-phys. Kl., 45–71 ; *Works*, vol. 1, 313–342.

1862a. ' Démonstration d'un théorème d'Abel . . . ', *J. math. pures appl.*, (2)7(1862), 253–255 ; *Works*, vol. 2, 303–306.

Dobrovolski, W. A.

1971a. ' Contribution à l'histoire du théorème fondemental des équations différentielles ', *Arch. int. hist. sci.*, (1969 : publ. 1971), 223–234.

Drabkin, I. E.

1950a. ' Aristotle's wheel : notes on the history of a paradox ', *Osiris*, 9(1950), 162–198.

Dubbey, J. M.

1963a. ' The introduction of differential notation to Great Britain ', *Ann. sci.*, 19(1963), 37–48.

du Bois Reymond, P. D. G. *see* Bois Reymond

Dugac, P.

1973a. ' Eléments d'analyse de Karl Weierstrass ', *Arch. hist. exact sci.* 10(1973), 41–176.

Dugas, R.

1940a. *Essai sur l'incompréhension mathématique* (1940, Paris).

Ecole Polytechnique

1819a. *Programme de l'enseignement de l'Ecole Royale Polytechnique arrêtes par le Conseil de Perfectionnement pour l'année 1818–1819* (1819 [?], Paris).

Euclid

Elements. [Edition used :] *The thirteen books of Euclid's Elements* (3 vols., ed. T. L. Heath : $_1$1905, $_2$1925, Cambridge = 1956, New York).

Euler, I.

Works. *Opera omnia* (29 + 31 + 12 + [about] 12 vols., ed. Societatis Scientarum Naturalium Helveticae : 1911–   , Leipzig, Berlin, Zurich and [now] Basel).

1736a. *Mechanica sive motus analytice exposita* (2 vols., 1736, St. Petersburg); *Works*, ser. 2, vols. 1–2.

1744a. *Methodus inveniendi lineas curvas maximi minimive proprietate gaudentes* (1744, Lausanne and Geneva); *Works*, ser. 1, vol. 24.

1748a. *Introductio ad analysin infinitorum* (2 vols., 1748, Lausanne); *Works* ser. 1, vols. 8–9.

1755a. 'Remarques sur les mémoires précédens de M. Bernoulli', *Mém. Acad. Roy. Sci. Berlin*, 9(1753 : publ. 1755), 196–222; *Works*, ser. 2, vol. 10, 233–254.

1755b. *Institutiones calculi differentialis* (1755, St. Petersburg=1787, Turin); *Works*, ser. 1, vol. 10.

1768–1770a. *Institutiones calculi integralis* (3 vols., 1768–1770, St. Petersburg); *Works*, ser. 1, vols. 11–13.

1777a. 'Disquisitio ulterior super seriebus . . .', *Nova acta Acad. Sci. Petrop.* 11(1793 : publ. 1798), 114–132; *Works*, ser. 1, vol. 16, pt. 1, 333–355. [Written in 1777.]

Fermat, P. de

Works. *Oeuvres* (4 vols., ed. P. Tannery and C. Henry: 1891–1912, Paris); *Supplément* (ed. C. de Waard: 1922, Paris).

1637a. 'Ad locos planos et solidos isagoge' (completed 1637, manuscript); *Works*, vol. 1, 91–103. [Revised version in *1679a*, 1–8.]

1638. 'Centrum gravitatis parabolici conoidis' (1638, manuscript); *1679a*, 64–66; *Works*, vol. 1, 136–139.

1638b. 'Méthode de maximis et minimis expliquée et envoyée par M. Fermat à M. Descartes' (1638, manuscript); Henry *1880a*, 184–189; *Works*, vol. 2, 154–162.

1643a. [Manuscript sent to P. Brûlart de Saint-Martin in 1643]; *Works*, *Supplément*, 121–125.

1662a. 'Methodus de maxima et minima' (1662, manuscript); *Works*, vol. 1, 170–173.

1679a. *Varia opera . . .* (ed. S. de Fermat: 1679, Toulouse).

De aequationum. 'De aequationum localium transmutatione et emendatione . . .' (about 1658, manuscript); *1679a*, 44–57; *Works*, vol. 1, 255–285.

Doctrinam. 'Doctrinam tangentium' [opening words of an undated, untitled manuscript, probably written around 1640]; *1679a*, 69–73; *Works*, vol. 1, 158–167.

Methodus. 'Methodus ad disquirendam maximam et minimam (about 1636, manuscript); *1679a*, 63–64; *Works*, vol. 1, 133–136.

Syncriseos. [Undated, untitled manuscript beginning 'Dum syncriseos et anastrophes Vietae', probably written around 1640]; Henry *1880a*, 180–183; *Works*, vol. 1, 147–153.

Flett, T. M.

1974a. ' Some historical notes and speculations concerning the mean value theorems of the differential calculus ', *Bull. Inst. Maths. Applics.*, 10(1974), 66–72.

Fourier, J. B. J.

1822a. *Théorie analytique de la chaleur* (1822, Paris) ; *Oeuvres* (2 vols., ed. G. Darboux : 1888–1890, Paris), vol. 1. E : *The analytical theory of heat* (trans. A. Freeman : 1878, Cambridge = 1955, New York).

Fraenkel, A. A. H.

1930a. ' Georg Cantor ', *Jber. Dtsch. Math.-Ver.*, 39(1930), 189–266 = (1930, Leipzig). [Abridged version in Cantor *Papers*, 452–483.]

1953a. *Abstract set theory*$_1$ (1953, Amsterdam).

Fraenkel, A. A. H., Bar-Hillel, Y., and Levy, A.

1973a. *Foundations of set theory*$_2$ (1973, Amsterdam).

Frege, F. L. G.

Writings. *Kleine Schriften* (ed. I. Angelelli : 1967, Hildesheim).

1884a. *Die Grundlagen der Arithmetik* . . . (1884, Breslau = 1964, Hildesheim). E : *The foundations of arithmetic* . . . (trans. J. L. Austin : $_2$1953, Oxford).

1890a. ' Entwurf zu einer Besprechung . . . ' (1890, manuscript) ; *1969a*, 76–80.

1892a. ' Resenzion von : Georg Cantor, Zum Lehre vom Transfiniten ', *Ztschr. Phil. philos. Kritik*, 100(1892), 269–272 ; *Writings*, 163–166.

1893a. *Grundgesetze der Arithmetik* . . ., vol. 1 (1893, Jena = 1962, Hildesheim). E : see under *1903a*.

1903a. [*1893a*], vol. 2 (1903, Jena = 1962, Hildesheim). Part E : *The basic laws of arithmetic* . . . (trans. M. Furth : 1964, Berkeley and Los Angeles).

1906a. ' Über die Grundlagen der Geometrie ', *Jber. Dtsch. Math.-Ver.*, 15(1906), 293–309, 377–403, 423–430 ; *Writings*, 281–323. E : *1971a*, 49–112.

1923a. ' Logische Untersuchungen. Dritter Teil. Gedankenfüge ', *Beiträge Phil. dtsch. Idealismus*, 3(1923), 36–51 ; *Writings*, 378–394. E : *Mind*, n.s., 72(1963), 1–17 ; E. Klemke (ed.), *Essays on Frege* (1968, Urbana), 537–558.

1960a. *Translations from the philosophical writings of Gottlob Frege* (ed. P. Geach and M. Black : $_2$1960, Oxford).

1969a. *Nachgelassene Schriften* (ed. H. Hermes, F. Kambartel and F. Kaulbach : 1969, Hamburg).

1971a. *On the foundations of geometry* . . . (trans. and ed. E.-H. W. Kluge : 1971, New Haven and London).

Galilei, G.

Works. *Opere* (20 vols., Edizione Nazionale : $_1$1890–1909, $_2$1929–1939, Florence).

Gericke, H.

1966a. ' Aus der Chronik der Deutschen Mathematiker-Vereinigung ', *Jber. Dtsch. Math.-Ver.*, 68(1966), 46–74.

Gillispie, C. C.

1971a. [Ed.] *Lazare Carnot, savant* (1971, Princeton).

Gödel, K.

1931a. ' Über formal unentscheidbare Sätze . . . ', *Monats. Math. Physik*, 38(1931), 173–198. E : van Heijenoort *1967a*, 596–616. [Also other translations.]

1944a. ' Russell's mathematical logic ', Schilpp *1944a*, 123–153. [Various reprints.]

Grattan-Guinness, I.

1970a. *The development of the foundations of mathematical analysis from Euler to Riemann* (1970, Cambridge, Mass.).

1970b. ' An unpublished paper by Georg Cantor : . . . ', *Acta math.*, 124(1970), 65–107.

1971a. ' The correspondence between Georg Cantor and Philip Jourdain ', *Jber. Dtsch. Math.-Ver.*, 73(1971), pt. 1, 111–130.

1971b. ' Towards a biography of Georg Cantor ', *Ann. sci.*, 27(1971), 345–391, plts. xxv–xxviii.

1972a. ' Bertrand Russell on his paradox . . . ', *J. phil. logic*, 1(1972), 103–110.

1974a. ' The rediscovery of the Cantor–Dedekind correspondence ', *Jber. Dtsch. Math.-Ver.*, 76(1974–75), pt. 1, 104–139.

1974b. ' The Russell Archives : some new light on Russell's logicism ', *Ann. sci.*, 31(1974), 387–406.

1975a. ' On Joseph Fourier : . . . ', *Ann. sci.*, 32(1975), 503–514.

1977a. *Dear Russell—dear Jourdain* . . . (1977, London).

Grattan-Guinness, I., and Ravetz, J. R.

1972a. *Joseph Fourier 1768–1830* . . . (1972, Cambridge, Mass.).

Grelling, K., and Nelson, L.

1907a. ' Bemerkungen zu den Paradoxien von Russell und Burali–Forti ', *Abh. Fries'schen Scule*, (2)2(1907), 301–334 ; Nelson *1959a*, 59–77.

Gudermann, C.

1838a. ' Theorie der Modular-Functionen und der Modular-Integralen ', *J. rei. ang. Math.*, 18(1838), 1–54, 142–175, 220–258, 303–364.

Guldin, P.

1635–1741a. *Centrobaryca seu de centro gravitatis trium specierum quantitatis continuae* (4 vols., 1635–1641, Vienna).

Gutzmer, A.

1904a. 'Geschichte der Deutschen Mathematiker-Vereinigung . . .', *Jber. Dtsch. Math.-Ver.*, 10(1904), 1–49 = (1904, Leipzig).

Haas, K.

1956a. 'Die mathematischen Arbeiten von Johann Hudde', *Centaurus*, 4(1956), 235–284.

Hankel, H.

1870a. *Untersuchungen über die unendlich oft oscillierenden und unstetigen Functionen* (1870, Tübingen); *Math. Ann.*, 20(1882), 63–112; Ostwald's Klassiker, no. 153, 44–102.

Hardy, G. H.

1904a. 'A theorem concerning infinite cardinal numbers', *Quart. j. pure appl. maths.*, 35(1904), 87–94.

1918a. 'Sir George Stokes and the concept of uniform convergence', *Proc. Cambridge Phil. Soc.*, 19(1918), 148–156.

Harnack, C. G. A.

1885a. 'Über den Inhalt von Punktmengen', *Math. Ann.*, 25(1885), 241–250.

Harriot, T.

1631a. *Artis analyticae praxis* (1631, London).

Harward, A. E.

1905a. 'On the transfinite numbers', *Phil. mag.*, (6)10(1905), 439–460.

Hawkins, T. W.

1970a. *Lebesgue's theory of integration* . . . (1970, Madison, Wis. = 1975, New York).

Heijenoort, J. van *see* van Heijenoort

Heine, E. H.

1870a. 'Über trigonometrische Reihen', *J. rei. ang. Math.*, 71(1870), 353–365.

1872a. 'Die Elemente der Functionenlehre', *J. rei. ang. Math.*, 74(1872), 172–188.

Henry, C.

1880a. *Recherches sur les manuscrits de Pierre de Fermat* (1880, Rome).

Hilbert, D.

Papers. *Gesammelte Abhandlungen* (ed. D. Hilbert and others: 1932–1935 = 1970, Berlin = 1966, New York).

1899a. 'Grundlagen der Geometrie'$_1$, *Festschrift zur Feier der Enthüllung des Gauss-Weber Denkmals in Göttingen* (1899, Leipzig), 1–92. [Later editions in book form.] E: *Foundations of geometry* (trans. L. Unger: $_1$1902, Chicago; $_2$1971, La Salle, Illinois).

1904a. ' Über die Grundlagen der Logik und Arithmetik ', *Verhandlungen des dritten Internationalen Mathematiker-Kongresses* (ed. A. Kraser: 1905, Leipzig=1967, Lichtenstein), 174–185; *1899a* [from third edition onwards]. E: van Heijenoort *1967a*, 129–138.

1922a. ' Neubegründung der Mathematik ', *Abh. Math. Seminar Hamburg Univ.*, 1(1922), 157–177; *Papers*, vol. 3, 157–177.

Hofmann, J. E.

1949a. *Die Entwicklungsgeschichte der Leibnizschen Mathematik* . . . (1949, Munich). E: *1974a.*

1956a. *Ueber Jakob Bernoulli's Beiträge zur Infinitesimalmathematik* (1956, Geneva).

1959a. ' Um Eulers erste Reihenstudien ', K. Schröder *1959a*, 139–208.

1974a. *Leibniz in Paris 1672–1676* . . . (trans. A. Prag and D. T. Whiteside: 1974, Cambridge).

l'Hôpital, G. F. de

1696a. *Analyse des infiniment petits* . . . (1696, Paris).

1697a. ' Solutio problematis . . . ', *Acta erud.*, (1697), 217–218; Johann Bernoulli *Correspondence*, 342–344.

Hudde, J.

1659a. ' Johannis Huddenii epistola secunda de maximis et minimis ', *Geometria a Renato Des Cartes anno 1637 Gallicè edita* (ed. F. van Schooten: $_2$1659, Amsterdam), 507–516.

1659b. [Letter in Dutch of 21 November 1659 to F. van Schooten]. French translations: *J. literaire de la Haye*, 1(1713: publ. 1715–16), 460–464; C. I. Gerhardt, *Der Briefwechsel von Gottfried Wilhelm Leibniz mit Mathematikern* (1899, Berlin=1962, Hildesheim), 234–237. [This last edition cited.]

Huygens, C.

Works. *Oeuvres complètes* (22 vols., ed. Sociéte Hollandaise des Sciences: 1888–1950, The Hague).

1691a. ' Christiani Hugenii . . . solutio . . . ', *Acta erud.*, (1691), 281–282; *Works*, vol. 10, 95–98.

Itard, J.

1947a. ' Fermat, précurseur du calcul différentiel ', *Arch. int. hist. sci.*, 1(1947) 589–610.

Jacoli, F.

1875a. ' Evangelista Torricelli ed il metodo delle tangenti detto metodo del Roberval ', *Bull. bibl. storia sci. mat. fis.*, 8(1875), 265–304.

Jordan, M. E. C.

1882a. *Cours d'analyse* . . .$_1$, vol. 1 (1882, Paris).

1892a. ' Remarques sur les intégrales définies ', *J. math. pures appl.*, (4)8(1892), 69–99 ; *Oeuvres* (4 vols., ed. G. Julia and others : 1961–1964, Paris), vol. 4, 427–457.

1893a. [*1882a*$_2$] (1893, Paris).

Jourdain, P. E. B.

1906–1909a. ' The development of the theory of transfinite numbers [pts. 1 and 2] ', *Arch. Math. Phys.*, 10(1906), 254–281, 14(1908–09), 287–311.

Keisler, H. J.

1976a. *Elementary calculus* [and] *Foundations of infinitesimal calculus* (2 vols., 1976, Boston).

Kennedy, H. C.

1975a. ' Nine letters from Giuseppe Peano to Bertrand Russell ', *J. hist. phil.*, 13(1975), 205–220.

Kepler, J.

Works$_1$. *Opera omnia* (8 vols., ed. C. Frisch : 1858–1871, Frankfurt).

Works$_2$. *Gesammelte Werke* (18 vols., ed. M. Caspar, F. Hammer and others : 1938–1969, Munich).

1615a. *Nova stereometria doliorum vinariorum* (1615, Linz) ; *Works*$_2$, vol. 9, 5–133.

Kline, M.

1972a. *Mathematical thought from ancient to modern times* (1972, New York and London).

Kneale, W. and M.

1962a. *The development of logic* (1962, Oxford).

Knopp, K.

1951a. *Theory and application of infinite series* (trans. R. C. H. Tanner : $_2$1951, London and Glasgow).

Koppelman, E.

1972a. ' The calculus of operations and the rise of abstract algebra ', *Arch. hist. exact sci.*, 8(1972), 155–242.

Kronecker, L.

Works. *Werke* (5 vols., ed. K. Hensel : 1895–1931, Leipzig = 1968, New York).

1886a. ' Über einige Anwendungen der Modulsysteme auf elementare algebraische Fragen ', *J. rei ang. Math.*, 99(1886), 329–371 ; *Works*, vol. 3, pt. 1, 145–208.

1887a. ' Über den Zahlbegriff ', *Philosophische Aufsätze, Eduard Zeller zu seinem fünfzigjährigen Doctor-Jubiläum gewidmet* (1887, Leipzig), 261–274 ; *J. rei. ang. Math.*, 101(1887), 337–355 ; *Works*, vol. 3, pt. 1, 249–274.

Lacroix, S. F.

1816a. *An elementary treatise on the differential and integral calculus* (trans. G. Babbage, G. Peacock and J. F. W. Herschel: 1816, Cambridge).

Treatise. *Traité du calcul différentiel et du calcul intégral* (3 vols., $_1$1797–1800, $_2$1810–1816, Paris).

Lagrange, J. L.

Works. *Oeuvres* (14 vols., ed. J. A. Serret and G. Darboux: 1867–1892, Paris = 1968, Hildesheim).

1759a. 'Recherches sur la nature, et la propagation du son', *Misc. Taur.*, 1(1759), cl. math., i–x, 1–112; *Works*, vol. 1, 39–148.

1762a. 'Essai d'une nouvelle méthode pour déterminer les maxima et les minima . . .', *Misc. Taur.*, 2(1760: publ. 1762), 173–195; *Works*, vol. 1, 333–362.

1772a. 'Sur une nouvelle espèce de calcul . . .', *Nouv. mém. Acad. Roy. Sci. Berlin*, (1772: publ. 1774), cl. math., 185–221; *Works*, vol. 3, 439–470.

1788a. *Mechanique analitique*$_1$ (1788, Paris). [$_2$(2 vols., 1811–1815, Paris); *Works*, vols. 11–12.]

Functions. *Théorie des fonctions analytiques* . . . ($_1$1797, $_2$1813, Paris); $_2$*Works*, vol. 9.

Laplace, P. S.

1809a. 'Mémoire sur divers points d'analyse', *J. Ecole Polyt.*, cah. 15, 8(1809), 229–265; *Oeuvres*$_2$ (14 vols., ed. Académie des Sciences: 1878–1912, Paris = 1966, Hildesheim), vol. 14, 178–214.

Lebesgue, H. L.

Works. *Oeuvres scientifiques* (5 vols., ed. G. Chatelet and G. Choquet: 1972–1973, Geneva).

1902a. 'Intégrale, longueur, aire', *Ann. mat. pura appl.*, (3)7(1902), 231–359 = (1902, Milan) = *Works*, vol. 1, 201–331.

1905a. 'Sur les fonctions représentables analytiquement', *J. math. pures appl.*, (6)1(1905), 139–216 = *Works*, vol. 3, 103–180.

1926a. 'Notice sur la vie et les travaux de Camille Jordan (1838–1922)', *Mém. Acad. Sci.*, 58(1926), xxxix–lxvi. [Also other editions.]

Leibniz, G. W. von

Writings. *Mathematische Schriften* (7 vols., ed. C. I. Gerhardt: 1849–1864, Berlin and Halle = 1961–1962, Hildesheim).

1684a. 'Nova methodus pro maximis et minimis . . .', *Acta erud.*, (1684), 467–473; *Writings*, vol. 5, 220–226.

1686a. 'De geometria recondita . . .', *Acta erud.*, (1686), 292–300; *Writings*, vol. 5, 226–233.

1691a. 'De linea in quam flexile se pondere propria curvat . . .', *Acta erud.*, (1691), 277–281; *Writings*, vol. 5, 243–247.

1697a. ' Communicatio . . . ', *Acta erud.*, (1697), 201–206 ; *Writings*, vol. 5, 331–336.

Lejeune-Dirichlet, J. P. G. *see* Dirichlet

l'Hôpital, G. F. de *see* Hôpital

Lhuilier, S. A. J.

1786a. *Exposition élémentaire des principes des calculs supérieurs* . . . (1786, Berlin).

1795a. *Principorum calculi differentialis et integralis* (1795, Tübingen).

Liouville, J.

1851a. ' Sur des classes très-étendues de quantités . . . ', *J. math. pures appl.*, (1)16(1851), 133–142.

Lipschitz, R. O. S.

1864a. ' De explicatione per series trigonometricas . . . ', *J. rei. ang. Math.*, 63(1864), 296–308. [French translation : *Acta math.*, 36(1913), 281–315.]

Loria, G.

1902a. *Spezielle algebraische und transcendente ebene Kurven* (2 vols., 1902, Leipzig).

MacLaurin, C.

1742a. *Treatise on fluxions* (2 vols., 1742, Edinburgh).

Mahoney, M. S.

1973a. *The mathematical career of Pierre de Fermat 1601–1665* (1973, Princeton).

Manning, K. R.

1975a. ' The emergence of the Weierstrassian approach to complex analysis ', *Arch. hist. exact sci.*, 14(1975), 297–383.

May, K. O.

1973a. *Bibliography and research manual of the history of mathematics* (1973, Toronto and Buffalo).

Medvedev, F. A.

1975a. *Ocherki istorii teorii funktsii* . . . [Essays in the history of the theory of functions of a real variable] (1975, Moscow).

1976a. *Frantsuzskaya shkola teorii funktsii i mnozhestv* . . . [The French school of the theory of functions and sets at the turn of the 19th and 20th centuries] (1976, Moscow).

Mersenne, M.

1644a. *Cogitata physico-mathematica* (1644, Paris).

Correspondence. *Correspondance du P. Marin Mersenne* (ed. C. de Waard and others : 1932– , Paris).

Meschkowski, H.

1964a. *Ways of thought of great mathematicians* ; . . . (trans. J. Dyer-Bennet : 1964, San Fransisco and London).

1965a. ' Aus den Briefbüchern Georg Cantors ', *Arch. hist. exact sci.*, 2(1965), 503–519.

1967a. *Probleme des Unendlichen. Werk und Lebens Georg Cantors* (1967, Braunschweig).

Mittag-Leffler, M. G.

1923a. ' Die ersten 40 Jahre des Lebens von Weierstrass ', *Acta math.*, 39(1923), 1–57.

1923b. ' Weierstrass et Sonja Kowalewsky ', *Acta math.*, 39 (1923), 133–198.

Moigno, F. N. M.

1840–1844a. *Leçons de calcul différentiel et de calcul intégral*, vols. 1 and 2 (1840–1844, Paris).

Mostowski, A.

1966a. *Thirty years of foundational studies* (1966, Oxford).

Nelson, L.

1959a. *Beiträge zur Philosophie der Logik und Mathematik* (1959, Frankfurt).

See also Grelling and Nelson

Neumann, J. von *see* von Neumann

Newton, I.

Papers. *The mathematical papers of Isaac Newton* (8 vols., ed. D. T. Whiteside : 1967– , Cambridge).

Works$_1$. *Opera quae extant omnia* (5 vols., ed. S. Horsley : 1779–1785, London).

Works$_2$. *The mathematical works of Isaac Newton* (2 vols., ed. D. T. Whiteside : 1964–1967, New York and London).

1666a. [The October 1666 tract on fluxions] (1666, manuscript) ; *Papers*, vol. 1, 400–448. [Also other editions.]

1669a. ' De analysi per aequationes numero terminorum infinitas ' (1669, manuscript) ; *Papers*, vol. 2, 206–247. [Also other editions.]

1671a. ' Methodus fluxionum et serierum infinitarum ' (about 1671, manuscript). [Many editions and translations ; see especially] *Papers*, vol. 3, 32–353.

1687a. *Philosophiae naturalis principia mathematica* (1687, London). [Many later editions.] E : *Mathematical principles of natural philosophy* (trans. F. Motte, ed. F. Cajori : 1934, Berkeley).

1693a. ' De quadratura curvarum ' (1693 [in final form], manuscript). [Many editions and translations.] E : *Works*$_2$, vol. 1, 141–160 [reprint of 1710 translation].

1697a. ‘ Epistola . . . ’, *Acta erud.*, (1697), 223–224 ; *Phil. trans. Roy. Soc. London*, 19(1695–97), 384–389 ; *Works*$_1$, vol. 4, 411–416.

Noether, M. *see* Brill and Noether

Osgood, W. F.

1897a. ‘ Non-uniform convergence and the integration of series term-by-term ’ *Amer. j. math.*, 19(1897), 155–190.

Paplauskas, A. B.

1966a. *Trigonometricheskie ryadi ot Eulera do Lebega* [Trigonometric series from Euler to Lebesgue] (1966, Moscow).

Pappus of Alexandria

Collections. *Mathematicae collectiones* (ed. F. Commandino : 1588, Pesaro) ; *La collection mathématique* (2 vols., ed. P. Ver Eecke : 1933, Paris).

Pascal, B.

Works$_1$. *Oeuvres* (14 vols., ed. L. Brunschvicg, P. Boutroux and F. Gazier : 1904–1914, Paris).

Works$_2$. *Oeuvres complètes* (ed. J. Mesnard : 1964– , Paris).

1654a. ‘ Traité du triangle arithmétique ’ (1654, manuscript) ; *1665a* ; *Works*$_1$, vol. 3, 445–503 ; *Works*$_2$, vol. 2, 1288–1323.

1658a. [First circular letter on the cycloid, written June 1658] ; *Works*$_1$, vol. 7, 343–347.

1658b. [Second circular letter on the cycloid, written July 1658] ; *Works*$_1$, vol. 8, 17–19.

1658c. *Lettre de A. Dettonville à Monsieur de Carcavy* (1658, Paris) ; *Works*$_1$, vol. 8, 331–384 and vol. 9, 3–133.

1665a. *Traité du triangle arithmétique*, . . . (1665, Paris) ; *Works*$_1$, vols. 3 and 11 *passim* ; *Works*$_2$, vol. 2, 1166–1332.

Postetatum. ‘ Postetatum numericarum summa ’ (about 1654, manuscript) ; *1665a* ; *Works*$_1$, vol. 3, 346–367 ; *Works*$_2$, vol. 2, 1259–1272.

Peano, G.

Works. *Opere scelte* (3 vols., ed. Unione Matematica Italiana : 1957–1959, Rome).

1884a. *Calcolo differenziale . . . pubblicatto con aggiunte dal Dr. Giuseppe Peano* (1884, Turin). [Main text written by Angelo Genocchi.]

1889a. *Arithmetices principia*, . . . (1889, Turin) ; *Works*, vol. 2, 20–55. E : *1973a*, 101–134. Part E : van Heijenoort *1967a*, 83–97.

1890a. ‘ Sur une courbe, qui remplit toute une aire plane ’, *Math. Ann.*, 36(1890), 157–160 ; *Works*, vol. 1, 110–114.

1892a. ‘ Dimostrazione dell’impossibilità di segmenti infinitesimi costanti ’, *Riv. di mat.*, 2(1892), 58–62 ; *Works*, vol. 3, 110–114.

1894a. ' Sur la définition de la limite d'une fonction . . . ', *Amer. j. math.*, 17(1894), 38–68 ; *Works*, vol. 1, 228–257.

1906a. ' Additione ', *Riv. di mat.*, 8(1906), 143–157 ; *Works*, vol. 1, 344–358. E : *1973a*, 206–218.

1908a. [Ed.] *Formulaire de mathématiques*$_5$ (1908=1960, Rome).

1973a. *Selected works of Giuseppe Peano* (trans. H. C. Kennedy : 1973, Toronto and London).

Pedersen, K. M.

1968a. ' Roberval's method of tangents ', *Centaurus*, 13(1968), 151–182.

Peters, M.

1961a. *Lied eines Lebens* (1961, Halle [privately printed]).

Pincherle, S.

1880a. ' Saggio di una introduzione alla teoria delle funzioni analitiche . . . ', *Giorn. mat.*, (1)18(1880), 178–254, 317–357.

Poincaré, J. H.

1883a. ' Sur les fonctions à espaces lacunaires ', *Acta Soc. Scient. Fennicae*, 12(1883), 343–350 ; *Oeuvres* (11 vols., ed. P. Appell and others : 1928–1956, Paris), vol. 4, 28–35.

1906a. ' Les mathématiques et la logique [pt. 3] ', *Rev. méta. morale*, 14(1906), 294–317. E : in *1913a*.

1913a. *The foundations of science* (trans. G. B. Halsted : 1913=1946, Lancaster, Penna.).

Poisson, S.-D.

1823a. ' Suite du mémoire sur les intégrales définies . . . ', *J. Ecole Polyt.*, cah. 19, 12(1823), 404–509.

Prag, A.

1929a. ' John Wallis ', *Quellen Studien Gesch. Math.*, *Abt. B*, 1(1929), 381–412.

Pringsheim, A.

1898a. ' Irrationalzahlen und Konvergenz unendlicher Prozesse ', *Enc. Math. Wiss.*, vol. 1, pt. 1 (1898, Leipzig), 47–146.

1899a. ' Grundlagen der allgemeinen Funktionentheorie ', *Enc. Math. Wiss.*, vol. 2, pt. 1 (1899, Leipzig), 1–53.

1900a. ' Zur Geschichte des Taylorschen Lehrsatzes ', *Bibl. math.*, (3)1(1900) 433–479.

Prony, G. C. F. M. Riche de

1798a. ' Introduction aux cours d'analyse . . . ', *J. Ecole Polyt.*, cah. 5, 2(1798), 213–218.

Ramsey, F. P.

1931a. *The foundations of mathematics* . . . (ed. R. B. Braithwaite : 1931, London).

Ravetz, J. R. *see* Grattan-Guinness and Ravetz

Richard, J.

1905a. 'Les principes des mathématiques . . .', *Rev. gen. sci. pures appl.*, 16(1905), 541. E: van Heijenoort *1967a*, 142–144.

Riche de Prony, G. C. F. M. *see* Prony

Riemann, G. F. B.

Works. *Gesammelte mathematische Werke* (ed. H. Weber with R. Dedekind: $_2$1892, Leipzig = 1953, New York).

1854a. 'Über die Darstellbarkeit einer Function durch eine trigonometrische Reihe', *Abh. Gesell. Wiss. Göttingen*, 13(1866–67: publ. 1868), math. Kl., 87–132; *Works*, 227–271. [Written in 1854.]

1854b. 'Über die Hypothese, welche der Geometrie zu Grunde liegen', *Abh. Gesell. Wiss. Göttingen*, 13(1866–67: publ. 1868), math. Kl., 133–152; *Works*, 254–269. [Written in 1854.]

Roberval, G. P. de

Works$_1$. *Divers ouvrages de mathématique et de physique par Messieurs de l'Académie Royale des Sciences* (1693, Paris).

Works$_2$*a*. *Mém. Acad. Roy. Sci. depuis 1666 jusqu'à 1699*, vol. 6 (1730, Paris). [Contains the works of Roberval.]

Works$_2$*b*. [*Works*$_2$*a*, in an edition containing also the works of the Abbé Picard.]

Observations. 'Observations sur la composition des mouvemens . . .' (undated manuscript); *Works*$_1$, 69–111; *Works*$_2$*a*, 1–67; *Works*$_2$*b*, 1–89.

Traité. 'Traité des indivisibles' (undated manuscript); *Works*$_1$, 190–245; *Works*$_2$*a*, 207–290; *Works*$_2$*b*, 247–359.

Robins, B.

1761a. *Mathematical tracts* (2 vols., 1761, London).

Robinson, A.

1966a. *Non-standard analysis* (1966, Amsterdam).

Royden, H. L.

1968a. *Real analysis*$_2$ (1968, London and New York).

Russell, B. A. W.

1901a. 'Recent work on the principles of mathematics', *Int. monthly*, 4(1901), 83–101; *1917a*, ch. 5.

1901b. 'Sur la logique des relations . . .', *Rev. de math.*, 7 (1900–01), 115–148. E: *1956a*, 1–38.

1903a. *The principles of mathematics* ($_1$1903, Cambridge; $_2$1937, London [with a new preface]).

1904a. 'The axiom of infinity', *Hibbert j.*, 2(1903–04), 809–812; *1973a*, 256–259.

1906a. ' On some difficulties in the theory of transfinite numbers and order types ', *Proc. London Math. Soc.*, (2)4(1906–07), 29–35 ; *1973a*, 135–164.

1906b. ' Les paradoxes de la logique ', *Rev. méta. morale*, 14(1906), 627–650. E : *1973a*, 190–214.

1908a. ' Mathematical logic as based on the theory of types ', *Amer. j. math.*, 30(1908), 222–262 ; *1956a*, 59–102 ; van Heijenoort *1967a*, 150–182.

1910a. ' Some explanations in reply to Mr. Bradley ', *Mind*, n.s., 19(1910), 373–378.

1910b. [Review of G. Mannoury, *Methodologisches und Philosophisches zur Elementar-Mathematik* (1909, Haarlem)], *Mind*, n.s., 19(1910), 438–439.

1911a. ' Sur les axiomes de l'infini et du transfini ', *C.r. séances Soc. Math. France*, (1911), no. 2, 22–35 = *Bull. Soc. Math. France*, 39 (1911) [1967 reprint], 488–501. E : Grattan-Guinness *1977a*, 162–174.

1917a. *Mysticism and logic* (1917, London).

1919a. *Introduction to mathematical philosophy* (1919, London).

1927a. *The analysis of matter* (1927, London).

1944a. ' My mental development ', Schilpp *1944a*, 1–20.

1956a. *Logic and knowledge* (ed. R. C. Marsh : 1956, London).

1959a. *My philosophical development* (1959, London).

1973a. *Essays in analysis* (ed. D. Lackey : 1973, London).

See also Whitehead and Russell

Schilpp, P. A.

1944a. [Ed.] *The philosophy of Bertrand Russell* (1944, New York).

Schneider, I.

1968a. ' Der Mathematiker Abraham de Moivre ', *Arch. hist. exact sci.*, 5(1968–69), 177–317.

Schönflies, A. M.

1900a. ' Die Entwicklung der Lehre von den Punktmannigfaltigkeiten '$_1$, *Jber. Dtsch. Math.-Ver.*, 8, pt. 2 (1899 : publ. 1900).

1927a. ' Die Krisis in Cantors mathematischem Schaffen ', *Acta math.*, 50(1927), 1–23.

Schröder, F. W. K. E.

1898a. ' Ueber zwei Definitionen der Endlichkeit und G. Cantorsche Sätze ', *Abh. Kaiserl. Leop.-Car. Akad. Naturf.*, 71(1898), 301–362.

Schröder, K.

1959a. [Ed.] *Sammelband der zu Ehren des 250. Geburtstages Leonhard Eulers* (1959, Berlin).

Seidel, P. L.

1848a. ' Note über eine Eigenschaft der Reihen, . . . ', *Abh. Bayer. Akad. Wiss. Munich*, 7(1847–49), math.-phys. Kl., 381–393 ; Ostwald's Klassiker, no. 116, 35–45.

Serret, J. A.

1868a. *Cours du calcul différentiel et intégral*$_1$, vol. 1 (1868, Paris).

Sierpinski, W.

1918a. ' L'axiome de M. Zermelo . . . ', *Bull. Acad. Sci. Cracovie, cl. sci. math. nat.*, (A), (1918), 97–152 ; *Oeuvres choisis* (3 vols., ed. S. Hartman and A. Schinzel : 1974– , Warsaw), vol. 2, 208–255.

1965a. *Cardinal and ordinal numbers*$_2$ (1965, Warsaw).

Smith, H. J. S.

1875a. ' On the integration of discontinuous functions ', *Proc. London Math. Soc.*, (1)6(1875), 140–153 ; *Collected mathematical papers* (2 vols., ed. J. W. L. Glaisher : 1894, Oxford = 1965, New York), vol. 2, 86–100.

Stevin, S.

Works. *The principal works* (5 vols., ed. E. Crone, E. J. Dijksterhuis and others : 1955–1968, Amsterdam).

1586a. *De Beghinselen der Weeghconst* (1586, Leyden) = *Works*, vol. 1, 35–285 [together with E].

Stokes, G. G.

1849a. ' On the critical values of the sums of periodic series ', *Trans. Cambridge Phil. Soc.*, 8(1849), 535–583 ; *Mathematical and physical papers* (5 vols., ed. G. Stokes and others : 1880–1905, Cambridge), vol. 1, 236–313.

Stolz, O.

1881a. ' B. Bolzano's Bedeutung in der Geschichte der Infinitesimalrechnung ', *Math. Ann.*, 18(1881), 255–279.

1884a. ' Über einen zu einer unendlichen Punktmenge gehörigen Grenzwerth ', *Math. Ann.*, 23(1884), 152–156.

Struik, D. J.

1969a. [Ed.] *A source book in mathematics, 1200–1800* (1969, Cambridge, Mass.). [Contains translations of parts of several works discussed in this book.]

Taylor, A. E.

1974a. ' The differential : nineteenth and twentieth century developments ', *Arch. hist. exact sci.*, 12(1974), 355–383.

Taylor, B.

1715a. *Methodus incrementorum directa et inversa* (1715, London).

Todhunter, I.

1861a. *A history of ... the calculus of variations* ... (1861, Cambridge and London).

Torricelli, E.

Works. *Opere* (3 vols. in 4, ed. G. Loria and G. Vassura : 1919–1944, Faenza).

1644a. *Opera geometrica* (1644, Florence) ; *Works*, vols. 1–2 *passim*.

Truesdell, C. A., III

1960a. ' A program toward rediscovering the rational mechanics of the age of reason ', *Arch. hist. exact sci.*, 1(1960–62), 3–36 ; *Essays in the history of mechanics* (1968, Berlin), 85–137.

1960b. *The rational mechanics of flexible or elastic bodies 1638–1788* (1960, Zurich [Euler *Works*, ser. 2, vol. 11, pt. 2]).

Tucciarone, J.

1973a. ' The development of the theory of summable divergent series from 1880 to 1925 ', *Arch. hist. exact sci.*, 10(1973), 1–40.

van Heijenoort, J.

1967a. [Ed.] *From Frege to Gödel* ... (1967, Cambridge, Mass.).

Ver Eecke, P.

1932a. ' Le théorème dit de Guldin considérée au point de vue historique ', *Mathesis*, 46(1932), 395–397.

Veronese, G.

1891a. *Fondamenti di geometria* (1891, Padua).

1894a. *Grundzüge der Geometrie* ... (trans. A. Schepp : 1894, Leipzig). [German translation of *1891a*.]

Viète, F.

Works. *Opera mathematica* (ed. F. van Schooten : 1646, Leiden = 1970, Hildesheim [with new introduction by J. E. Hofmann]).

1591a. *In artem analyticen isagoge* (1591, Tours) ; *Works*, 1–12. E. J. Klein, *Greek mathematical thought* ... (1968, Cambridge, Mass.), 315–353.

1973a. *Einführung in die neue Algebra* (trans. and ed. K. Reich and H. Gericke : 1973, Munich). [German edition of *1591a*.]

Vitali, G.

1904a. ' Sui gruppi di punti ', *Rend. Circ. Mat. Palermo*, 18(1904), 116–126.

Vivanti, G.

1891a. ' Sull'infinitesimo attuale ', *Riv. di mat.*, 1(1891), 135–153, 248–256.

Volterra, V.

Works. *Opere mathematiche* (5 vols., ed. Academia Nazionale dei Lincei : 1954–1962, Rome).

1881a. ' Alcune osservazioni sulle funzioni punteggiate discontinue ', *Giorn. mat.*, 19(1881), 76–87 ; *Works*, vol. 1, 7–15.

1881b. ' Sui principii del calcolo integrale ', *Giorn. mat.*, 19(1881), 333–372 ; *Works*, vol. 1, 16–48.

von Neumann, J.

1925a. ' Eine Axiomatisierung der Mengenlehre ', *J. rei. ang. Math.*, 154(1925), 219–240 [and correction at 155(1926), 128] = *Collected works* (6 vols., ed. A. Taub : 1961–1963, New York), vol. 1, 34–56. E : van Heijenoort *1967a*, 393–413.

Voss, A. E.

1899a. ' Differential- und Integralrechnung ', *Enc. Math. Wiss.*, vol. 2, pt. 1 (1899, Leipzig), 54–134.

Wallis, J.

Works$_1$. *Operum mathematicorum*, *Pars prima/altera* (2 vols. in 1, 1656–1657, Oxford).

Works$_2$. *Opera mathematica* (3 vols., ed. J. Wallis : 1693–1699, Oxford = 1972, Hildesheim [with new introduction by C. J. Scriba]).

1956a. *Arithmetica infinitorum* (1656, Oxford) ; *Works*$_1$, pars altera ; *Works*$_2$, vol. 1, 355–478.

1659a. *Tractatus duo. Prior de cycloide . . . Posterior . . . de cissoide* (1659, Oxford) ; *Works*$_2$, vol. 1, 489–541.

Weierstrass, K. T. W.

Works. *Mathematische Werke* (7 vols., ed. K. Weierstrass and others : 1894–1927, Berlin = 1967, Hildesheim and New York).

1841a. ' Zur Theorie der Potenzreihen ' (1841, manuscript) ; *Works*, vol. 1, 67–74.

1872a. ' Über continuerliche Functionen eines reellen Arguments, . . . ' (1872, manuscript) ; *Works*, vol. 2, 71–74.

1880a. ' Zur Functionenlehre ', *Monatsb. Preuss. Akad. Wiss. Berlin, math.-phys. Kl.*, (1880), 719–743 ; *Works*, vol. 2, 201–233.

1885a. ' Über die analytische Darstellbarkeit sogenannte willkürlicher Funktionen einer reellen Veränderlichen ', *Sitz.-Ber. Preuss. Akad. Wiss. Berlin*, (1885), 633–639, 789–805 ; *Works*, vol. 3, 1–37.

Weyl, C. H. H.

Papers. *Gesammelte Abhandlungen* (4 vols., ed. K. Chandrasekharan : 1968, Berlin).

1910a. ' Über die Definitionen der mathematischen Grundbegriffe ', *Math.-naturw. Blätter*, 7(1910), 93–95, 109–113 = *Papers*, vol. 1, 298–304.

Whewell, W.

1838a. *The doctrine of limits* . . . (1838, Cambridge and London).

Whitehead, A. N.

1907a. *The axioms of projective geometry* (1907, Cambridge = n.d., New York).

Whitehead, A. N., and Russell, B. A. W.

PM. *Principia mathematica* (3 vols., $_1$1910–1913, $_2$1925–1927, Cambridge).
See also Russell

Whiteside, D. T.

1961a. ' Patterns of mathematical thought in the later seventeenth century ', *Arch. hist. exact sci.*, 1(1960–62), 179–388.

1966a. ' Newton's marvellous year : 1666 and all that ', *Notes records Roy. Soc. London*, 21(1966), 32–41.

Wilbraham, H.

1848a. ' On a certain periodic function ', *Cambridge Dublin math. j.*, 3(1848), 198–201.

Woodhouse, R.

1803a. *The principles of analytical calculation* (1803, Cambridge).

1810a. *A treatise on isoperimetrical problems and the calculus of variations* (1810, Cambridge = 1964, New York [under the title *A history of the calculus of variations during the eighteenth century*]).

Wren, C.

1659a. [Appendix to ' De cycloide '] ; in Wallis *1659a.*

Young, W. H.

1905a. ' On the general theory of integration ', *Phil. trans. Roy. Soc. London*, 20(1905), 221–252.

Yushkevich, A.-A. P.

1959a. ' Euler und Lagrange über die Grundlagen der Analysis ', K. Schröder *1959a*, 224–244.

1971a. ' Lazare Carnot and the competition . . . on the mathematical theory of the infinite ', Gillispie *1971a*, 147–168.

1971b. ' Notes à la " Dissertation " de Carnot ', Gillispie *1971a*, 263–267.

1973a. ' J. A. da Cunha et les fondements de l'analyse infinitésimale ', *Rev. d'hist. sci.*, 26(1973), 1–22.

1974a. ' La notion de fonction chez Condorcet ', *For Dirk Struik* . . . (ed. R. Cohen and others : 1974, Dordrecht), 131–139.

1976a. ' The concept of function up to the middle of the 19th century ', *Arch. hist. exact sci.*, 16(1976), 37–85.

Zermelo, E. F. F.

1901a. ' Über die Addition transfiniter Kardinalzahlen ', *Nachr. Gesell. Wiss. Göttingen*, (1901), math.-phys. Kl., 34–38.

1904a. ' Beweis, dass jede Menge wohlgeordnet werden kann ', *Math. Ann.*, 59(1904), 514–516. E : van Heijenoort *1967a*, 139–141.

1908a. ' Neuer Beweis für die Möglichkeit einer Wohlordnung ', *Math. Ann.*, 65(1908), 107–128. E : van Heijenoort *1967a*, 183–198.

1908b. ' Untersuchungen über die Grundlagen der Mengenlehre I ', *Math. Ann.*, 65(1908), 261–281. E : van Heijenoort *1967a*, 199–215.

Zlot, W. L.

1960a. ' The principle of choice in preaxiomatic set theory ', *Scripta math.*, 25(1960), 105–123.

Name Index

I. Grattan-Guinness

Where known, the names and dates of the persons cited are included. Editorial additions in square brackets indicate parts of names, or alternative names, or alternative forms of the name, by which the person was or is widely known but which have not been used in the text.

Subject Index

I. Grattan-Guinness

The subject matter covered by this book constitutes an intricate and multiply connected collection of topics. In order that the reader can retrieve as much information as possible, it is essential to have a subject index which both is comprehensive in coverage and comprehensible in structure. Students are insufficiently trained in the use of indexes in textbooks (and also of books in libraries, and of catalogues and abstracting services), so that this index may be useful practice. Older textbooks are worse ; often they were not indexed at all.

The structure of the index is provided primarily by the use of relatively few principal headings under which many sub-headings are supplied. Some of these headings are used by their sub-headings in the singular and/or the plural, and the sub-headings either read after or before the heading, or describe a particular case of that general category. For example, the heading ' Set(s) ' has these five among its sub-headings : 'covering of a ', ' decomposition of ', ' of first *or* second category ', ' denumerable ' and ' multiplicities '. The same points apply to the (rarely used) sub-sub-headings relative to their sub-heading, and also to the other headings which have a sub-heading structure.

The principal headings are :

Calculus	Differential(s)	Limit(s)
Cardinal(s)	Fourier series	Numbers
Convergence	Function(s)	Ordinal(s)
Curve(s)	Infinitesimals	Series
Definitions	Integral	Set(s)
Derivative(s)	Integration	Variable(s)

' Calculus ' covers entries applying to either the differential or the integral (or other) forms, which themselves are (largely) covered respectively by ' Derivative(s) ' and ' Differential(s) ', and ' Integral ' and ' Integration '. A similar relationship applies between ' Numbers ', ' Cardinal(s) ' and ' Ordinal(s) '. In addition, ' Definition(s) ' contains references only to general features of definitions ; particular definitions are located at the appropriate (sub-)heading elsewhere in the index. This kind of point applies also to the headings ' Axiom(s) ', ' Mathematics ', ' Proof ' and ' Variable(s) '.

In order to help the reader further around the index, a large number of cross-references are provided. ' *See* ' directs *all* citations to the other place(s) ;

it takes the form '*See under*' when the recommendation is to the appropriately worded sub-heading (which is not itself always explicitly given) of the heading(s) then named. '*See above*' and '*See below*' direct attention from one sub-heading to another under the same heading. When closely related topics are named *in addition* to the locations given by the (sub-)heading, then the word '*also*' is added to the above instructions. On occasion we see combinations such as '*See below under*' or '*See also above*'; and quite often more than one type of cross-reference is used for a (sub-)heading, when the word '*See*' is used only at the beginning of the instructions, and link-words such as '*and*' and '*or*' are also sometimes employed.

The names of well-known theorems or techniques are usually entered under those names rather than under a broader heading to which the result relates. All journals named in the text are listed, as are several books or other writings which were cited several times. Some cities and institutions have been included on similar grounds. Square brackets are used once, to fill out editorially the name of a journal which was abbreviatively named 'Crelle's *Journal*' in the text.